GLASS TECHNOLOGY

GLASS TECHNOLOGY

Developments Since 1978

Edited by J.I. Duffy

NOYES DATA CORPORATION

Park Ridge, New Jersey, U.S.A.

1981

Published in the United States of America by
Noyes Data Corporation
Noyes Building, Park Ridge, New Jersey 07656

Library of Congress Cataloging in Publication Data

Duffy, Joan Irene, 1950-
   Glass technology.

   (Chemical technology review ; no. 184)
   Includes index.
   1. Glass manufacture--Patents.   I. Title.
II. Series.
TP858.D83   666'.1'0272   80-26045
ISBN 0-8155-0838-7

FOREWORD

The detailed, descriptive information in this book is based on U.S. patents, issued beginning with January 1979, that deal with glass technology.

This book is a data-based publication, providing information retrieved and made available from the U.S. patent literature. It thus serves a double purpose in that it supplies detailed technical information and can be used as a guide to the patent literature in this field. By indicating all the information that is significant, and eliminating legal jargon and juristic phraseology, this book presents an advanced commercially oriented review of recent developments in the field of glass technology.

The U.S. patent literature is the largest and most comprehensive collection of technical information in the world. There is more practical, commercial, timely process information assembled here than is available from any other source. The technical information obtained from a patent is extremely reliable and comprehensive; sufficient information must be included to avoid rejection for "insufficient disclosure." These patents include practically all of those issued on the subject in the United States during the period under review; there has been no bias in the selection of patents for inclusion.

The patent literature covers a substantial amount of information not available in the journal literature. The patent literature is a prime source of basic commercially useful information. This information is overlooked by those who rely primarily on the periodical journal literature. It is realized that there is a lag between a patent application on a new process development and the granting of a patent, but it is felt that this may roughly parallel or even anticipate the lag in putting that development into commercial practice.

Many of these patents are being utilized commercially. Whether used or not, they offer opportunities for technological transfer. Also, a major purpose of this book is to describe the number of technical possibilities available, which may open up profitable areas of research and development. The information contained in this book will allow you to establish a sound background before launching into research in this field.

Advanced composition and production methods developed by Noyes Data are employed to bring these durably bound books to you in a minimum of time. Special techniques are used to close the gap between "manuscript" and "completed book." Industrial technology is progressing so rapidly that time-honored, conventional typesetting, binding and shipping methods are no longer suitable. We have bypassed the delays in the conventional book publishing cycle and provide the user with an effective and convenient means of reviewing up-to-date information in depth.

The table of contents is organized in such a way as to serve as a subject index. Other indexes by company, inventor and patent number help in providing easy access to the information contained in this book.

16 Reasons Why the U.S. Patent Office Literature Is Important to You

1. The U.S. patent literature is the largest and most comprehensive collection of technical information in the world. There is more practical commercial process information assembled here than is available from any other source. Most important technological advances are described in the patent literature.

2. The technical information obtained from the patent literature is extremely comprehensive; sufficient information must be included to avoid rejection for "insufficient disclosure."

3. The patent literature is a prime source of basic commercially utilizable information. This information is overlooked by those who rely primarily on the periodical journal literature.

4. An important feature of the patent literature is that it can serve to avoid duplication of research and development.

5. Patents, unlike periodical literature, are bound by definition to contain new information, data and ideas.

6. It can serve as a source of new ideas in a different but related field, and may be outside the patent protection offered the original invention.

7. Since claims are narrowly defined, much valuable information is included that may be outside the legal protection afforded by the claims.

8. Patents discuss the difficulties associated with previous research, development or production techniques, and offer a specific method of overcoming problems. This gives clues to current process information that has not been published in periodicals or books.

9. Can aid in process design by providing a selection of alternate techniques. A powerful research and engineering tool.

10. Obtain licenses—many U.S. chemical patents have not been developed commercially.

11. Patents provide an excellent starting point for the next investigator.

12. Frequently, innovations derived from research are first disclosed in the patent literature, prior to coverage in the periodical literature.

13. Patents offer a most valuable method of keeping abreast of latest technologies, serving an individual's own "current awareness" program.

14. Identifying potential new competitors.

15. It is a creative source of ideas for those with imagination.

16. Scrutiny of the patent literature has important profit-making potential.

CONTENTS AND SUBJECT INDEX

INTRODUCTION

The term "glass" means an inorganic product of fusion which solidifies to a rigid, noncrystalline condition upon cooling. Most of the commonly used glasses are silicate glasses. These include container glass, plate glass, borosilicate glass, fused silica, special high-melting glasses, glasses designed specifically for subsequent devitrification, sodium silicates, fiber glass, glass wool, slag wool, and rock wool.

Various techniques are presently being used to manufacture flat sheet glass. Typically, premixed glass-forming materials are fed onto the surface of a bath of molten glass contained in a furnace. In the fuel-firing of the regenerative tank type furnaces, the materials are melted by hot gases from flames playing across the furnace above the glass surface. In the more modern electric furnaces, heat is produced by passing electric current through the bath of molten glass between electrodes immersed in the glass. Also, a combination of both heating methods is sometimes employed.

The furnaces described above assume various shapes. Early regenerative, fuel-fired, tank type furnaces were generally horizontal and rectangular in shape with raw material received in one end and molten glass formed in a continuous sheet on the opposite end. This furnace at one time enjoyed considerable popularity in view of the abundance of relatively cheap natural gas energy resources. However, as natural gas fuel became scarce and therefore expensive, the energy consumption deficiencies of the regenerative furnace soon became apparent.

In particular, the horizontal regenerative furnace experienced considerable heat loss because of its relatively large exposed cross-sectional areas. Therefore the trend in recent years has been to employ vertical furnaces. These furnaces are characterized by smaller cross-sectional area, and therefore less heat loss. However, these furnaces likewise have not been without problems. A perennial problem with electric furnaces has been heat localization around the electrodes, and the integrity of the furnace wall surrounding the localized electrode heat pockets. Furthermore, normal electrode wear requires regular replacement, which has resulted in shutdown of the furnace.

A glass sheet is tempered by a two-step process in which the glass is first heated to an elevated temperature and then is cooled very rapidly to a temperature below the strain point. Tempering provides glass sheets with a stress pattern in which the glass sheet develops a thin skin of compression stress surrounding an interior stressed in tension. Such a stress distribution makes the glass sheet much stronger than untempered glass so that tempered glass is less likely to shatter when struck by an object. Contained in this volume are a number of processes dealing with improved methods of strengthening and toughening glass, as well as apparatus designed to handle the glassware more efficiently and with less breakage.

The significance of the role to be played by optical fibers in information transmission systems is no longer in dispute. The emphasis of research and development programs in this field has shifted from that of proving practicality to one of improving transmission efficiency. An active area, which has been particularly fruitful in yielding such improvements, involves the reduction of losses in optical fibers so that they may be used for long distance transmission. The lower the optical losses in such fibers the less frequent the need for multiple optical repeaters and, consequently, the cheaper the cost of the total system.

Cladded core fibers generally consist of a fiber core and cladding composed of materials which have been selected so that the refractive index of the core is higher than the refractive index of the material forming the cladding.

In self-focusing fibers, the index of refraction decreases from the center of the core (again cylindrical) to the periphery thereof. The refractive index along the radius of the cylindrical cross section of the core is often pseudoparabolic. If the radial gradient is sufficiently large in its absolute value, all the light rays (visible or invisible) are refocused and, because they are unable to escape from the fiber, are propagated by it without any losses.

This book describes the syntheses and treatment of glasses, glass fibers and glass-ceramics and presents formulation and evaluation data for a variety of processes presented in the U.S. patent literature from January 1979 through mid-1980. The processes are grouped according to their major use, but it should be recognized that many of these formulations may be used for other applications as well.

GLASSMAKING

MELTING

Glass Batch Wetting and Mixing Apparatus

Glass batch in its usual form is a mixture of finely-divided solids which are thoroughly mixed and delivered to a refractory furnace by a system of hoppers, gravity flow chutes and other positive displacement conveyors. Since the batch is a finely-divided material, severe dusting conditions are commonly encountered when the batch is exposed to the high velocity hot gases of the melting furnace.

Additionally, the glass batch is extremely abrasive and will erode even the hardest of materials in a relatively short time where it frictionally contacts the moving parts of conventional positive displacement conveyance means, such as screw conveyors, augers or the like.

A.D. Heller; U.S. Patent 4,172,712; October 30, 1979; assigned to Dart Industries, Inc. describes a structural arrangement wherein the typical hopper of a glass furnace charger is enlarged to such an extent that it will appropriately accommodate a mixing arrangement. The principal mode of movement through the mixer, therefore, continues to be that which is common to the glass furnace charger, i.e., gravity.

Mixing action within the hopper is achieved by means of two rotating shafts having uniquely structured agitating means positioned therearound. These rotary agitators tend not only to mix the batch, but to move same toward the center of the hopper during the mixing action.

Furthermore, responsive to the active feeding action of the charger itself, suitable circuitry is designed to provide a uniform and constant flow of fluid to spray heads positioned above the hopper.

Accordingly, as new batch enters the inlet opening of the hopper, such is thoroughly wetted to a degree that will assure a desired wetness level as same exits the charger into the glass furnace.

Referring to the figures in which like numerals have been used to designate like components, Figure 1.1a shows the relationship between this apparatus and a typical glass melting furnace **10**.

**Figure 1.1: Glass Melting Furnace and Furnace Charger
with Mixing Apparatus**

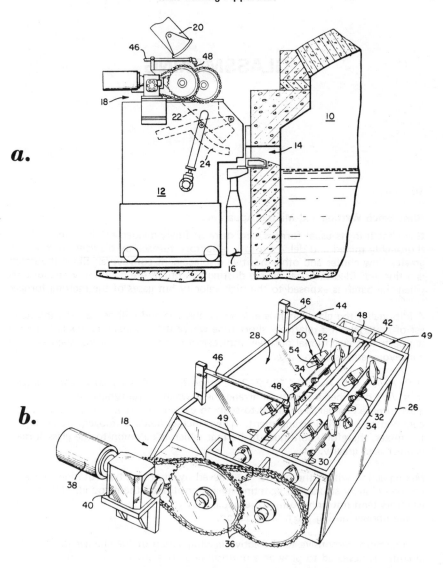

(a)   Side elevation view
(b)   Top view of mixing apparatus

Source: U.S. Patent 4,172,712

Adjunct to the melting furnace **10** is the glass furnace charger **12** which is movably mounted for easy positioning at the furnace opening **14**. Also provided is an air duct **16** that conducts cooling air to both the glass furnace opening and the furnace charger. Positioned atop the glass furnace charger **12** is the glass batch mixing apparatus **18** and positioned immediately adjacent such mixing apparatus is the batch delivery means **20**.

As is apparent, glass batch is delivered to the mixing apparatus **18** through the delivery means **20** and subsequent to being wetted and mixed, exits into the furnace charger hopper **22**. Thereafter, the batch is intermittently delivered to the furnace opening **14** by the pusher element or feeding means **24**. In effect, therefore, the mixing apparatus **18** becomes an extension of the charger hopper **22** and, as has been typical, glass batch that is delivered to the hopper moves therethrough basically as a result of gravity flow.

The glass batch mixing apparatus **18**, as can best be seen in Figure 1.1b, is composed of a receptacle **26** having a top inlet opening **28** and a bottom outlet opening **30** which will accommodate the gravity flow of glass batch through the receptacle **26**. Also, it should be noted that the wall construction of the receptacle **26** is designed such that the bottom outlet opening **30** is of approximately the same size as the inlet opening of the furnace charger hopper **22**. Accordingly, upon attachment of the mixing apparatus **18** to the furnace charger **12**, these respective openings **22, 30** are aligned and the mixing apparatus thereby effectively becomes an enlargement of the basic hopper construction **22** of furnace charger **12**.

The receptacle **26** has positioned therein a rotatably mounted mixing means **32** which is comprised of two main rotatable shafts **34**, each being driven through a suitable chain and sprocket assembly **36**, and gear reducer **40** by motor **38**. Also positioned within the receptacle **26** is a deflection baffle **42** which is adapted to redirect glass batch as it is delivered from delivery means **20** to a position above each of the mixing means **32**.

It is preferred that the sprocket assemblies **36** be of the type that incorporates a slip clutch arrangement. Such arrangement will minimize the possibility of damaging the various driving means **36, 38, 40** in the event of the mixer jamming or plugging.

Similarly, there is affixed to the receptacle **26** a fluid delivery system **44**, comprised of fluid conduits **46** and delivery means **48**, the latter being typical full cone fluid spray nozzles. These nozzles **48** are similarly positioned above the deflection baffle **42** and adjacent to the batch delivery means **20** so that as batch is delivered therefrom, it may be fully wetted in the preparation for mixing and passage through the receptacle **26**.

Full cone nozzles **48**, as opposed to hollow cone or flat spray nozzles, are preferred because such tend to provide the most uniform wetness to the surface of the batch upon its initially entering the mixer **18**.

The driving means **36, 38, 40** is adapted for constant operation so that the rotatably mounted mixing means **32** within the receptacle **26** is in constant motion, thereby agitating the glass batch within that receptacle even in the absence of batch movement through same.

The particulate materials in the noted receptacle accordingly are not afforded any opportunity to cake or otherwise solidify because of their wetted condition. The speed of rotation will, of course, be determined by the batch consistency, its speed of movement through the receptacle and the degree of wetness of the batch itself. Accordingly, such will be subject to experimentation and adjustment based upon operating conditions that may be encountered in any particular situation.

Each of the rotatably mounted mixing means **32** incorporates upon shaft **34**, agitating means **50** which is composed of radically protruding paddle members **52** affixed to pins **54**. Each of the shafts **34** is suitably bored to accept the pins **54**. These bore holes are positioned along and around the shafts in a symmetrical relationship that provides for a slight overlap of the area swept by paddle members **52** during shaft rotation and at approximately a 90° offset with respect to each adjacent bore hole. The agitating means **50** is affixed to the shafts **34** as is shown.

Melting Apparatus Using Gas-Free Materials

E.T. Strickland; U.S. Patent 4,138,238; February 6, 1979 describes a method of melting glass-forming materials comprising:

  (a) Establishing a bed of particulate substantially gas-free, glass-forming materials;

  (b) Urging the particulate materials into close proximity with a resistance heating member, the heating member having at least one outlet for molten glass;

  (c) Melting the glass-forming materials with heat transmitted from the resistance heating member to form molten glass;

  (d) Maintaining only a thin film of molten glass on the heating member;

  (e) Flowing the molten glass through the outlet; and

  (f) Collecting the molten glass in a heated reservoir having a gas space over the molten glass.

Unlike the common prior art glass melting processing, the raw materials for this process should be substantially gas-free. Most raw materials can be readily rendered substantially gas-free merely by preheating or calcining. Calcination temperatures are well-known to the art. The raw materials are substantially gas-free in order to avoid a rapid and unmanageable generation of gas during melting.

In the process, the charge moves toward the melter essentially as a unified body. Excessive gas formation at or near the melter can disrupt the integrity of the batch and can greatly reduce the efficiency of the melting process. Small amounts of gas, including the gas in the interstices between the particles of the charge, can be tolerated without disrupting the process. The resistance heater should achieve a temperature of at least about 2600°F, with temperatures of from about 2900° to 3100°F being particularly preferred.

In Figure 1.2a, the apparatus **10** includes a hopper **2** overlying and in communication with resistance heater **1**. The resistance heater is supported by ceramic supporting means **5** and overlies heated reservoir **6**.

Figure 1.2: Glass Melting Apparatus

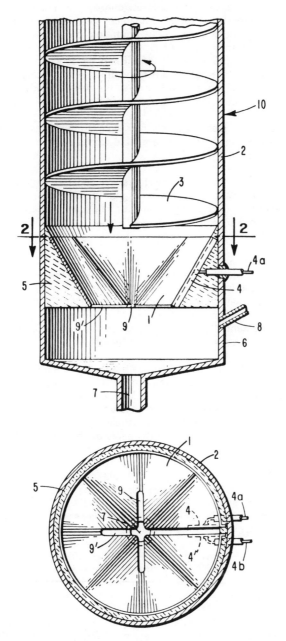

a.

b.

(a) Cross-sectional view
(b) Resistance heater cross section along line **2-2** of Figure 1.2a

Source: U.S. Patent 4,138,238

In order to maintain a close proximity between the particulate glass-forming materials and the resistance heater, a screw conveyor is schematically shown as **3**. It will be understood that other mechanical means for maintaining close contact can also be used including, without limitation, a plunger, piston or the like. If desired, the hopper may include heating means to preheat the outer surface of the charge, for example, to about 2000°F. If a screw conveyor is used and the sides of the hopper are heated, flutes or other arrangements may be used to avoid having the batch merely rotate with the screw conveyor. Such situation would not be encountered if a plunger or piston is used.

Since the hopper itself does not reach very high temperatures, a variety of materials, all well-known in the art, are available for its construction. Steel or stainless steel can be used, if desired.

The resistance heater **1**, as shown in Figures 1.2a and 1.2b, contains flanges **4** and **4'** for electrical power connections **4a** and **4b**. The space between the flanges is filled with electrical insulating material, such as a ceramic, or the flanges may be cooled with a cooling tube or the like so that the space between the flanges becomes filled with solid glass.

The resistance heater may be formed from a strip of metal which is rolled and then has one end crimped without drawing to provide the slots **9** for passage of molten glass as shown in Figure 1.2b. In order to avoid hot spots, the current flow should be maintained substantially uniform throughout the resistance heater. This may be accomplished, as shown in Figures 1.2a and 1.2b, by using a strip having a substantially uniform thickness and width (for example, 0.03" x 4" x 14") and forming the heater therefrom.

In the embodiment shown, the current path adjacent the outlet slots **9** and **9'** is essentially the same length as the current flow around the upper periphery of the resistance heater. If the length of the path of current flow varies along the resistance heater as encountered, for example, in a smooth cone, substantially constant resistance paths may, nevertheless, be maintained substantially uniform by appropriately varying the thickness of the resistance heater.

The material chosen for the resistance heater will vary depending upon the glass to be formed. For E glass, an alloy of 80% platinum and 20% rhodium is entirely satisfactory and is well-known in the art. The thickness of the resistance heater again will vary somewhat, depending upon the specific application, but thicknesses ranging from about 0.005" to 0.05" generally will be used and thicknesses ranging from about 0.01" to 0.03" are preferably used.

Since the resistance heater will frequently be operated near its melt temperature and must withstand the force of the glass-forming materials being urged toward it, the heater, as shown, is supported by a ceramic support member **5**. Other support means may also be used.

A variety of high temperature ceramic materials are known in the art and may be used. Zirconia casting cement, for example, has been successfully used. Casting is particularly useful since an intimate fit results between the heater and its support.

The heated reservoir or atrium **6** is formed of a metal which is suitable for use with molten glass. Desirably, it will have a sloped bottom and be fully supported since vertical walls at the temperatures prevalent in the reservoir may collapse under their weight. Once again, the common 80:20 platinum:rhodium alloy may be used.

The atrium itself may be used as a resistance heater or separate heating means may be located within or around the atrium. Since the glass entering the atrium is already molten and very hot, the atrium heating means need supply only a very small amount of maintenance heat.

Since the atrium is used, inter alia, to permit gas bubbles entrained in the molten glass from the resistance heater to escape, it is provided with a gas outlet schematically shown as **8** in Figure 1.2a.

The outlet is desirably heated to avoid condensation of off-vapors. In addition, a liquid outlet schematically shown as **7** is provided from the atrium. The glass outlet means may, of course, be heated and provided with flow control means if desired.

In operation of the apparatus of Figure 1.2a, particulate gas-free, glass-forming materials in hopper **2** are urged toward resistance heater **1** by screw conveyor **3**. The resistance heater **1** is maintained at a temperature of at least 2600°F by electrical power connected to flanges **4** and **4'**. The glass-forming materials which come into close proximity with resistance heater **1** become molten and a thin film of molten material flows downwardly along resistance heater **1** through slots **9** and **9'** into atrium **6**.

Atrium **6** is heated by a heating means (not shown) to maintain a temperature of at least about 2600°F. The atrium is partially filled with molten glass which, during operation of the apparatus, will have an overlying thin layer of foam. As the foam breaks up, the gas exits from reservoir **6** through gas exit **8**. The residence time in the reservoir, generally from about 10 to 60 minutes, can be achieved by proper selection of reservoir size according to principles well-known in the art. Molten glass is removed through glass outlet **7**. The apparatus can be operated either on a continuous basis or a batch basis.

Energy-Efficient Fuel-Fired Glass Furnace

T.D. Erickson and C.M. Hohman; U.S. Patent 4,184,861; January 22, 1980; assigned to Owens-Corning Fiberglas Corporation describe a method of recycling the heat generated during glass manufacture by indirectly extracting heat from the flue gases in the chamber during the heating of the agglomerated glass batch. In this way, dusting is not increased, the amount of unused energy emitted to the atmosphere in the flue gases is significantly minimized and the heat transfer medium used in the indirect extraction of heat is in a condition conducive to the recovery of energy or heat therefrom.

Because this heat transfer medium is heated and the flue gases cooled, through an indirect heat exchange process, that medium does not contain any undesirable pollutants or contaminants and is ideally suited for the beneficial recovery and utilization of its energy.

Referring to Figure 1.3, it will be seen that glass-forming batch materials and
water are converted into individual agglomerates, preferably pellets, on a rotating
disc pelletizer. The free water content of the pellets may be about 10 to 20%
by weight and, while not shown, the pellets preferably are subjected to a screening operation to select pellets of a nominal size of about 3/8" to 5/8" diameter.

Figure 1.3: Energy-Efficient Glass Furnace

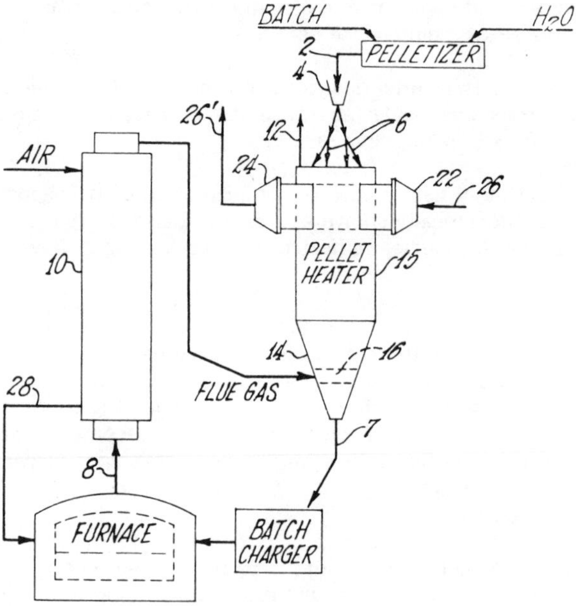

Source: U.S. Patent 4,184,861

These pellets are then transported by suitable means **2**, such as a belt conveyor,
to a feed hopper **4** and then, in turn, the pellets, through a spider-like feeding
arrangement **6**, are fed to a pellet heater which maintains a bed of pellets (not
shown) therein.

Pellets generally move downwardly in the bed of the pellet heater and are discharged therefrom as hot, individual pellets and supplied by a duct member **7**
to a batch charger which conveys them to a fossil-fuel-fired glass melting furnace.

The combustion gases, or flue gases, from the melting furnace are conveyed by
suitable means **8**, for example a duct, to a recuperator **10** where they are indirectly cooled with air, for example, from a temperature of about 2600°F to a
temperature on the order of about 1400° to 1500°F. The heated air **28** is then
supplied to the furnace as combustion makeup air.

The cooled flue gases are then conveyed by suitable duct means to the pellet
heater where they flow in direct contact with the pellets, in countercurrent flow
fashion, to dry the pellets and preheat them.

The flue gases leave the pellet heater by a suitable outlet, generally designated **12.** Preferably, the flue gases will be supplied to the pellet heater by a manifold type arrangement with entrances into the heater being on diametrically-opposed sides of a lower frustoconical portion **14.**

In accordance with sound engineering practices, the gases will be distributed generally uniformly across the heater as by using an inverted V-shaped member **16,** which spans frustoconical portion **14.**

The heat exchanger contemplated herein is positioned in the pellet bed of the cylindrical portion **15** of the pellet heater. As generally illustrated in Figure 1.3, the heat exchanger comprises an inlet manifold **22** to which is supplied a suitable heat transfer medium via a duct **26** and disposed on the opposite side, externally of the pellet heater, is an outlet manifold **24** from which the heated heat transfer medium is removed via a duct **26'.**

This duct can convey the heat transfer medium to any location for the beneficial use of its energy. In sealed, fluid communication with manifolds **22** and **24,** in accordance with a preferred embodiment of this process, is the heat exchanger in the form of a plurality of hollow, generally rectilinear duct members which are located in the pellet bed. Desirably, the system will be operated such that the uppermost level of pellets in the heater will be disposed upwardly of these ducts and, generally, the ducts will be located in the upper half of the pellet bed.

The above arrangement is ideally suited for manufacturing a wide variety of glasses, but is especially well adapted for the manufacture of fiberizable textile glasses. Typically, these glasses are low-alkali metal oxide-containing glasses, for example, glasses containing, if at all, less than 3% by weight of alkali metal oxides and, more typically, less than 1% by weight.

Exemplary of such glasses are the alkaline earth aluminosilicates where, for example, the cumulative amount of the alkaline earth oxides, plus alumina, plus silica, is in excess of about 80% by weight and quite commonly between in excess of about 90% up to, in some instances, virtually 100% by weight.

Energy-Saving Electric Glass Melting Furnace

H. Pieper; U.S. Patent 4,184,863; January 22, 1980; assigned to Sorg GmbH & Co. KG, Germany, describes a method for melting glass comprising:

(a) Feeding batch material into a melting zone to form molten glass;

(b) Passing the molten glass from the melting zone to a horizontally contiguous refining zone without a reduction in cross section and then vertically downwardly therein in a nonturbulent stream;

(c) Heating the stream in the refining zone by passing electrical energy therethrough until final gas removal is obtained;

(d) Thereafter passing the refined glass to a homogenizing zone below the refining zone in a nonturbulent stream without passing electrical energy therethrough; and

(e) Withdrawing the homogenized glass from a lower portion of the homogenizing zone.

The glass melting furnace, according to the process, includes a shallow melting section **1** and an adjacently situated, substantially deeper, hexagonal or rectangular refining section **2**, the sections being interconnected such that molten glass flows from section **1** into section **2** without a reduction in cross section as is clearly shown in Figure 1.4.

Figure 1.4: Glass Melting Furnace Having Hexagonal Refining Section

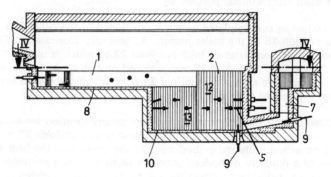

Source: U.S. Patent 4,184,863

In this case, the depth of the tank or molten glass bath in section **2** is about two or three times greater than that in section **1**, i.e., the bottom **10** of the refining section **2** is arranged two to three times deeper than the bottom **8** of the melting section.

From the refining section **2**, an outlet leads to a riser **7**, wherein energy is supplied to the stream of the discharged glass in conventional manner by auxiliary electrodes **9** such that the temperature of this stream is maintained or prevented from decreasing excessively.

The refining section **2** comprises an upper refining zone **12** and a lower refining zone **13** with a plurality of electrodes **5** being positioned within such refining zone **12**. These electrodes **5** may be mounted in the sides, both of the hexagonal and of the rectangular refining section **2**, and according to a modified embodiment, but also as shown in Figure 1.4, the electrodes may be positioned in two or more levels.

The vertical position of the electrodes is selected such that these electrodes are in a plane below the bottom **8** of the melting section **1** or approximately at the level of bottom **8**, respectively.

The total surface of the molten glass bath may be covered by a blanket of batch material; however, it is also feasible to supply batch material to the surface of the melting section **1** only.

In most instances, the melting section **1** has a greater surface area than the refining section and two-thirds of the total surface area may be constituted by the melting section, while the refining section comprises one-third of such surface area.

Depending on the type of glass to be molten and on the specific efficiency of the furnace, these ratios may be varied as well; for example, the ratio of the surface areas of melting and refining sections also may be equal to 3:1 or 1:1, respectively.

The method is carried out as follows. The batch material supplied onto the total surface area of the molten glass bath is contacted from above with the combustion gases of the burners such that it is heated and molten both at the surface and at the interface between the batch material and the molten glass bath.

The molten, unrefined glass flows from the melting section 1 into the refining section 2 where it is mixed with the glass molten in the latter section. Thereupon, the glass flows uniformly downwards through the refining zone 12, including the electrodes 5, whereby the glass is further heated and refined to descend into the homogenizing zone 13 disposed therebelow, so as to become homogenized therein.

In zone 13, the temperature of the glass is slightly reduced, while the flow is without turbulence. A descent of unmolten batch material particles into zone 13 is positively prevented from occurring by the refining zone 12, including the electrodes 5. The thus-refined glass flows to a transferring glass channel or passage 7 which may be formed as a riser.

Alternatively, the molten glass batch in the region of the refining section 2 may be left uncovered by the batch material also. Although the specific efficiency of the furnace is reduced by this measure, it is thereby rendered possible to melt glass of a particularly high quality or to produce special types of glass which are difficult to melt.

In general, however, this measure will be unnecessary since the arrangement of the melting zone (of the electrodes in one level of the refining zone) and of the homogenizing zone, one below the other, provides for a sufficiently satisfactory refining action.

Obviously, this glass melting furnace, depending on the cost of the various types of energy, may be operated in even more economical fashion than fully electrically operated glass melting furnaces.

Addition of Barium Carbonate in Electric Furnace Process

The practice of producing and melting glass in electric furnaces is well-known. In this type of glass furnace, which may be either rectangular or hexagonal in shape, electrodes are present in either the bottom or the sides of the furnace and these electrodes, when electrolyzed, melt glass batch which is present between the electrodes to form glass.

In normal operation, a uniform, coherent layer of glass batch, normally termed a "batch blanket", is placed over the melted glass to serve both as a feed for the glass furnace and also to insulate the furnace against loss of heat from the molten glass present in the bottom of the furnace.

Despite the advantages of electric glass melting furnaces, the operation of these furnaces has encountered some serious problems. One major problem that occurs is that the batch blanket becomes stiff and rigid and does not permit the free passage and escape of gases through such blanket.

The gases are generated during the fusion and reaction of the ingredients of the glass batch and must be removed from the furnace. When this blanket stiffening occurs, numerous hot spots appear on the surface of the batch blanket in the form of large bubbles with heavy foaming and violent eruptions. These violent and clearly visible breaches in the batch blanket result in rapid drops in furnace temperatures, the exposure of molten glass surface and loss of large amounts of heat which disrupt the integrity of the batch blanket and which interfere seriously with any chance for an efficient, completely controlled operation of the electric furnace.

W.H. Manring; U.S. Patent 4,135,905; January 23, 1979; assigned to FMC Corporation has found that operation of electric glass furnaces can be improved so as to maintain a controlled and thermally-insulating batch blanket with a minimum of hot spots by adding from 0.1 to 7.0% by weight, and preferably 0.5 to 1.5% by weight, expressed as BaO, of barium carbonate in the glass batch.

The barium carbonate is simply mixed with the glass batch in the appropriate proportions desired and the glass batch is then fed uniformly to the top of the furnace so as to maintain a blanket of glass batch over the surface of the electric furnace. In general, the added glass batch, which is uppermost in the batch blanket, should remain cool, indicating that little of the furnace heat is being evolved through the top of the blanket.

The general method of feeding such glass furnaces is by the so-called distributing chargers, which use a long arm with a moving belt mounted thereon, to distribute the charge uniformly over the surface of the blanket on a continual basis.

The location of the batch spill end of the moving belt can be varied by extending or contracting the support arm on which the belt rests and, in this way, a uniform charge can be distributed equally and uniformly over the entire surface of the glass furnace.

Differential Extraction of Heat

G.A. Dickinson; U.S. Patent 4,200,448; April 29, 1980; assigned to Pilkington Brothers Limited, England, describes a method of producing molten glass comprising feeding glass-forming materials into an inlet end of a glass melting tank, melting the material in a melting zone of the tank adjacent the inlet end of the tank, refining the molten material at a position downstream of the melting zone and conditioning the molten glass adjacent an outlet end of the tank so that the molten glass is ready for use in a forming process.

The method further comprises applying heat to the contents of the tank so as to cause forward and return flows of molten glass, stirring the molten glass which is flowing in a substantially forward direction by a plurality of stirrers spaced laterally apart across the tank, extracting heat from the forward flowing glass through the stirrers and regulating the amount of heat extracted by the stirrers so as to achieve a differential extraction of heat from the glass which flows along a return path towards the inlet end after stirring, and that glass which continues along a forward path to the outlet end of the tank.

In normal operation, it is desirable to regulate the heat extraction by the stirrers so that less heat is extracted from the glass which flows along a return path towards the inlet end after stirring than that which is extracted from the glass which continues along a forward path to the outlet end of the tank.

Preferably, stirring is effected on the glass passing from the refining zone to the conditioning zone. In this way, the stirring is effected at a position downstream of the hot spot. The stirring may be effected in the refining zone and/or immediately adjacent the inlet of the conditioning zone.

Preferably, the differential heat extraction is such that there is a variation in heat extraction across the width of the tank. Commonly, the outlet of the tank is arranged in line with a central axis along the length of the tank and in such a case, it is preferable that the differential heat extraction is arranged to extract heat from the central region of the forward flowing glass than from the glass adjacent the edges of the tank.

It has been found that by stirring the glass and cooling with a differential heat extraction so that the minimum of cooling occurs on that glass which subsequently returns in a return flow towards the melting zone, it is possible to improve the homogeneity of the glass, not only because the glass passing the stirrers is attenuated, causing thinning of the layers of glass of differing composition and thereby increasing the diffusion between layers, but also by reducing the temperature within the body of the molten glass and reducing the need for the high surface cooling which can give rise to inversion flow in the conditioning zone.

That glass which returns towards the melting zone does so in a more homogeneous form and is made even more homogeneous on its eventual return to the conditioning zone. Furthermore, the stirrers in this particular instance do not cool the glass to the same extent as those placed in the glass flowing directly to the forming process and, consequently, more heat is preserved in those areas where it is mostly required, namely upstream of the conditioning zone.

Reduction of Sulfur Emissions

W.H. Turner; U.S. Patent 4,138,235; February 6, 1979; assigned to PPG Industries, Inc. describes a method of operating a continuous flat glass melting furnace so as to reduce sulfur-containing emissions. This is achieved without sacrificing glass quality and without the use of substitute fining agents, and yet permits the use of high batch-to-cullet ratios while operating at normal furnace temperatures and throughputs.

It has been discovered that a relatively small, specifically defined amount of salt cake used in the batch results in reduced emissions without sacrificing glass quality. More specifically, it has been found that the sulfur content of the batch (expressed as SO_3), which includes both the salt cake in the batch as well as SO_3 in the cullet, should be less than about 2.25 (preferably less than 2.0) times the amount of SO_3 retained in the final glass product. In other words, for a given tonnage of throughput, an optimum amount of salt cake has been found which is independent of the batch-to-cullet ratio. This is contrary to the prior art belief that inclusion of salt cake in an amount proportional to the sand in the batch was necessary.

It has been found that this prior art practice of maintaining a fixed amount of salt cake for a given amount of sand in the batch can lead to excessive sulfur-containing emissions and to the production of excess foam on the surface of the melt and associated defects, particularly at high batch-to-cullet ratios, such as 70 pbw batch to 30 pbw cullet or higher. However, by utilizing a batch with a total SO_3 content of no more than about 2.25 times that of the outgoing glass stream, it has been found that batch-to-cullet ratios higher than 70:30 may be used with reduced sulfurous emissions, while at the same time maintaining the usual low defect densities of flat glass.

It has additionally been discovered that even greater improvements can be obtained if the conventional inclusion of coal in the glass batch is reduced or essentially eliminated. Coal has long been considered an essential ingredient in the glass batch for the purpose of aiding melting.

It was believed that the coal serves to break down salt cake into sodium oxide and sulfur dioxide and that the sodium then serves to dissolve sand grains. It has been discovered, however, that such a reaction results in premature volatilization of the sulfur dioxide, with the result that less of the salt cake is available for fining at the subsequent fining stage of the continuous melting process and that excess foam is produced on the surface of the melt, particularly at high batch-to-cullet ratios.

By eliminating coal, the salt cake is not so rapidly broken down, so that a greater portion of the salt cake may be retained in the glass melt so as to act as a fining agent in the fining zone of the melting furnace. While the fining ability of a given amount of salt cake is thus enhanced, it has been found quite surprisingly that no difficulty in melting sand grains and producing homogeneity is caused thereby in a flat glass melting operation.

As a result, considerably less salt cake is needed to produce glass of a given quality and the amount of sulfurous emissions is reduced. Additionally, even though less salt cake is used, it has been found that fining is improved and that the final glass composition includes slightly more SO_3 than with previous practice. Also, the amount of foam in the furnace is reduced.

Glass-Contacting Member of Platinum-Coated Refractory

M.G. Chrisman; U.S. Patent 4,192,667; March 11, 1980; assigned to Owens-Corning Fiberglas Corporation describes a process for forming a thin, fused layer consisting essentially of at least one platinum group metal directly and strongly adhered on the surface of a refractory substrate which comprises depositing a solid, small particle size form of the metal in a thin layer on the refractory substrate and heating the layer at a temperature and for a time sufficient to fuse the deposited layer into a thin, fused, uniform layer.

The solid small particle size form of the metal is produced by a process which comprises substantially quantitatively reducing a complex of the metal in an aqueous solution and precipitating the elemental form of the metal therefrom as a solid small particle size form of the metal and then drying the precipitated metal under conditions of temperature and time so as to avoid sintering the particles. Suitably, the complex will be a halide, for example, chloride, or nitrite complex.

Quite outstanding results will be obtained when the metal is an alloy of platinum and rhodium where the platinum desirably will be in a major (greater than 50% by weight) amount and the rhodium will be present in a minor amount. Outstanding results will be obtained by using an alloy consisting essentially of about 70 to 80 wt % platinum and about 20 to 30% rhodium.

Desirably, the reducing and precipitating step will comprise adding hydrazine or a hydrazine hydrate to an aqueous solution of nitrite or halide complexes of platinum and rhodium and precipitating platinum and rhodium under alkaline conditions.

According to a further feature of this process, prior to heating the deposited layer to fusion, a sheet consisting essentially of a platinum group metal or an alloy of platinum group metals is positioned on the particulate mass and the heating is effected so as to bond the sheet to the refractory. Such a structure may be used as a glass-contacting member in conventional glass handling apparatus.

The solid, small particle size form of the platinum group metal or alloy thereof may be deposited onto the refractory substrate in any conventional manner. For example, the material may be applied as a dry powder, as for example, by brushing a layer onto the refractory substrate or using a doctor blade technique. If desired, these particles may be slurried in a suitable volatile carrier, for example, alkanols, ketones, esters, ethers, or the mono- and dialkyl ethers of ethylene glycol and their derivatives, in which case the slurry may, for example, be applied by flow coating, spraying and the like.

Generally, the small particle form material will be applied in a thickness sufficient that upon fusing the resulting thickness will be in the range of about 0.0005" to 0.125" and preferably, at least for bonding, about 0.005" to 0.001".

In the embodiment where the solid small particle size material is to be used to bond a sheet of a platinum group metal to the refractory or a sheet of an alloy of a platinum group metal, such sheet will generally be on the order of at least about 0.003" in thickness.

Various heating cycles may be used to effect the fusion and/or bonding, but generally the intensities of such cycles are, surprisingly, quite low. For example, the fusion and/or bonding may be effected without any significant pressure, i.e., at atmospheric pressure, at temperatures in the range of about 2000° or 2200° to 2700°F. The time for heating, of course, will vary with the temperature, but even at temperatures between 2200° and 2400°F, times on the order of about one-half to one hour will be quite satisfactory.

No particular care or any special equipment is required for the fusing and/or bonding step, but, of course, in those instances where a volatile carrier is used, that carrier will have to first be evaporated with suitable precautions taken depending on the nature of the material used.

Molten Glass Homogenizer

H.L. Penberthy; U.S. Patent 4,195,981; April 1, 1980 describes an improved arrangement for physically stirring glass to reduce or eliminate the cordiness of molten glass and the problems it causes, no matter what its source.

The stirrer is located in the forehearth where the glass is cooler and, hence, less aggressive in attack on the refractories. It can be installed with only a few hours of shut-down and impedes the flow of glass by only a very small and perfectly acceptable amount.

In the process, a blade is provided of refractory material and is vertically oriented and aligned substantially parallel to the direction of flow of the glass. In this way, its damming effect against the flow of glass is small. While the blade preferably is a flat plate, the blade could be of rounded configuration as a segment of a cylinder or a shallow V-shape.

The blade is reciprocated across the width of the forehearth to accomplish mixing. It traverses close to each side wall in turn to mash away the glass, which is so to speak, held there normally by viscous drag. That glass moves along slowly as is well-known in canal theory.

Figure 1.5 shows a modified drive linkage construction, particularly designed to give a clawing action to the blade **22** for improved mixing. In this arrangement, a swing arm **50** is hung from a shaft **52** held in pillow blocks **54**, attached to a rigid frame **56**. Blade shaft **24** is held at its outer end in a clamp **58**. The assembly is reciprocated by a double-acting air cylinder **60** pivoted to a rigid frame **62** and to the arm **50** as indicated at **64**.

Figure 1.5: Drive Linkage Constructed for Glass Batch Stirrer

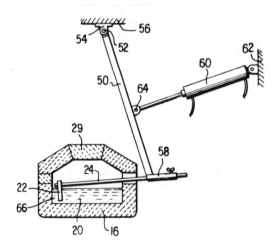

Source: U.S. Patent 4,195,981

In the arrangement illustrated in Figure 1.5, the blade **22** travels in an arc determined by the height of the shaft **52** above the floor of glass channel **20**. When this distance is eight feet for a forehearth twenty-six inches wide, the bottom of blade **22** rises at both ends of the stroke about one-half inch above its level at the lowest point, which is when the blade is in the middle of its stroke directly under the shaft **52**.

This rise is beneficial in allowing glass **66**, trapped between the blade **22** and the side wall of the forehearth, to escape under the blade. This accomplishes some extra mixing when part of the glass **66** is caught on the return stroke and carried toward the center of the forehearth.

Inclusion Melting with Radioactive Components

H. Piper; U.S. Patent 4,139,360; February 13, 1979; assigned to Sorg GmbH & Co. KG, Germany, describes a method of inclusion melting or remelting glass with radioactive components, especially from the reprocessing of nuclear fuel (burner) elements, and a glass melting furnace for carrying out such a method.

The radioactive components are introduced into the furnace from above in the form of an aqueous suspension or of a slurry in combination with a mixture suitable for the melting or melt-forming of glass; the molten mass comprising glass and radioactive components (melt) is heated by passing electrical current directly therethrough. The flow of the mass takes place from above to below in the vertical direction only.

Advantageously, the power supply to the molten mass may be effected through electrodes in contact with the molten mass and in order to increase the operational life of the electrodes and to substantially reduce corrosion thereof, a slight convection current may be present in the upper portion of the tank adjacent the electrodes. Measures may be taken so that in the upper portion of the central part of the furnace a higher concentration of energy is present than adjacent the edges and the electrodes.

Owing to this low energy concentration, corrosion of the electrodes is reduced by the lesser amount of heat produced at the electrodes, while the energy concentration in the central part of the furnace provides the high temperatures required for the melting of the mixture in this region without allowing a caking of material to the furnace walls or increased corrosion of the walls of the furnace. The glass melting furnace for carrying out the above method is made of ceramic materials.

Furthermore and preferably, the furnace is provided with electrodes made of tin-oxide or molybdenum, and the furnace has, in its upper portion below the deposit layer of the mixture, a greater cross-sectional area than in at least one portion therebelow in which the glass has a higher temperature.

In order to prevent stronger convection currents from occurring in the region of the electrodes, the furnace has a greater cross-sectional area in the region adjacent the upper electrodes, and these electrodes are preferably positioned within niches or recesses and/or above projections, whereby the formation of convection currents passing through the furnace is prevented.

The greater cross-sectional area in the upper portion of the furnace is advantageous when the furnace is operated in a discontinuous fashion whereby the batches are thusly supplied discontinuously. The level of the molten material is thereby not raised to any substantial degree even by greater batches of the mixture, and the quantity supplied may be measured by placing the furnace upon a plurality of pressure gauges, the load conditions of which provide distinct information with respect to the loading of the furnace.

The discontinuous mode of operation of the furnace is advantageously promoted by the fact that the glass outlet or discharge spout includes an electrode in the form of sealing brick adapted to be pressed thereagainst from the outside, which electrode is electrically connected to another electrode positioned interiorly of the furnace, whereby the current flowing from the sealing brick electrode to the inner electrode acts to positively keep open and unobstructed the outlet or discharge passage, even when no material flows therethrough.

Prevention of freezing during pauses of operation is augmented by radiator heater elements arranged exteriorly in front of the outlet brick, the latter being adapted to be replaced if wear is observed which would affect the operation.

In order to provide for sufficient heating of the mixture, including the quantity of aqueous suspension or of aqueous slurry within the upper part of the furnace, this portion of the furnace may, additionally, include upper furnace heating means and shield or cover means to protect the heating means against the entry of steam which would destroy the heater elements.

GLASS MANUFACTURE

Electric Glass Sheet Manufacturing Process

F. Anderson; U.S. Patent 4,162,907; July 31, 1979 describes an all-electric glass manufacturing process and apparatus having a vertical air-cooled electric furnace and a transverse air-cooled refiner section.

The furnace and the refiner are provided with a plurality of molybdenum electrode cartridges housed in lock-in cartridge casings designed to facilitate electrode removal. Molten glass is removed from the furnace refiner section by means of a plurality of basin cylinders symmetrically disposed within the transverse refiner. The basin cylinders are provided with a plurality of open faces which receive molten glass from the furnace refiner and which, upon rotation of the basin cylinder, deliver a prescribed amount of molten glass to an extruder mechanism.

The extruder mechanism accepts molten glass from the basin cylinder and applies the molten glass to a pair of extrusion rolls. The extrusion rolls are eccentrically pivoted off-center such that the separation distance between them can be varied in accordance with the desired thickness of the sheet glass being formed.

The entire extruding mechanism, including the extruder rolls, is mobile such that as molten glass is extruded through the extruder rolls, it is deposited on a molten tin bath which imparts an ideal smoothness to the surface of the newly-formed glass sheet.

A cutting frame is then lowered over the tin bath such that the molten glass sheet is held firmly between a flange on the cutting frame and a flange on the tin bath. The molten glass sheet is, thereafter, cut into lite sizes by a cutting mechanism having a plurality of discrete glass cutting block assemblies, each of which is independently adjustable to a resolution of $1/32$". These cutting block assemblies are housed in two cutting carriages which alternately sweep the length and width of the newly-formed glass sheet while the sheet is stationary on the tin bath.

After cutting, a transfer unit having a plurality of suction orifices connected to a reversible vacuum source is lowered over the glass sheet. The transfer unit lifts the newly-cut glass sheet off the tin bath and transfers the glass sheet either to an annealing stage or to a tempering stage. Notably, the transfer unit is also used for tempering, with the reversible vacuum source applying cold air to the glass sheet.

After temperature treatment, the glass sheet cut to lite sizes undergoes electronic inspection and packaging. The entire process is under the control of a minimum of operator personnel located in a central control center.

Production of Glass in a Rotary Furnace

K. Kiyonaga; U.S. Patent 4,185,984; January 29, 1980; assigned to Union Carbide Corporation describes an improvement over prior art processes for producing glass in a rotary furnace by increasing thermal efficiency without melting all or part of the batch mixture prior to its introduction into the rotary furnace and without preferential carry-over of fines.

The process for producing molten glass in a generally cylindrical, continuously-rotating chamber comprises the following steps:

(a) Feeding inorganic raw materials, including silica in major proportion, into the chamber;

(b) Providing a flame of high intensity heat produced by the combustion of fuel with a gas containing about 50 to 100% by volume oxygen and directing the flame into the chamber in such a manner that the raw materials are melted;

(c) Rotating the chamber at a sufficient speed and cooling the exterior of the chamber with a liquid coolant in such a manner that the inner surface of the chamber is coated with a layer of molten glass; the layer is solidified and a solidified layer of glass is maintained throughout the process whereby the solidified layer essentially prevents impurities from the inner surface of the chamber from entering the melt; and

(d) Withdrawing molten glass.

The improvement comprises heating the silica prior to step (a) to a temperature in the range of about 500° to 2500°F.

Manufacture of Flat Glass by the Float Process

The float process is typically characterized by the delivery of a sheet of molten glass onto the surface of a bath of molten metal such as tin, where the glass is normally delivered from the broad surface of a continuous glass tank over a refractory lip and onto the central one-third or one-half of the width of the molten bath. The ribbon of molten glass flows outwardly upon the molten bath until the force tending to cause the spreading, represented by the thickness and density of the glass, and the force resisting the spreading, represented by the surface tension and the radius of curvature of the glass edge, have reached an equilibrium.

The theoretical thickness of the glass at such equilibrium is, therefore, about 0.28", which is much thicker than required for normal architectural window glass and automobile windows. Accordingly, in order to produce the required thinner sheet, edge portions of the expanded sheet flowing outwardly on the molten bath are engaged by knurled rollers and the velocity of the sheet downstream from such rollers is increased to draw down and produce thinner sheets as shown in U.S. Patent 3,853,523.

Unless the glass which is supplied to the molten bath has been well stirred so as to be completely homogeneous and free of cord, surface streaks or ridges may appear in the drawn glass as a result of the attenuation of nonhomogeneous glass containing cord or striae.

G.C. Shay; U.S. Patent 4,203,750; May 20, 1980; assigned to Corning Glass Works describes a process which relates to the delivery of a sheet of homogeneous molten glass onto the surface of a bath of molten metal. Molten glass is delivered through a conventional narrow forehearth where it is well stirred, and homogeneous glass is delivered therefrom to the inlet end of a sheet-forming overflow trough device. The trough is positioned above and extends laterally across an inlet end of a molten metal bath for receiving a sheet of molten glass to be drawn therealong in the production of float glass.

A sheet-forming overflow trough device which may be utilized for delivering the molten sheet to the surface of the metal bath is shown in U.S. Patent 3,338,696, wherein molten glass, which is delivered to the inlet end of an overflow channel, wells up and evenly overflows both sides in molten sheet-like paths which converge at the bottom of the forming member into a molten sheet.

In view of the fact that the molten glass delivered to the overflow channel is homogeneous, having been thoroughly stirred immediately prior to delivery of the overflow channel, and further in view of the fact that the only free surface which the glass is exposed to is, in fact, the outer surface of the drawn sheet, it is possible to deliver a sheet of molten glass to the molten metal bath of the float process with improved optical qualities through the virtual elimination of cord and striae.

Loading Containers into an Annealing Lehr

E.H. Mumford; U.S. Patent 4,193,784; March 18, 1980; assigned to Owens-Illinois, Inc. describes a method of loading containers into an annealing lehr wherein a plural section bottle-forming machine supplies ware to a machine conveyor and the ware moves from the machine conveyor by way of a ware transfer to a cross-conveyor from which the ware is pushed onto the lehr in which the improved method comprises changing the normal interval between sweep-out operations in relation to the speed of the machine conveyor and increasing the speed of the ware transfer and maintaining normal speed at the cross conveyor such that groupings of ware on the cross conveyor will be serially arranged in relation to the cycle of the forming machine and separated by a predetermined gap.

Referring to Figure 1.6, an 8-section, double-gob, glass-forming machine generally designated **10** is schematically shown as a series of rectangular boxes **11** which represent individual sections of the glass-forming machine.

Figure 1.6: Glassware-Forming System

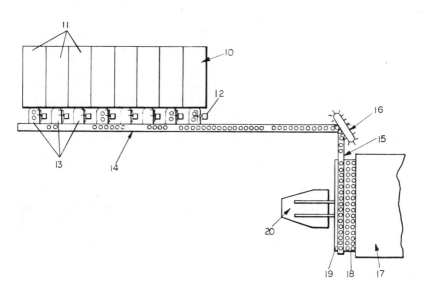

Source: U.S. Patent 4,193,784

Each of the boxes **11** will have a sweep-out mechanism **12** associated therewith. The sweep-out mechanisms **12** are of conventional design and may be similar to those shown, for example, in U.S. Patent 3,812,690. Each sweep-out mechanism **12** is associated with a machine section dead plate **13**. The dead plate, while being attached or mounted to the side of a machine conveyor **14**, generally is thought of as being part of the forming machine section. The dead plate serves as a cooling site for the ware after it has been taken from the blow molds.

The ware is placed upon the dead plate preparatory to being swept off onto the surface of the machine conveyor which will be moving to the right, as viewed in the figure. Each of the eight sweep-out mechanisms will be mechanically operated through 90°, or slightly more, by an individual cam carried by a common shaft which extends the full length of the machine **10**.

The machine conveyor **14** will carry the ware away from the forming machine to a cross conveyor **15** which may be positioned, for example, at right angles with respect to the machine conveyor. At this point of juncture of the two conveyors, a ware-transfer device **16** is provided for transferring the ware from the machine conveyor to the cross conveyor while maintaining the same spacing of the ware.

The cross conveyor may be similar to that shown in U.S. Patents 4,067,434 and 2,547,791. In both patents, the cross conveyors are positioned to travel across the width of a lehr belt or mat and in front of a pusher mechanism or lehr loader, whose function is to push the ware from the cross conveyor onto the lehr mat.

One of the major problems confronting manufacturers of glass containers is the inability to smoothly move the ware from the cross conveyor onto the lehr mat by a pusher mechanism that will not interfere with the leading article on the cross conveyor which will form the next row in the lehr.

With increased production, speeds of the cross conveyors and of the machine conveyor have necessarily increased. In the process there is illustrated a lehr **17** having a moving lehr mat **18**. A pusher bar **19**, operated by a lehr loader mechanism **20**, is positioned in alignment with the lehr mat and, as illustrated, is co-extensive with the width of the lehr mat.

When using a push bar loader to transfer bottles from the cross conveyor to the lehr mat, the loader bar **19** must pass through the row of bottles with enough speed to clear the oncoming bottles on the cross conveyor and provide a clean cut-off.

Previously, a side shift mechanism was developed which caused the push bar to move with the flow of bottles on the cross-conveyor during the push-in stroke. This proved to be an adequate solution to the problem at production speeds which were then in use. Since that time, speeds have nearly doubled and shifting the bar at these speeds has become impractical.

The method overcomes this clearance or cut-off problem by putting the bottles in groups on the cross conveyor such that there is a gap between the bottles on the cross conveyor which will provide for the push bar clearance. This gap is produced by proper selection of the machine conveyor speed and the setting of the sweep-out cams. As a specific example, an IS 8-section, double gob machine produces sixteen bottles in one machine cycle.

It is normal to have a conveyor speed which results in evenly-spaced bottles with the sweep-out cams set so that a pair of bottles is swept out for each 45° of cam rotation. If the cams are set a few degrees less than 45°, the bottles will be equally spaced in groups of eight, then a gap, then another group of eight, a gap, etc. By proper selection of the cam setting angles, the effect on the machine conveyor could be the same as having all bottles equally spaced, with one bottle missing, then a group of eight, one missing, a group of eight, etc.

At the point where the bottles transfer from the machine conveyor to the lehr conveyor, the transfer device **16** is geared to operate at a speed which will be seventeen paddles or seventeen transfer fingers per machine cycle instead of sixteen as is normal. One paddle will be empty for each gap in the line of bottles on the machine conveyor, thereby leaving a gap in the row of bottles on the cross conveyor which will be the space for cut-off clearance.

Gob Weighing System

J.A. Kwiatkowski and C.L. Wood; U.S. Patent 4,165,975; August 28, 1979; assigned to Ball Corporation describe a system for determining the mass of a moving molten glass gob. The gob is directed along a curved channel having a force-measuring device over which the gob passes for measuring the force exerted by the gob normal to its path.

The velocity of the gob is determined both before and after the normal force is measured by means of two sets of two-position sensors, one set located on each side of the force-measuring device, and a timer which determines the time period for the gob to move between the position sensors of each set. The timer also determines the time between velocity measurements.

The data is used to calculate a value of acceleration which is substantially proportional to the acceleration normal to the path. The acceleration value and the output of the force-measuring device are then interrelated in order to calculate the mass of the gob.

Referring to Figure 1.7, gob **10** is produced by a source of molten glass (not shown) and falls vertically onto the upper portion of scoop **12**. Gob **10** then follows the curved contour of scoop **12**, during which time gob **10** is accelerating.

Figure 1.7: Gob Weighing Device

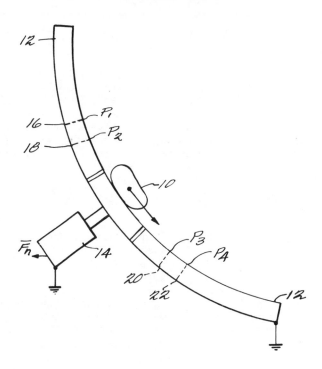

Source: U.S. Patent 4,165,975

Note that the acceleration along the curved section of scoop **12** has two components, one tangential and one normal, to the path. The velocity of gob **10** as it first touches scoop **12** is:

(1) $$\frac{ds}{dt} = gt_1 = V_0$$

where t_1 is the time it took gob **10** to fall from the source to scoop **12**.

In the curved portion of scoop **12**, the velocity of gob **10** is:

(2) $\dfrac{ds}{dt} = V_o + (g \cos \theta)t$

and the acceleration of gob **10** along the curved section is:

(3) $a = \dfrac{d^2s}{2}e_t - \dfrac{1}{r}\left[\dfrac{ds}{dt}\right]^2 e_n = (g \cos \theta)e_t - \dfrac{[V_o + (g \cos \theta)t]^2}{r}e_n$

where e_t and e_n are unit vectors in the tangential and normal directions to the path of gob **10** and where r is the radius of the curved portion of scoop **12**.

The normal force of gob **10** over the curved section of scoop **12** can be expressed as:

(4) $F_n = ma_n$

where:

(5) $a_n = \dfrac{[V_o + (g \cos \theta)t]^2}{r}$

Force sensor **14**, inserted as a part of scoop **12**, detects the normal component of the force exerted by gob **10**, that is F_n. Force detector **14** can be any such detector known in the art, but in the preferred embodiment, is a load cell.

The output of force detector **14**, F_n, is supplied to a computer which can be any computer well-known in the art, capable of measuring periods of time, storing data applied to it and performing basic mathematical functions. Thus, if the normal component of acceleration of gob **10** can be determined, the mass of gob **10** can be determined. It is known that:

(6) $a = \dfrac{dv}{dt} \approx \dfrac{\Delta V}{\Delta t}$

Position indicators **16**, **18**, **20** and **22** determine the acceleration of gob **10**. Although the position indicators such as indicator **16** could be any of the indicators known in the art, the preferred embodiment uses photoelectric cells that detect the moment gob **10** passes along scoop **12**.

The photoelectric cells **16** and **18** are positioned relatively close together on one side of force detector **14**, while photocells **20** and **22** are positioned relatively close together on the other side of force detector **14**. P_1, P_2, P_3 and P_4, the respective output of photocells **16**, **18**, **20** and **22**, are also connected to this computer.

As gob **10** passes cell **16**, a clock in the computer is started, the output of which is used to increment a counter. As gob **10** passes cell **18**, the counter is stopped. The information stored in the counter is the time for gob **10** to transverse the distance between cells **16** and **18**. By knowing the distance between cells **16** and **18**, the velocity between cells **16** and **18**, hereafter referred to as V_2, can be determined.

The same technique is used with cells **20** to determine the velocity between cells **20** and **22**, hereafter referred to as V_1. ΔV in equation (6) then is $V_2 - V_1$. Δt is the time for gob **10** to traverse the distance between cell **18** and cell **20**, which is also determined by computer.

Thus, by determining the time it takes gob **10** to traverse the distance between cells **16** and **18**, cells **18** and **20**, and cells **20** and **22**, the computer has the data necessary to calculate the acceleration. The computer can be used to calculate the mass:

$$(7) \qquad m = \frac{F_n}{a} = \frac{F_n}{(\Delta V/\Delta t)} = \frac{F_n \Delta t}{V_2 - V_1}$$

The computer can determine whether the mass of gob **10** is less or greater than the ideal and can control a motor to vary the mass of future gobs.

Thus, the process offers the advantage of a closed feedback system that continually corrects for any changes in the mass of a gob without destroying the gob that is measured. The mass is also measured without any interference to the glassware-forming machines in use. Since the mass of the gobs is continually adjusted, no containers need to be discarded because of incorrect weight. Since the mass of the gobs varies slowly with time, the mass of every gob need not be determined.

Experimentation reveals that the measurement of every sixth or tenth gob provides sufficient information in order to maintain the gob mass within conventional standards.

High Silica, Metal-Oxide-Containing Granules

P.P. Bihuniak; L.H. Brandes and D.L. Guile; U.S. Patent 4,200,445; April 29, 1980; assigned to Corning Glass Works describe a method for producing glassy granules of high silica glass (>70% by weight SiO_2) containing at least one metal oxide from Groups III and IV of the Periodic Table selected from the group of Al_2O_3, B_2O_3, GeO_2, HfO_2 and TiO_2, which consists of the steps:

(a) A silica-producing component and the Group III and/or Group IV metal oxide-producing component are passed through a flame burner to form a composite-fumed product of uniform composition;

(b) The composite-fumed product is mixed with a polar liquid to form a flowable sol containing up to 45% solids content;

(c) The polar liquid is removed from the sol to produce a fragmented solid; and

(d) The fragmented solid is then calcined within the temperature range of 1150° to 1500°C to produce glassy granules.

Example: In this example, the sol was prepared in a closed bowl mixer known as a high shear mixer. The mixer is of the type described in U.S. Patent 2,945,634 and characteristically embodies two sets of flat blades vertically spaced and set at right angles to each other.

1610 g of MS-7 fumed-silica were combined with 3750 g of water in the mixer to produce s suspension containing about 30% solids. The silica was predampened by hand mixing with water to reduce its fluffy nature and thus enable mixing a larger batch at one time. The dampened silica was placed in the mixer and the mixer rotated at 1,800 rpm for 30 seconds. At this point, the material suddenly became a flowable suspension, as indicated by a sudden decrease in power to operate the mixer.

Normally, the batch is mixed for a full minute to render the suspension viscosity more uniform as well as to thoroughly disperse any additive or dopant that is present.

The suspension, having a viscosity on the order of a few hundred centipoises, was poured onto drying trays. Each tray had a peripheral barrier about $\frac{1}{8}$" thick within which was provided about two square feet of flat surface. The tray was filled with suspension and set over a flat, radiant industrial heater of approximately the same surface geometry as the tray. The heater was electrically powered and was regulated to heat the suspension just below boiling.

As the water was removed, the pool of silica sol gradually rigidified and fragmented into a pattern resembling cracks in a dried mud flat. Further drying caused the rigid material to dice into relatively uniform size particles about $\frac{1}{8}$" in cross section.

The $\frac{1}{8}$" particle size has been found particularly suitable for subsequent milling to form a casting slip. Prior to grinding, however, the granules are calcined for 30 minutes at 1350°C.

Moldable Material Containing Crystallizable Glass

K.K.K. Kroyer; U.S. Patent 4,157,907; June 12, 1979 describes a process based on using particles of crystallizable glass obtained by fusing sand, chalk and dolomite in a rotary kiln and possible subsequent crystallization of the formed crystallizable glass by heating in a rotary kiln and grinding the crystallizable or crystallized glass in a ball mill to a desired particle size.

British Patent 992,782 describes a method of making such a glass. The product is also known as synopal particles. Such particles contain a substantial number of microbubbles deriving from the formation of carbon dioxide during the melting process, and in the process described, these microbubbles remain in the material.

Synopal particles have been used as an aggregate for road surfaces and as fillers in a variety of products such as coating materials and molding compositions. Hence, it is known to mix synopal particles with a binder or an organic resin and to use such mixtures as molding compositions for producing various articles or semiproducts.

For the production of the subject molding composition there is preferably selected a proportion of particles to binder whereby the amount of crystallized glass constitutes from 50 to 95%, based on the weight of finished composition. Such proportions are suitable for manufacturing articles to be shaped by compression, extrusion or injection molding. The preferred content of crystallized glass is from 70 to 90% by weight, based on moldable material.

For special purposes, the content of crystallized glass may be higher, 95% or more, based on the weight of molding composition. This, therefore, applies where it is desired to manufacture porous articles such as filter boards.

The preferred binder is thermosetting binder(s) as these impart high strength and resistance to heat to the finished product after setting. If desired, however, it is also possible to use thermoplastic binders which result in products that are easy to shape by injection molding or extrusion. Examples of suitable thermosetting binders are epoxy resins, melamine resins and urea-formaldehyde resins. Other curable plastics that may be used are polyurethane resins, phenol-formaldehyde resins and unsaturated polyesters.

Among suitable thermoplastic resins for the subject molding composition should be mentioned polyethylene, polypropylene, polyvinyl acetate, polyacrylates, polystyrene, polyamides and polycarbonates.

The binder may be introduced to the ball mill as a dry powder. As a result of grinding, it will be uniformly distributed on the surface of the particles. The distribution may be further facilitated by adding a solvent, or the binder may be introduced as a solution of the organic resin in a suitable solvent. The type of solvent depends on the nature of the binder used. Hence, water may be used in combination with melamine resins and urea-formaldehyde resins. Besides, it is possible to use conventional solvents such as acetone, benzene, alcohols or esters. Also, hydrocarbons such as turpentine or petroleum are suitable. The solvent may also, if desired, be added before introducing the binder. Such solvents as a rule evaporate on account of the frictional heat generated during the grinding operation.

If desired, it is possible to add minor amounts of various adjuvants to achieve desired effects or as additional fillers. Hence, silicon oxides or hydroxides, such as Aerosil, may be added in amounts of from 0.01 to 1% and this will result in increased fluidity, improved density and uniformity of the finished product.

The moldable material produced according to the process makes it possible to manufacture products having good mechanical characteristics such as great strength. The material further possesses good electrical properties such as high dielectric constant and good leakage current characteristics. The material is, therefore, suitable for manufacturing electrical articles such as switches.

Among applications of the subject material should also be mentioned building materials, such as boards, partitions and roofing tiles; household articles, such as cutlery, plates, cups and goblets; packaging; filters; porous tiles; materials for filling teeth; and industrial articles.

Example 1: A ball mill having a volume of 10 m³, lined with aluminum oxide blocks and partly filled with flint nodules, was charged with 4 tons of synopal glass having a grain size of from 0 to 2 mm, crystallized by heating in a rotary kiln and thereafter cooled.

The ball mill was run for about 3 hours, resulting in the particle size being reduced to from 0 to 200 μ. This was followed by adding 800 kg of Uredana, which is a urea formaldehyde resin.

After an additional 3 hours of running, there was obtained a material having a particle size of from 0 to 150 μ, which material was excellent for compression in conventional heated molds for manufacturing cups and similar articles.

Example 2: A conventional rotary kiln having a length of 10 m and a contents of 6 tons of grinding material was charged on an hourly basis with 2 tons of synopal glass having a particle size of from 0 to 10 mm. The material was continuously introduced to one end of the kiln, and after setting to a stationary position, 2 tons of ground product were removed from the other end of the kiln.

Through a pipe inserted into the inlet of the kiln to terminate 2 m from the inlet were introduced 2 kg of Aerosil per hour. Through a pipe inserted into the discharge to terminate 3 m from same were introduced 400 kg of melamine resin per hour, and through a pipe inserted 2 m into the discharge were introduced 2 kg of Aerosil per hour. Frictional heat was generated during the grinding process, causing the temperature to rise to 50° to 60°C. The discharged material coated with resin has a grain size of from 0 to 150 μ.

The product displayed excellent fluidity and by compression in molds having a temperature of 180°C there were obtained articles of particularly high density and uniformity.

Blistered, Crystallizable Glass from Waste Materials

K.K.K. Kroyer; U.S. Patent 4,191,546; March 4, 1980 describes a process of making a blistered, crystallizable glass material, whereby calcareous and siliceous minerals are introduced into a rotary kiln, melted and then discharged, cooled and possibly crystallized. The process is characterized by introducing as raw materials a mixture of materials originating from domestic or industrial waste products; waste from purification of wastewater or any type of ashes, for example, power works and refuse disposal plants, with calcareous materials and; optionally, other glass-forming materials.

Example: A rotary kiln having a length of 70 m and a diameter which was 2 m over the first approximately three-quarter length and about 2.8 m over the remaining length was used. The kiln inclination was about 2° and the kiln made one revolution for every 65 seconds.

Iron chains are hanging from the walls over the first 10 to 15 m of the kiln to produce good heat transfer to the raw materials introduced and afford a larger surface contributory to retention of dust, especially when the wet process is applied:

(a) A mixture of raw materials in slurry form was prepared from the following raw materials: 65 parts of sand from Limfjorden and North Jutland; 45 parts of chalk from the same region, namely, Hillerslev and; additionally, 15 parts of dolomite from Hammerfall in the northern part of Norway; the above materials were mixed with water to form a slurry having a moisture content of 24%.

With a capacity equal to 13 tons of dry raw materials per hour, the slurry was introduced to the upper end of the rotary kiln. The materials were heated in the kiln by firing with heavy fuel oil from the lower end thereof.

As the heat in the upper end of the kiln partially desiccated the raw materials, the latter formed nodules. Further down in the kiln there was formed a pasty material with large lumps by partial caking of the nodules.

Passage through the kiln takes about 3 to 4 hours, and at the lower end of the kiln where the temperature was about 1450°C, the mixture melted to form a blistered liquid product which was discharged from the lower end of the kiln.

This product could be cooled to form crystallized blistered glass suitable for a large number of purposes, such as aggregate for road surfaces, building materials, etc.

(b) The composition of raw materials was continuously modified by using 80 pbw of fly ash (Studstrup Power Works at Aarhus) and 20 pbw of chalk from the above location were added. This fly ash is composed as follows:

SiO_2	55.9 % by wt
Al_2O_3	4.16 % by wt
Fe_2O_3	10.6 % by wt
CaO	19.1 % by wt
SO_3	1.17 % by wt
P_2O_5	0.23 % by wt
MgO	3.31 % by wt
TiO_2	0.18 % by wt
Na_2O	0.17 % by wt
K_2O	0.36 % by wt
Li_2O	96.00 ppm
Ignition loss	17.93 % by wt

These raw materials were likewise admixed with water to obtain a consistency permitting them to be introduced in slurry form to the upper end of the kiln. The raw materials traveled down through the kiln in the same manner as described above. A crystallizable material, similar to that described above, was formed, except that this material had a dark color, almost olive black and partly transparent.

(c) In a third step, carried out in continuation of step (b), a portion of the same raw materials, that is, fly ash and chalk (in the same proportions), in a dry state, was introduced at the lower end of the kiln. An amount of raw materials equal to about 35% of the total amount of raw materials was introduced at the lower end of the kiln. It was found that they could be introduced both together with the primary air and through a separate tube next to the intake of primary air.

When introducing the raw materials in the stated amount at the lower end of the kiln, it was possible to reduce the oil consumption from about 1,300 ℓ/hr in step (a) to about 950 ℓ/hr in step (c). This proves that the carbon content of the fly ash (Studstrup) was utilized in heating the kiln.

The amount of finished material as blistered crystallizable glass discharged from the lower end of the kiln was about 220 tons per day. A sample fraction with a granulometry of from 5 to 7 mm had a bulk density of 850 g/ℓ indicating a high blister content.

From the upper end of the kiln there was deposited in the associated filter, an amount of chalk and fly ash of about 10% of the amount introduced, and an analysis showed that practically all of the carbon of the fly ash had been removed and thereby utilized during the stay in the rotary kiln. The filter content was recycled to the slurry tank.

High Sodium Oxide Composition

W.H. Manring, P.M. DeBello and E.G. Imperato; U.S. Patent 4,142,907; March 6, 1979; assigned to FMC Corporation describe a process whereby a high sodium oxide composition, suitable for use in the manufacture of glass and fiber glass, is prepared by heating a sodium carbonate ore with a silicate or other source of glass-making oxides, such as clay or oil shale.

The product obtained is rich in sodium, calcium and magnesium oxides and contains less than 35 wt % SiO_2. It may be substituted for soda ash in a glass batch and will reduce the fuel required in glass manufacturing.

The chemical analysis of the high sodium oxide composition will vary depending on the amount and chemical composition of the particular silicate or other source of glass-making oxide and sodium carbonate ore that are mixed prior to heating.

Preferred compositions are those containing the aluminum, magnesium and calcium oxides that are useful in the manufacture of glass, characterized by a sodium content (determined as sodium oxide) of from about 25 to as much as 58 wt %. The ratio of sodium oxide to aluminum, magnesium and calcium oxides is preferably in the range of from about 0.67:1 to about 1.13:1.

Particularly preferred are sodium oxide compositions containing from about 25 to 40 wt % sodium (determined as sodium oxide) and a ratio of sodium oxide to magnesium and calcium oxides of from about 0.76:1 to about 1.44:1.

Economically advantageous high sodium oxide compositions are those containing minor amounts (less than 35 wt %) of silicon dioxide and a ratio of silicon dioxide to sodium, potassium, aluminum, calcium and magnesium oxides in the range of from about 0.35:1 to about 0.57:1.

Example 1: This example illustrates the use of oil shale in the production of a high sodium oxide composition. One thousand pounds of raw trona ore ($Na_2CO_3 \cdot NaHCO_3 \cdot 2H_2O$) containing 12.5% by weight of water-insoluble material are intimately mixed with 1,350 lb of mine run floor oil shale (analyzing 4.48 gal of water and 31.96 gal of oil per ton).

The mixture is charged into a melting furnace and heated to 2600°F (1427°C). The oil burns and molten residue is run out into a water Bosch. One thousand two hundred twenty pounds of dry-granulated high-sodium oxide composition is obtained.

The analyses of the starting materials, and final product (high sodium oxide composition) are summarized in the following table. The yield, based on the total weight of trona and oil shale charged to the furnace, is 52%. Most of the heat required by the reaction was supplied by the oil in the shale.

	.........Pounds of Oxide in			
		1,350 lb		Oxides in
	1,000 lb	Floor	1,222 lb	Product
	Trona	Shale	Product	(wt %)
SiO_2	32	280	312	25.5
Al_2O_3	6	77	83	6.8
Fe_2O_3	3	29	32	2.6
CaO	23	183	206	16.9
MgO	9	91	100	8.2
Na_2O	364	78	442	36.2
K_2O	3	44	47	3.8

Example 2: This example illustrates the use of spent oil shale in the production of a high sodium oxide composition. 1,400 lb of low-grade oil shale (analyzing 5.91 gal of water and 2.61 gal of oil per ton) is heated at 2000°F (1093°C) for 1½ hours to remove the oil, decompose the carbonate materials and burn off any residual oil or other carbonaceous residue.

The burned spent shale residue (950 lb) is intimately mixed with 1,000 lb of raw trona ore (containing 12.5% insoluble material) and heated at 2700°F (1482°C) for 1 hour until the evolution of carbon dioxide ceases. The residual product (high sodium oxide composition) weighs 1,390 lb and the analytical data is summarized in the following table. The yield, based on the total weight of trona and spent shale charged to the furnace, is 71%.

	.........Pounds of Oxide in			
		950 lb		Oxides in
	1,000 lb	Spent	1,390 lb	Product
	Trona	Shale	Product	(wt %)
SiO_2	32	416	448	32.2
Al_2O_3	6	54	60	4.3
Fe_2O_3	3	18	21	1.5
CaO	23	206	229	16.5
MgO	9	180	189	13.6
Na_2O	364	42	406	29.2
K_2O	3	34	37	2.7

GLASS COATINGS

Glass-Coated Polycarbonate Article

W.L. Hall and J.S. Humphrey, Jr.; U.S. Patent 4,200,681; April 29, 1980; assigned to General Electric Company describe a shaped, nonopaque polycarbonate article having improved mar-, abrasion-, scratch- and organic solvent-resistance comprising a polycarbonate substrate having deposited thereon:

    (a)   A primer coating layer comprised of the photoreaction product of at least one polyfunctional acrylic monomer represented by the general formula:

$$\left[H_2C{=}CH{-}\overset{\displaystyle O}{\overset{\displaystyle \|}{C}}{-}O \right]_n R$$

where n is an integer having a value of from 2 to 4, and R
is a n valent hydrocarbon radical, n valent substituted hydrocarbon radical, n valent hydrocarbon radical containing at
least one ether linkage, and a n valent substituted hydrocarbon radical containing at least one ether linkage; and

(b)   A thin top layer of vapor-deposited glass on the primer
coating layer.

Example 1: An aromatic polycarbonate is prepared by reacting 2,2-bis(4-hy-
droxyphenyl)propane and phosgene in the presence of an acid acceptor and a
molecular weight regulator and having an intrinsic viscosity of 0.57. The product
is then fed into an extruder, which extruder is operated at about 265°C and the
extrudate is comminuted into pellets.

Example 2: An intermediate coating composition is prepared by blending together 50 pbw of ethylene glycol diacrylate, 50 pbw of pentaerythritol triacry-
late, 2 pbw of α,α-diethoxyacetophenone, 5 pbw of resorcinol monobenzoate
and 0.5 pbw of a silicone oil-type surface-active agent [BYK-300 (Mallinckrodt
Chemical Co.)].

A film of about 12.5 μ thickness of this coating composition is applied to the
polycarbonate panels prepared substantially in accordance with Example 1,
using a wire-wound drawdown bar.

The coated polycarbonate panels are then passed through a Linde photocuring
apparatus, which apparatus consists of a variable-speed conveyor running through
a chamber containing germicidal-type mercury vapor lamps which emit light
mainly at 2537, 3150 and 3605 Å operating in air, where the nitrogen pressure
is 25 psi nitrogen and the speed of the conveyor is 50 ft/min. The resulting
primer coating is hard and tack-free.

Example 3: A polycarbonate test panel prepared substantially in accordance
with Example 1 and precoated substantially in accordance with Example 2 is
placed into a vacuum deposition chamber containing a crucible about which is
disposed a radio frequency induction coil, the coil being connected to a
Pachydyne 50 Induction Heating Power Supply operating at between 15 and
30 kW.

The crucible, which contains quartz, is located 10" below the polycarbonate
test panel. The crucible is positioned so that its longitudinal axis is transverse
to the longitudinal axis and the direction of travel of the polycarbonate test
panel, thereby enabling the volatilized quartz stream to evenly impinge upon
the side of the polycarbonate panel facing the crucible and resulting in a silicon
dioxide layer which is evenly and uniformly deposited across the entire width
of the test panel as it passes over the crucible.

The vacuum deposition chamber is maintained at a pressure of approximately
1×10^{-4} mm Hg and the polycarbonate test panel is transported across the
crucible at a rate of 1 ft/min. A coating of silicon dioxide 3 μ thick is evenly
and uniformly deposited on the precoated polycarbonate test panel.

The glass coating which is vapor-deposited on a polycarbonate panel primed in accordance with the process results in a coating having a much superior durability after exposure to humidity and provides a much greater degree of solvent-resistance than does a glass coating which is vapor-deposited on a polycarbonate panel which is not primed. This results in a glass-coated polycarbonate panel which can be utilized successfully in many commercial applications.

Corrosion-Resistant Coating for Metals

M. Berretz; U.S. Patent 4,196,004; April 1, 1980 describes a composite glass coating and a method of preparing same, wherein a host or parent glass formulation is provided with one or more additive substances which impart superior mechanical strength, impact- and abrasion-resistance to the host or parent glass, with no impairment or reduction in its corrosion-resistance. The composite glass coating is used in accordance with the established enameling techniques of spraying and dusting.

Uniformly satisfactory results have been obtained in those instances in which the host or parent glass comprised glass frits formulated from the ingredients set forth in Tables 1 and 2. Table 1 sets out the glass formulation for the transparent acid- and alkali-resistant species, whereas Table 2 contains the formulation for the opacified glass that is particularly resistant to acid and alkaline etch.

Table 1

	Percent
SiO_2	48–70
TiO_2	1.5–10
ZrO_2	2–10
Li_2O	0.5–12
Na_2O	5–20
K_2O	1–5
MgO	1–4
CaO	1.5–6
CaF_2	1–6
SrO	1–5

Table 2

	Percent
SiO_2	45–72
TiO_2	12–20
ZrO_2	5–16
Li_2O	0.5–5
Na_2O	6–12
K_2O	0–6
MgO	0.5–4
CaO	1–4
P_2O_5	2–8
ZnO	0–4
Al_2O_3	0–3
B_2O_3	2–8

The abovementioned batch compositions are smelted between 1900° and 2400°F (1038° to 1316°C) and then water-quenched after which the resultant frit is combined with selected, insoluble additives which will give the desired improved physical and chemical properties to the final product. The glass frit and selected additives are ground into a fine powder or a slip, and both the additive and the slip must be of the appropriate fineness to insure consistent application.

The selected additives that are added to the parent glass composition of Tables 1 and 2 fall into two categories: (1) those added to compositions from Table 1 and (2) those added to compositions from Table 2.

The additives to the optically-clear frit derived from Table 1 are one or more of the following: lithium titanate, lithium silicate, lithium aluminum silicate, silica and magnesium titanate, in proper ratios to give the desired physical and chemical properties as determined by the service conditions and the type of base metal to which the coating is applied.

The additives to the opaque frit derived from Table 2 are one or more of the following: magnesium silicate, lithium silicate, silica and lithium aluminate.

The chemical constituents used as mill additives in both cases above are obtained commercially and a typical mill additive may have the following composition:

Host glass (Tables 1 and 2)	1,000 g
Lithium silicate	100 g
Bentonite (clay)	25 g
Syloid (electrolyte)	5 g
Water	500 cc

The additives are ground in a ball mill if necessary to a range of −100+325 mesh. They should comprise 5 to 80% of the mill formulation, the quantities depending on the desired properties of the coating. Lithium titanate, lithium silicate and silica substantially enhance the anticorrosion properties of the parent glass. The lithium aluminum silicate, lithium aluminate and magnesium silicate enhance the temperature resistance of the composite glass; magnesium titanate enhances its impact-resistance.

The objects or equipment to be coated with the above formulations comprise fabricated mild steel, stainless steel, or any other suitable metal. The metal objects are first prepared using standard techniques that have been developed for porcelain enameling.

FOAMED GLASS

High Silica Borosilicate Composition

S.B. Joshi; U.S. Patent 4,192,664; March 11, 1980; assigned to Pittsburgh Corning Corporation describes a method of making a cellular body whereby a mixture of alumina, boric oxide, alkali metal carbonate, a cellulating agent and a portion of the scrap trimmings from previously cellulated material is introduced into a ball mill and subjected to comminution for a sufficient period of time to intimately mix the constituents and form a first mixture containing a preselected amount of comminuted scrap trimmings.

Amorphous silica is added, thereafter, to the first mixture in the ball mill to form a second pulverulent mixture. The second pulverulent mixture is thereafter positioned in suitable covered molds and introduced into a cellulating furnace. The molds are immediately subjected to a temperature of 1200°C and maintained at that temperature until the carbon is fixed in the mixture.

The temperature of the furnace is then increased at a preselected rate to about 1390°C, where it is maintained for a sufficient period of time to coalesce the mixture and gasify the cellulating agent and form a cellular body that has a substantially uniform cell structure.

The temperature to which the cellular body is subjected is then rapidly reduced to a temperature of about 760°C where it is maintained for a sufficient period of time to permit the cellular body to cool. The cellular body is thereafter slowly cooled from 760°C to ambient temperature and trimmed and shaped to form a cellular body having a desired configuration.

A suitable composition for making the precellulated material for the process consists of the following range of constituents expressed in percent by weight. The cellulating agents, i.e., the carbon black and Sb_2O_3 are expressed in weight percent of the other constituents:

SiO_2	80-88
K_2O	1-3
B_2O_3	5-13
Al_2O_3	4
Carbon black	0.5
Sb_2O_3	0.5

Example: The high silica borosilicate batch contained the following constituents expressed as oxides in weight percent:

SiO_2	88.0
Al_2O_3	4.0
B_2O_3	6
KOH	1
Carbon black	0.5
Sb_2O_3	0.5

The boric acid was dissolved in hot water to form a boric acid solution. The potassium hydroxide was dissolved in water. The SiO_2, KOH, Al_2O_3, C and Sb_2O_3 were mixed in a high shear mixer and the boric acid solution was added thereto.

After thorough mixing, the material was dried in a drum drier until it contained about 2% by weight water. The dried material was ground in a ball mill until the pulverulent glass batch had an average particle size of about 2.5 μ.

The ground material was positioned in a graphite tray and then compacted by mechanical shaking. The tray was covered with a graphite plate and introduced into an electrically-fired cellulating furnace. The tray with the pulverulent glass batch was subjected to the following heating schedule in the cellulating furnace.

The material was maintained for about one-half hour at an initial temperature of 1200°C; thereafter, the temperature within the furnace was raised to 1450°C and held at that temperature for about 90 minutes. The temperature of the furnace was then reduced to 1200°C and the tray was removed from the furnace at that elevated temperature.

After cooling, the foamed mass within the graphite tray was then placed in a crusher mixer and 0.4% by weight carbon and 0.5 % by weight antimony trioxide, based on the weight of the foamed mass, were added thereto to form a precellulated material. The precellulated material was crushed and mixed in the ball mill until an average particle size of about 2.5 μ was obtained.

The precellulated material was then placed in a graphite tray and compacted by mechanical shaking. The tray was introduced into a cellulating furnace and subjected to substantially the same heating schedule. After removal from the furnace, the graphite tray was insulated to permit slow cooling of the cellulated body.

The resultant cellular glass body was tested and compared with products made from the same glass batch by melting and also the same glass batch by calcining. The tests indicated that the product obtained by the above-discussed process had substantially the same properties as material obtained by either melting or calcining.

Oxygen Acids as Bonding Agents

C. Wüstefeld; U.S. Patent 4,178,163; December 11, 1979 describes a method for the manufacture of foamed glass comprising the steps of:

(a) Providing a mixture of a finely-divided glass and a bonding agent of 100 parts glass to 5 to 10 parts bonding agent, the bonding agent being an agent selected from the group of (1) the aqueous solutions of the oxygen acids of beryllium, boron, aluminum, silicon, germanium, arsenic, antimony, tellurium, polonium, astatine and phosphorus; and (2) the aqueous solutions of the salts formed by the oxygen acids neutralized with basic oxides, or basic hydroxides of beryllium, boron, aluminum, silicon, germanium, arsenic, antimony, tellurium, polonium, astatine and of the transition metals having a variable oxidation number;

(b) Drying the mixture at a temperature from 20° to 600°C to thereby transform the bonding agent into a gel having water bound thereto; and

(c) Heating the dried mixture to a temperature from 800° to 1000°C to thereby melt the mixture and release the bound water from the gel, the released water forming a vaporous cellulating agent which effects the foaming of the molten glass.

Example 1: 100 g glass powder with additions of 6.12 g sodium metaphosphate $[(NaPO_3)_6]$, 0.36 g soot and 0.135 g iron sulfate ($FeSO_4 \cdot 2H_2O$) are ground in the ball mill to a fineness of 1 m^2/g and mixed with sufficient water to yield a workable mass.

A plate, which is dried and hardened at 200°C, is formed from this mass. The solid plate is heated to foam at approximately 880°C and kept at this temperature for 15 to 30 minutes. Subsequently follows cooling down to approximately 500°C in the air and further slow cooling down for expansion in the expansion oven.

Example 2: A mixture of 85% glass powder and 15% ground basalt is treated, respectively, with 1 g boron phosphate, silicon carbide and talcum. Silicon carbide has a particle size from 0 to 1 μm; talcum has particle sizes 50% of which were above 0.1 mm. The mixture was ground in a ball mill to a BET surface of 1 m^2/g; its product foamed glass weighed 0.15 g/cm^3.

Example 3: A mixture of 99.5 pbw of glass powder and 0.5 pbw manganese black was ground to a fineness with a BET surface of 1 m^2/g and subsequently heated to 875°C in a metal mold. After cooling down, a foamed glass body with closed cells has been obtained in which the cell diameter was approximately 1.5 mm.

With the weight per unit volume of 150 kg/m^3, the compressive strength was approximately 15 kg/cm^2. The coefficient of thermal conduction determined as per DIN 52 612 was approximately 0.040 kcal/mh °C at 25°C.

Production Without Prolonged Heating Schedule

J.D. Kirkpatrick; U.S. Patent 4,198,224; April 15, 1980; assigned to Pittsburgh Corning Corporation describes a process for making cellulated material where a formulated glass is used as one of the constituents and which includes obtaining glass cullet of a preselected composition, such as glass cullet of conventional soda lime glass, and introducing the glass cullet into a ball mill.

A cellulating agent, such as carbon black, is introduced into the ball mill thus admixed with the glass cullet and cocomminuted to obtain an average particle size of between 3 and 10 μ. The bulk density of the pulverulent batch is between about 50 to 60 pcf (0.80 to 0.96 g/cc).

The comminuted batch mixture is then sintered in any suitable manner as, for example, in a mold pan or as preagglomerated pellets in a rotary kiln under oxygen deficient conditions.

In sintering, the pulverulent batch is subjected to a sintering temperature of between about 1200° and 1400°F (650° and 760°C) and maintained at that temperature for a sufficient period of time for the vitreous material in the glass batch to become soft and coalesce. The material during sintering does not, however, cellulate and the carbonaceous cellulating agent does not react with the other materials in the glass batch to cellulate the material. The sinter itself has a true density of about 140 pcf (2.24 g/cc). The sintered material is then cooled and comminuted.

The comminuted material is screened so that it has a size capable of passing through a –10 mesh U.S. Standard screen and exhibits a bulk density of about 80 to 100 pcf (1.28 to 1.6 g/cc).

Where pellets of the batch are sintered, the comminution and sizing of the sintered material is not necessary since the pellets can be formed of a preselected size so that after sintering, the pellets will reduce in volume and have a size capable of passing through a –10 mesh screen. Because of the increase in bulk density, the volume of the material in a mold pan decreases substantially.

The sintered, sized material which contains the unreacted cellulating agent is placed in a mold pan and the mold pan is introduced into a cellulating furnace where the sintered material is subjected to cellulating temperatures and maintained at these cellulating temperatures until the desired cellular material is attained. The cellular material in the form of blocks is then cooled and quenched to stop the cellulating process and the blocks are then removed from the mold.

The cellular blocks are annealed to slowly cool the cellular blocks through the thermal range between the annealing and strain point of the vitreous material and thereafter cooled to ambient temperature. The cellular material is trimmed to form generally rectangular blocks of cellular material useful as insulation.

The heating schedule for the sintered material is more versatile than the heating schedule for the conventional cellulating process in which the pulverulent batch materials are continuously heated while the material first sinters and then cellulates.

In the conventional cellulation process, the pulverulent batch is slowly heated to the sintering temperature because of its pulverulent nature and the difficulty of transferring heat into a fine powder. After the pulverulent material has softened and coalesced, the temperature is slowly increased from the sintering temperature to the cellulating temperature and then maintained at this cellulating temperature for a sufficient period of time to cellulate the material.

With the described process, the pulverulent batch material may be sintered in separate facilities and after being properly sized and positioned in a mold pan, the mold pan can be introduced into a cellulating furnace which is at the cellulating temperature without using the prolonged slowly increasing heating schedule.

With this versatility, it is possible to substantially reduce the time required to maintain the material in the furnace. It is also possible to vary the peak cellulating temperature to thus control the density of the resulting cellulating material.

Example 1: A batch was prepared from a glass cullet having the following composition in percent by weight:

SiO_2	72.7
Na_2O	12.0
CaO	5.2
MgO	3.8
K_2O	0.7
Al_2O_3	4.5
Fe_2O_3	0.2
SO_3	0.3

The cullet was admixed with 0.2% carbon black and cocomminuted in a ball mill to an average particle size of between 4 and 5 μ to form a pulverulent glass batch.

The glass batch was placed into a sintering pan and introduced into a sintering furnace. The furnace was at 1400°F (760°C) and had an oxygen deficient atmosphere to prevent the oxidation of the carbonaceous cellulating agent. The batch remained in the furnace for a period of about 30 minutes during which period the batch formed a sinter product.

The sinter product was removed from the furnace and cooled. The sinter was then comminuted and screened on a U.S. 10 mesh standard screen. The sintered material passing through the screen had a size of less than 10 mesh and was placed in a stainless steel mold pan and introduced into a cellulating furnace having an initial temperature of 1490°F (810°C).

The temperature was slowly increased so that a temperature of 1600°F (870°C) was attained in 15 minutes. The temperature was held at 1500°F (816°C) for 30 minutes during which time the pulverulent sintered product cellulated and formed a cellular material.

The cellular material was quenched and annealed and exhibited properties comparable to a cellular material cellulated under a conventional cellulating process. The density of the cellular material was 10.4 pcf (0.17 g/cc).

Example 2: The same glass cullet was prepared and sintered as described in Example 1. The cellulating temperature was 1650°F (899°C) and the cellular material had a density of 9.45 pcf (0.15 g/cc). Thus, by increasing the peak temperature, the density of the cellular material is reduced.

Example 3: The same process as described in Example 1 with the same batch was sintered as abovedescribed and the sintered material in the stainless steel mold was introduced into a cellulating furnace having a temperature of 1640°F (893°C) and was held at this temperature for 35 minutes. The cellular material had a density of 10 pcf (0.16 g/cc).

Thus, with the process described herein, it is possible to introduce the sintered material into a furnace at or above the cellulating temperature and obtain a cellular material that has desirable physical properties and is free of folds and other types of flaws.

The cellular material in Examples 1 through 3 was flaw-free and was cellulated utilizing heating schedules that could not be used in the conventional cellulating process. If the rapid heating schedule discussed above in Examples 1 through 3 were used on a raw batch, the product would be full of folds and other inhomogeneities.

Ash-Coated Pellets

E. Tseng and M. Bassin; U.S. Patent 4,143,202; March 6, 1979; assigned to Maryland Environmental Service describe cellular glass pellets having an ash coating bonded thereto which exhibit superior properties as lightweight aggregate for addition to various inorganic and organic matrices.

Such pellets are lightweight and have a closed- or open-cell structure which renders them particularly useful as lightweight aggregate. Also, the lightweight pellets have particular utility as bulk insulation for hollow building walls and the like.

Processing: Waste glass, which is primarily bottle glass of a soda-lime-silica composition, is collected from a waste disposal plant wherein the glass is relatively free of organic matter and has been crushed to a rough size of particles having less than about a half-inch maximum diameter, then placed in a ball mill or other device for reducing to a fine size.

After crushing the glass to a fine size, it is screened over a 200 mesh screen. The material passing through the 200 mesh screen is retained for further processing, while the over-sized material is returned for further size reduction.

The –200 mesh glass particles are mixed on an inclined revolving pan or other pelletizing device with about 6 to 10% by weight water, and preferably from about 7 to 9% water.

The green glass agglomerates are then fed into the forward portion of a kiln or similar firing device wherein the agglomerates are thoroughly mixed with ash as a release agent. The ash-coated agglomerates are then passed through the firing portion of the kiln wherein the temperature is raised to about 750° to 900°C for a period of about 15 to 30 minutes. At temperatures less than about 750°C, the pellets expand very little; at temperatures greater than 900°C, soda-lime-silica softens too much and the pellets are destroyed.

The fired, foamed pellets having a size range of about 0.5 to 20 mm are preferably directed through a cooler before being passed over a screen whereby the very fine material, that is, material having a particle size less than about 0.5 mm, is returned to be further pelletized and foamed.

The pellets produced from this process have a very fine and substantially uniform pore size with the pores having a diameter of about 1 μ for the small pellets to about 10^4 μ for the larger pellets. The cellulated pellets have a minimum bulk density of about 8 pcf, which is obtained by maintaining the firing temperature and the residence time near the maximum. The density may range from about 8 pcf up to substantially the density of solid glass, depending upon firing temperatures and residence time.

The pellets further have a refractory coating of fine ash particles physically bonded to the surface. The ash coating is substantially one particle in thickness.

Example: A waste glass, primarily of a soda-lime-silica bottle glass composition, was crushed, ground and sized to have a particle size less than about 200 mesh. It was then mixed with sufficient water to bring the moisture content up to about 8% by weight and then agglomerated to an inclined pelletizing pan operated at an angle of about 30° and revolved at about 60 rpm.

Fine agglomerates having a median size of about 3 to 5 mm were formed. These agglomerates were then passed into a kiln and mixed with fly ash having a median particle size of about 40 μ. The forward portion of the kiln was unheated. The fly-ash-coated agglomerates are then passed into a fired kiln portion.

In the firing portion of the kiln, the pellets had an average residence time of about 10 minutes at 840°C. The foam pellets were discharged into a rotating water cooler and from there were discharged at a temperature less than about 200°F.

The foam glass pellets had an average diameter of about 4 to 7 mm and had a thin coating of fly ash substantially bonded thereto. A portion of the pellets were returned to the cooler and rotated for a sufficient period of time to substantially abrade away the fly ash coatings so that a comparison could be made between foam glass pellets with a fly ash coating and foam glass pellets with substantially no coating.

The foam glass pellets with the fly ash coating generally appeared to be stronger and more rigid than the pellets without a coating. Also, the moisture absorption and permeability of the ash-coated pellets were much less than that of the uncoated pellets.

Ash-coated pellets and uncoated pellets were exposed to a moist atmosphere for a period of time sufficient to cause some disintegration of the uncoated pellets, while the ash-coated pellets showed disintegration from the humid atmosphere.

The compressive strength of the individual ash-coated foamed pellets varies exponentially with the density of the pellets, as illustrated in the following table:

Specific Gravity	Compressive Strength (psi)
0.2	100
0.3	300
0.4	700
0.5	1,200
0.6	1,800

The above data were obtained for foamed glass samples having up to about 15% by weight ash contained in the glass core.

GLASS PROCESSING

MOLDING AND SHAPING

Injection Molding of Hydrosilicates

R.W. Eolin, G.F. Foster, J.F. Mach and R.O. Maschmeyer; U.S. Patent 4,132,538; January 2, 1979; assigned to Corning Glass Works describe a process capable of producing articles of hydrosilicate glasses having complex, intricate shapes with excellent dimensional accuracy and repeatability and surface quality, and which process is not subject to the limitations inherent in compression molding. This objective can be achieved utilizing an injection molding process.

First, contrary to the situation with plastics, the injection molding system must be fully sealed and pressurized to prevent water loss to the ambient environment from the hydrosilicate material when it is at flowing temperatures. Accordingly, the injection cylinder must be sealed and pressurized. In practice, a valve will be required to separate the hot, pressurized hydrosilicate material in the injection cylinder from ambient pressure when the mold is open. Inert gas and/or steam is utilized in the mold to apply pressure to the hydrosilicate material so the mold must seal against gases. The mold cavity must be so designed that gas can be vented into the pressurized system as the mold fills to avoid trapping bubbles or pockets of gas.

Second, inasmuch as hydrosilicate materials when heated to temperatures to cause flowing adhere to most common materials, the mold cavity must be constructed of materials which are capable of good release of hydrosilicates plus provide an acceptable combination of formability, toughness, and thermal resistance making them useful construction materials. Such mold materials permit intimate contact of the fluid hydrosilicate with the cavity such that the geometry, dimensions, and surface quality thereof are faithfully transferred to the shaped hydrosilicate article.

Third, hydrosilicates exhibit virtually no flow orientation and anisotropy. Those phenomena are prevalent in plastic products and result in preferential stresses with consequent differential shrinkage and warpage.

Fourth, hydrosilicates display a certain inherent brittleness which allows large gates, particularly if designed as stress concentrators. The gate is that part of the molding apparatus constituting a flow restriction immediately before the cavity which permits the molded shape to be separated cleanly.

Apparatus suitable for injection molding hydrosilicate materials must be designed with the above four characteristics of such materials in mind. Figure 2.1 schematically depicts an operable apparatus.

Figure 2.1: Injection Molding Assembly

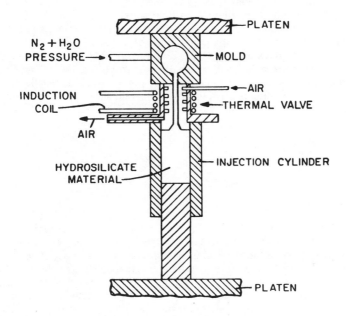

Source: U.S. Patent 4,132,538

The apparatus illustrated in Figure 2.1 contemplates three fundamental elements: (1) an injection cylinder which performs as a high pressure pump to inject the hydrosilicate; (2) a thermal valve which separates the hot, pressurized hydrosilicate in the injection cylinder from ambient pressure when the mold is open; and (3) a mold which must incorporate such features as pressurization capability, constructed of materials demonstrating ready release of hydrosilicates, and having the capability of being heated.

Two materials which are capable of good release of hydrosilicates and so are useful in the construction of injection molding equipment are: first, three commercially available, nickel-molybdenum alloys viz., Hastelloy B, Hastelloy C, and Hastelloy C-276; and second, thin films of noble metals and nobel metal alloys.

Hydrosilicate glasses having the base anhydrous compositions disclosed in U.S. Patent 3,912,481, have been found to be particularly suitable for the injection molding technique. Those glasses consist essentially, in mol % on the oxide basis, of about 3 to 25% Na_2O and/or K_2O and 55 to 95% SiO_2, the sum of those constituents comprising at least 55 mol % of the total composition. Al_2O_3, BaO, B_2O_3, CaO, MgO, PbO, and ZnO are observed as being useful additions with MgO being operable up to 35%, BaO being suitable up to 20%, and Al_2O_3, B_2O_3, CaO, and ZnO having utility up to 25%. PbO can be present up to 45%. The inclusion of other compatible metal oxides is noted, but individual additions of such will preferably be limited to less than about 10%.

Laboratory experience has demonstrated a particular group of glass compositions within the above ranges which exhibits excellent capability for injection molding and homogenization via mixing in the fluid state. Those glasses, which exhibit good chemical durability after hydration, have anhydrous compositions consisting essentially, in mol % on the oxide basis, of about 72 to 82% SiO_2, 10 to 17% Na_2O and/or K_2O, and 5 to 15% ZnO and/or PbO. Up to 5% Al_2O_3 and up to 3% B_2O_3 and/or BaO and/or MgO may also be included. One glass from this latter group which, after hydration, has proven to be especially amenable to the injection molding process has the following approximate anhydrous composition, reported both in terms of mol percent and weight percent.

	Mol Percent	Weight Percent
SiO_2	77.1	73
Na_2O	10.8	10.5
K_2O	3.0	4.5
Al_2O_3	1.3	2
ZnO	7.8	10

In preparing the hydrated glasses for the injection molding process, the general practice set forth in U.S. Patent 3,912,481 will commonly be followed. Thus, utilizing the abovementioned exemplary compositions as illustrative examples, the batch ingredients, therefore, will be blended together, a ball mill customarily being employed to assist in achieving a homogeneous melt, and the mixture then placed into a platinum or silica crucible. The actual batch constituents can consist of any materials, either the oxides or other compounds which, when melted together, will be converted into the desired oxide composition in the proper proportions.

The batch will be melted at 1450° to 1600°C and the melt formed into bodies of fine dimensions. For example, thin ribbon can be rolled which is subsequently fractured into pieces suitable for introduction into the apparatus, or the melt can be immediately formed into glass granules through such conventional methods as passing a thin stream through a hot flame, or running a stream thereof into water or over patterned rolls. Of course, shapes of larger bulk can be formed and then pulverized, but such practice is not economically desirable.

Compression Molding of Hydrosilicates

A.R. Olszewski and D.R. Parnell; U.S. Patent 4,142,878; March 6, 1979; assigned to Corning Glass Works describe a method for transforming a mixture of anhydrous silicate materials and water with optional additives into solid hydrosilicate bodies via a single step process of short duration.

This can be achieved through combining the anhydrous silicate materials with water and/or other protonic reagent capable of depolymerizing silica and then forming a body therefrom at elevated temperatures and under high pressures. The high pressure molding, denominated hydromolding, will be conducted in a closed system at temperatures of at least 100°C and at pressures of at least about 500 psi. The process is useful with compositions such as are discussed in U.S. Patent 3,912,481. Thus, the hydromolding can be carried out on actual glasses or on the batch ingredients from which a glass can be derived. Customarily, however, a more homogeneous body will be produced where glass constitutes the charge for compression molding.

Compositions operable in the process consist essentially, in mol % on the oxide basis, of about 3 to 25% Na_2O and/or K_2O and 50 to 95% SiO_2, the sum of those components constituting at least 55 mol % of the total composition. PbO, CaO, ZnO, and B_2O_3 can be included in amounts up to 25%; BaO and Al_2O_3 can be useful in quantities up to 20%; and MgO may advantageously be present in amounts up to 35%. In general, additions of other compatible metal oxides, e.g., CdO, TiO_2, ZrO_2, WO_3, P_2O_5, and SnO_2 will be held below 10% in individual amounts. Li_2O appears to impede hydration and hazard devitrification so, if present, will be maintained below 5%.

It is also possible to modify the glass composition by including inorganic additives during the hydromolding process. Hence, such additives can be blended homogeneously with the glass charge to the compression mold which will become an integral part of the pressed solid body.

Water or other protonic reagent content of at least 5% by weight has been found necessary to achieve homogeneous bodies. Amounts in excess of 50% by weight may be employed but such amounts generally afford weak bodies.

The preferred molding temperatures range between about 200° and 350°C. At temperatures approaching 500°C, cracking and/or delaminating of the body may occur. Also, higher temperatures tend to cause increased wear on the mold and sealing system required to prevent leaks. A molding pressure of at least about 500 psi has been found necessary. Pressures above about 25,000 psi can be employed but with no evident property advantage.

The process is also advantageous in permitting the ready shaping of glass bodies from difficultly hydratable glasses, i.e., glasses containing less than about 10% Na_2O and/or K_2O.

Shaping of Glass Sheets by Roll-Forming

R.S. Johnson and T.L. Wolfe; U.S. Patent 4,139,359; February 13, 1979; assigned to PPG Industries, Inc. describe the production of shaped, tempered sheets of glass.

Heat-softened glass sheets are shaped to either simple or compound bends while being continuously conveyed by means of forming rolls having transverse curvature, wherein, in at least the first portion of the bending station, the force of gravity alone causes the glass sheets to sag into substantial conformity with the transverse curvature of the forming rolls. Support for each glass sheet is gradually transferred from straight conveyor rolls to curved forming rolls without vertically reciprocating the rolls. Optionally, at the exit end of the forming station, a set of upper forming rolls having transverse curvature complementary to that of the lower forming rolls may be provided to assure close conformance to the desired curvature. Compound bends may be formed by providing the path defined by the lower forming rolls with a concave curvature in the direction of glass travel.

Shaping of Sheet Glass by Pressure-Molding

C.G. Bradbury; U.S. Patent 4,191,039; March 4, 1980; assigned to Leesona Corporation describes a shaping method and apparatus comprising, providing a sheet heated to a temperature sufficient to enable it to be shaped by pressure application thereto, placing such heated sheet in proximity to a mold shaping member and thereafter directing liquid coolant under pressure into direct contact with such sheet so as to force the sheet into contact with the member and thereafter vaporizing at least a part of the coolant upon contact with the sheet material to quickly form at least a solidified skin of cooled material on the sheet material surface by reason of the heat of vaporization effect.

Referring to Figure 2.2, in operation, the coolant enters the mold **28** through bore or bores **38** so as to force sheet S into immediate and intimate contact with the walls of the cavities **34**. Thus, the sheet is shaped to the configuration of the mold member **30**.

Figure 2.2: Sectional View of Mold in Open Position

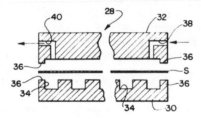

Source: U.S. Patent 4,191,039

Thereafter, the mold is either vented to atmosphere through the outlet bore **40** or otherwise depressurized so that at least a portion of the coolant immediately evaporates. Such change of phase from liquid to gas serves to immediately withdraw that amount of heat equal to the heat of vaporization from those surfaces of the material in contact with the coolant. Such action, i.e., the shaping and depressurizing steps are sequential or to some extent simultaneous; however, as a result thereof, at least those surfaces of the glass which the coolant comes in contact with, are solidified at least to the extent of the formation of a skin thereon which enables the sheet to be self-supporting. Once the mold members **30** and **32** are separated from each other, the shaped portion of the sheet is withdrawn therefrom for continued downstream processing.

Glass-Drawing Method

In the past, it has been a problem during the drawing of glass to find an acceptable combination of furnace temperature, drawing tension and drawing speed.

For example, a high temperature may be undesirable since the glass may become too fluid in the forming zone to maintain its desired shape, or because exotic or highly refractory materials of construction may be required when using such high temperatures. However, by reducing the glass temperature during the drawing process, the viscosity increases which results in an increase in the required drawing tension. Due to the increase in the glass viscosity, the drawing tension is often increased to the point where the freshly formed glass is overstressed resulting in breakage. In order to correct the breakage, the drawing speed may be reduced which automatically reduces tension; however, extremely low and uneconomical production rates may result, and further devitrification may occur in some glass compositions.

G.D. Lipp; U.S. Patent 4,175,942; November 27, 1979; assigned to Corning Glass Works describes the drawing of glass utilizing conventional method and apparatus but with the additional provision of means for introducing steam or water vapor into the atmosphere surrounding the glass being formed. This change in atmosphere materially improves the strength of freshly drawn glass and allows it to be drawn at low temperatures and high speeds where it would otherwise fail or break during such forming operations.

Referring now to Figure 2.3, a support member **20** is shown having a clamp **22** for controllably feeding a master blank **10** into an electrically powered redraw furnace **24**. The furnace has an open portion extending vertically therethrough which is lined with electric heating elements so as to heat and soften the blank into a flowable semimolten condition for attenuation as it passes through such opening.

Figure 2.3: Glass-Drawing Apparatus

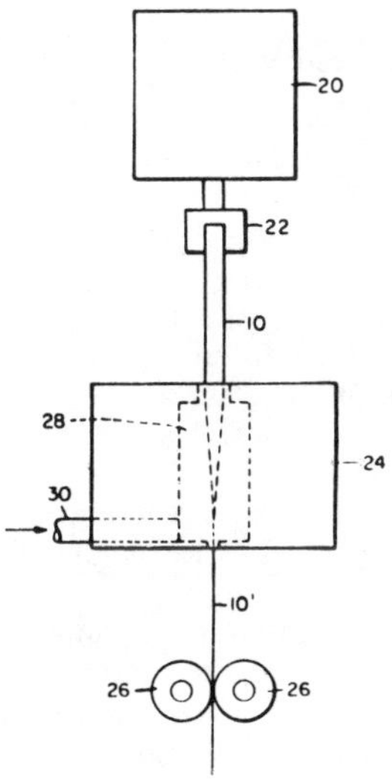

Source: U.S. Patent 4,175,942

Suitable pulling means such as driven tractor belts or pull rollers **26** are driven to draw down and attenuate the blank **10** into a drawn article **10'** of desired size and shape and withdraw the newly formed glass from the furnace for cooling into a rigid article **10'**.

The furnace **24** provides a chamber **28** about the glass being formed and an inlet conduit **30**, connected to a suitable source of water vapor such as steam, communicates with chamber **28**. Accordingly, steam or water vapor is introduced into the chamber via the inlet conduit so as to permeate the atmosphere surrounding the glass **10-10'** being formed within the chamber, and the spent water vapor is then exhausted from the furnace through the openings provided therein for the glass.

As a specific example, a master blank of a borosilicate glass having a width of about 5½" and a thickness of about 0.2" was heated to a semimolten state and drawn down and attenuated into a ribbon having a width of about 0.3" in a conventional redraw furnace. When the furnace atmosphere was saturated with low pressure steam, the drawing process was successfully accomplished with a draw rate of about 37" per minute in a continuous manner without experiencing breakage. However, when the source of steam was removed from the furnace, the glass ribbon broke and further attempts to restart the draw were unsuccessful due to continual breakage.

The temperature within the furnace was then raised to 900°C and the draw was again restarted and maintained with only occasional breakage. When the temperature was reduced to about 895°C breakage of the newly formed ribbon again occurred and continued during attempts to restart the draw. However, upon the reintroduction of steam into the atmosphere within the furnace and surrounding the glass while it was being formed, the temperature of the draw was able to be reduced to 880°C and a continuous draw maintained without breakage.

Bending of Glass Sheets to Curved V-Bends

D.L. Thomas and T.J. Reese; U.S. Patent 4,157,254; June 5, 1979; assigned to PPG Industries, Inc. describe a special glass sheet bending mold of outline metal rail construction capable of bending a glass sheet sharply about a curved line of sharp bending. This mold incorporates a heating ribbon of electroconductive material that is supported in electrically insulated relationship to the metal rail or rails that comprise the outline mold and is held in tension by weights but is supported intermediate its ends on a plurality of insulator guides disposed along a curved line facing a surface of the glass sheet to be bent along a curved line of sharp bending.

Special means are provided for holding the elongated ribbon in tension in closely spaced relation to the plurality of insulator guides disposed along the curved line so that the tensioned ribbon is longitudinally curved to face an elongated curved line of sharp bending to be imparted along a dimension of a glass sheet. The insulator guides are preferably composed of sectionalized ceramic spools comprising an intermediate ceramic roller of relatively small diameter flanked by larger diameter ceramic rollers and a special ribbon supporting element is provided for each ribbon supporting insulator guide. Preferably, U-shaped wire is wrapped around the circumference of the relatively small diameter ceramic roller in such a manner that the closed end of the U slidably supports the ribbon in closely spaced relation to the roller about which the U-shaped wire is wrapped. The wire is flexible enough to wrap around the roller so that its closed end is closely adjacent thereto and yet is sufficiently rigid after it is bent around the roller to maintain a slidable support for the ribbon between each roller and the closed end of the U-shaped wire.

According to a method of making a bent and tempered glass sheet having a sharp line of bending that is curved, a flat glass sheet is supported on an edge-wise disposed outline metal rail that forms an outline mold and a support structure containing the plurality of insulator guides disposed along a curved line is used to support an electroconductive ribbon along a curved line conforming to the curved line of bending desired in closely spaced relation to a surface of the glass sheet. While the specific embodiment described supports the curved ribbon above the glass sheet, it is understood that the ribbon may also be employed in facing relation to the lower surface of the glass sheet if desired.

The glass laden mold is heated to a temperature approaching the softening point of the glass and at an appropriate moment in the heating cycle, additional heat is supplied through the elongated ribbon along a curved line of sharp bending that faces the line of bending desired for the glass sheet. The glass sheet sags by gravity so that its outline conforms to the outline of the outline shaping surface of the metal mold rail while the line of the glass sheet that faces the curved elongated ribbon sags below the level of the remainder of the glass sheet along a curved line of sharp bending to provide a shape conforming in elevational contour and in outline to the shape desired for the glass sheet.

After the sheet is bent to the desired curvature, it is immediately removed from the hot atmosphere where it is shaped, and, while still mounted on the mold, is transported to a quenching position between upper and lower nozzles where air or other tempering medium is applied against the upper and lower surfaces of the glass at a relatively cold temperature sufficiently rapidly to cause the glass sheet to develop a stress profile which is associated with tempered glass.

Bending of Glass Plate Using Electrical Current

Y. Ebata, T. Ueno, N. Kataoka and A. Akao; U.S. Patent 4,173,461; November 6, 1979; assigned to Agency of Industrial Science and Technology, and Central Glass Company, Limited, both of Japan describe a method of bending a glass plate along a line traversing the plate, which method features the bending to be achieved with a small radius of curvature of the bend and that a glass to be bent needs no pretreatment; the product is free from defects such as stains and has a good appearance.

According to the process, the glass plate is provided with a pair of electrodes brought into contact with the glass plate respectively at two terminals of the line along which the bending is intended, and the glass plate is heated in the following manner.

First, a substantially entire portion of the glass plate is heated so as not to cause softening of the glass plate, and at the same time the glass plate is heated more intensely only in a zone which traverses the glass plate with a narrow width and contains the entire length of the bending line by the use of a heater arranged at a short distance from the glass plate to radiate heat along the bending line

such that only this zone of the glass plate is heated to a temperature at which
the glass plate does not yet soften but exhibits an appreciable lowering of its
electrical resistivity.

Then, a voltage is applied to the electrodes to cause an electric current to flow
between the electrodes through the intensely heated zone of the glass plate,
while the abovementioned heating is continued, such that only this zone of the
glass plate is further heated by the Joule effect of the current and softened.
While the aforementioned zone remains softened, the glass plate is bent so as to
give a bend along the line which is contained in the softened zone.

Extrusion Apparatus

*A. Coucoulas and J.R. Nis; U.S. Patent 4,195,982; April 1, 1980; assigned to
Western Electric Company, Incorporated* describe a gas head extrusion apparatus
10 (Figure 2.4) having upper **12** and lower **13** chambers, arranged to feed glass
particulate **60** into the upper chamber.

Figure 2.4: Extrusion Apparatus

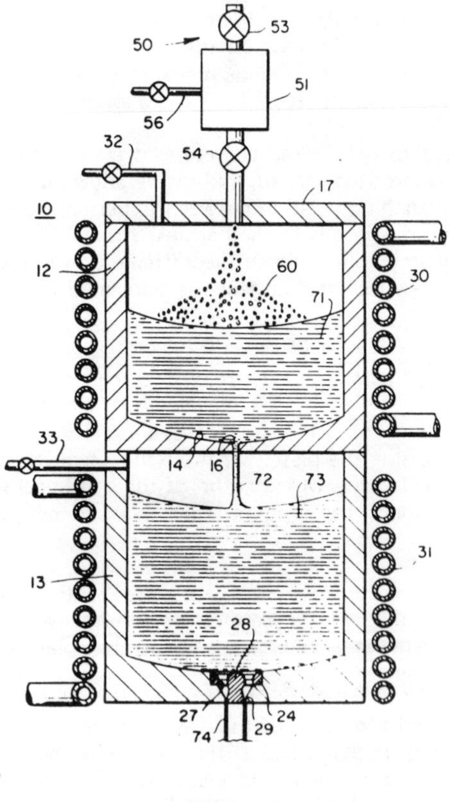

Source: U.S. Patent 4,195,982

The particulate **60** is heated to consolidate the particulate into a first viscous melt **71** as an elevated gas pressure is simultaneously applied within the chamber **12**. The first viscous melt is urged through a narrow opening **16** in the bottom **14** of the upper chamber **12** into the lower chamber **13** to form a second viscous melt **73**. The lower chamber **13** is pressurized to a second pressure, P_2, where $P_1 > P_2$. Gas bubbles, having an internal pressure, P_1, formed during the consolidation process in the upper chamber, will expand in the lower chamber to facilitate removal thereof from the melt. The lower pressure, P_2, also causes the melt in the lower chamber to be extruded through an annular opening **29**, in the bottom of the lower chamber, to form a substantially bubble-free hollow glass tube **74**.

Apparatus for Producing Narrow-Necked Containers

This process relates to the manufacture of blown glass articles such as bottles, jars, flasks, etc. According to the "narrow-neck blow and blow" method used and as described in U.S. Patent 1,911,119, a charge of glass is delivered to and compacted or caused to settle in the cavity of an inverted or neck-down, blank or parison mold. The glass of the charge will be extending from the neck portion of the mold cavity, part of the way up the sides thereof. A baffle is placed on the uppermost end of the inverted blank or parison mold and air under pressure is applied to the interior of the glass in the mold through the neck opening to counterblow such glass into conformity with the internal configuration of the blank or parison mold and against the baffle.

Thereafter, the counterblown blank or parison is transferred to an upright final blow mold in which the blank or parison is disposed in an upright or neck-up position and air under pressure is applied to the interior thereof. The counterblown blank or parison is thus expanded to the configuration of the final blow mold cavity, thereby forming an article of the final shape and size desired.

This method of forming articles of glassware has been practiced since the 1920s. Certain faults and shortcomings have been known and such defects as "settle waves" in the sidewalls of the article, marking the juncture of wall portions of two different thicknesses are common. Furthermore, articles of generally circular cross-sectional configuration have experienced the formation of excessively thick shoulders and relatively thin bottoms when produced by the abovedescribed method.

Those articles which would have a generally rectangular cross-sectional configuration or are of a flask shape usually have excessively thick sides or panels and relatively thin corner sections. As a matter of fact, different portions of practically all articles produced by the aforementioned method vary substantially in the thickness of the walls thereof. Therefore, for most articles of a given size and intended use, it has been necessary to use a glass charge of undue size and weight so as to insure that the article produced will be thick, and therefore strong enough at its thinnest wall portion, to enable the article to withstand the normal filling line abuse and later handling in the service for which the article is intended. A smaller charge could be used if better glass distribution were effected in the forming process resulting in a saving of material and energy.

J.M. Fortner and R.T. Kirkman; U.S. Patent 4,191,548; March 4, 1980; assigned to Owens-Illinois, Inc. describe a method and apparatus for forming glass containers by a process wherein a gob of glass is delivered to a parison mold and specially designed neck mold cavity. The gob is almost immediately settled in the neck mold about a low-heat removal plunger by vacuum. After the vacuum settle has been completed, the plunger is withdrawn and after a brief corkage reheat period, air under pressure is used to counterblow the charge in the parison mold to form a parison. The parison is then inverted and transferred from the parison mold to a final or blow mold where the parison is expanded into final article form. The special design of the neck-forming equipment permits the efficient use of vacuum to accomplish settling the glass in the neck-forming area and counterblow is effected sooner, resulting in a glass container of a selected volumetric capacity being formed with less glass and of a more uniform wall thickness and improved strength.

Turning now specifically to Figure 2.5, details of the construction of the neck ring and plunger mechanism will be described.

Figure 2.5: Neck Ring and Plunger Mechanism of Molding Apparatus

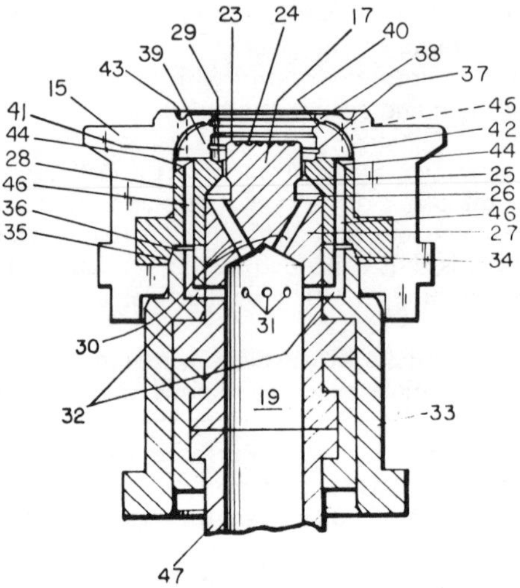

Source: U.S. Patent 4,191,548

In the particular arrangement shown in Figure 2.5, the plunger **17** is in its elevated position and as can be seen, the plunger has a generally horizontal top surface **23**. The surface is provided with a series of circular indentations **24**, thus providing a generally "bull's eye" top view of the concentric rings formed by the pattern of the indentations. The circumference of the plunger is generally cylindrical and extends downwardly to a point where it becomes relieved inwardly at **25**. At the bottom of the relieved area, the plunger presents a horizontal face **26** which extends to an enlarged barrel portion **27**.

The barrel portion is positioned concentrically with respect to a guide ring **28**.

The upper end of the plunger 17 extends through a central opening in the guide ring 28 with relatively close clearance therebetween. The guide ring presents a generally rounded annular surface 29 which, in cooperation with the plunger, and the internal surface of the neck mold 15, defines the "finish" or neck of the container to be produced. The barrel portion 27 of the plunger has a plurality of angularly extending, circumferentially spaced passages 30 formed therein which connect the passage 19 to the area immediately above the horizontal face 26 of the plunger.

In addition to the passages 30, a plurality of horizontally extending passages 31 (in actual practice 16 in number) extend horizontally through the wall of the lower portion of the plunger. These passages 31 align with an equal number of passages 32 formed in a thimble 33 which surrounds and guides the plunger, and at its upper annular end 34 presents an inwardly tapered surface 35. The tapered surface seats within a complementary tapered surface 36 formed at the lower inner circumference of the guide ring and effectively isolates and seals the system interconnecting chambers as well as aligns the parts to provide guidance for the plunger.

The specific internal configuration of the neck mold carries a spiral, thread-forming groove 37 and a groove 38 which serves to form a bead around the neck or finish that will provide a support for the parison at the time the parison is enclosed within and suspended in the blow mold prior to the inflating of the parison.

In view of the fact that the apparatus is to be used in the process wherein vacuum serves as the means by which the glass is settled around the neck pin and within the neck mold, diametrically opposed faces 39 and 40 of the neck mold are relieved, by milling to a depth sufficient to pass vacuum but not so large as to create a defect in the surface of the glass, in the one neck mold half only. A pair of chipped grooves 41 and 42 of greater depth than the relieved areas extend from within close proximity of the bead-forming groove 38, in surrounding relationship to the surfaces 39 and 40 and provides an area through which vacuum will be applied along the face of the neck mold half 15.

It should be understood that the opposite or complementary neck mold half will be made without comparable reliefs 39 and 40 and the chipped grooves. While it would be possible to divide the relieved portions between the two neck mold halves, this obviously would add to the expense in construction of the molds and sufficient relief and access to vacuum is provided by having the grooves formed in only one of the mold halves.

In addition to the abovementioned grooves, a circular groove 43 surrounds the central cavity formed in the neck mold. This semiannular groove 43 is in communication with an area 44 formed by the opening between the upper part of the guide ring and the lower surface of the interior of the neck mold. Annular groove 43 is connected to the area 44 by a series of vertical passages 45. The guide ring also is provided with a plurality of vertical passages 46. Thus, when the source of vacuum is connected to the passage 19 centrally of the plunger, the area surrounding the areas at the sides of the neck-forming cavity will be evacuated through the passages 31, 32 and 46 via the chipped grooves and milled faces 39 and 40 of the neck ring.

At the same time, the annular groove **43** will also be connected to the source of vacuum through the passages **45**. The upper surface of both halves of the neck mold **15** are relieved in the area encompassed by the groove **43** sufficiently to provide a vacuum to the match line between the neck ring and the parison mold that is, however, small enough to avoid glass intrusion during vacuum settle. Thus the upper area surrounding the neck mold is also subject to vacuum conditions when the vacuum is on.

With the plunger in the position shown in Figure 2.5, and with the application of vacuum to the passage **19**, the glass will be drawn into and settled about the plunger **17** and within the confines of the cavity of the neck mold. The particular configuration of the plunger face **23**, with the concentric grooves **24**, will contact the molten glass at the time that the glass is drawn into the neck mold cavity, only at those raised portions. The glass will not be drawn into the bottom of the grooves **24** and only the limited contact area of the tops of the grooves will be chilling the glass. In this manner a less chilled surface is provided in the neck area, centrally of the parison.

After retraction of the plunger by downward movement of a tubular member **47** which is coupled to the plunger, and a short corkage reheat period, air under pressure may be connected to the interior of the passage **19** resulting in air under pressure passing through the passages **30** and into the area previously occupied by the upper end of the plunger. This application of air results in the counterblowing of the glass into the completed parison form.

Production of Large Glass Containers

A. Cruccu; U.S. Patent 4,200,447; April 29, 1980; assigned to Industria Macchine Impianti, Italy describes a process and an apparatus for the automatic production of large glass containers.

The term, "large containers," is to be understood here to mean containers having a capacity generally between 5 and 50 liters (in Italy, for example, the familiar carboys). These containers cannot be produced with the conventional machines for the production of bottles and thus are still being produced today mostly by mainly manual methods.

The process and apparatus for the manufacture of large glass containers comprises preparing a glass gob of a predetermined weight in an automatic mechanical feeder and feeding it to a first preparatory mold at a predetermined frequency. A parison is formed in the preparatory mold at the same frequency, the parison comprising a finished mouth of the glass container to be formed. The parison is then transferred to a second, prefinishing mold and, at the same frequency, a prefinished parison is formed. The prefinished parison is transferred to a third, finishing mold and, at a frequency in a ratio of from 1:1-1:4 of the predetermined frequency, the finished glass container is formed. The expansion of the prefinished parison is controlled by the action of compressed air conveyed to the interior of the parison from its mouth.

Mold Lubricant and Method

G.I. Goodwin, J.L. Margrave and R.E. Wagner; U.S. Patent 4,165,974; Aug. 28, 1979 describe a glass mold lubricant which is particularly useful in providing permanent lubrication for the parts of the mold, including the neck ring, bottom plate and mold cavity. Its use results in longer mold life, no pollution, improved productivity and eliminates a potential fire hazard.

The lubricant may be described as graphite fluoride $(CF_x)_n$. In the preferred form, it is a stable carbon monofluoride having a maximum super stoichiometric fluorine to carbon ratio prepared by the method described in U.S. Patent 3,674,432. In the preferred form, the fluorine to carbon ratio is greater than 0.99/1.00 or 1.12/1.00, which provides for a more stable composition capable of withstanding high heat while retaining its lubricative properties.

In prior lubricating techniques, both the blank and blow molds, as well as the neck ring and the gob chute receive liberal applications of a carbonaceous material, which may be a petroleum product or oil-hydrocarbon mixture. It is sometimes referred to in the trade as "dope." The operator of the machine will, at predetermined periods, lubricate the mold by using a cotton swab or the like, swabbing the mold surfaces with liberal applications of dope. Any other part of the machine which comes into contact with the hot glass is similarly swabbed, such as the neck ring, chute, etc.

The dope, when applied to the hot surfaces, gives off large quantities of smoke and in some cases even causes the swab to burst into flames. The swabbing technique for lubrication is somewhat haphazard and, as might be expected, the distribution is nonuniform. If not carefully applied, the lubricant will drip onto other parts of the machine, creating a fire hazard as well as being unsightly and untidy.

In one method of practicing the process, the cavity surfaces were cleaned by sandblasting and a small quantity of carbon monofluoride in powdered form was placed in the cavity. A cotton cloth, preferably a twill, was used to burnish the carbon monofluoride into the cavity surfaces. Thereafter, the mold was heated to a temperature of between 700° to 800°F.

The treated mold halves were then placed in the machine with the result that no additional lubrication was needed. The glass gob contacting surfaces on the funnel were lubricated in a similar fashion by cleaning through gritblasting and burnishing the carbon monofluoride into the surfaces with a cotton cloth. The ring was heated in the same manner as the molds. Carbon monofluoride was applied by the same method to the neck ring with similar good results in that the neck ring ran for more than eight hours without any problems of sticking and no additional lubrication. The test was terminated for reasons other than sticking of the neck ring.

It is expected that the carbon monofluoride lubricant would last for several hours. Good results were obtained in using carbon monofluoride having a fluorine to carbon ratio of 0.7/1.00, which is grey. Better results were obtained with carbon monofluoride having a ratio of above 0.99/1.00 and preferably in the range of about 1.12/1.00. The temperature resistance of the latter lubricant is increased and the powder is practically pure white in the higher ratio so that any decomposition will result in a colorless, odorless and nontoxic gas.

In another method, the surfaces which come into contact with the hot gob of glass, i.e., the funnel, mold cavities and neck ring are cleaned with a degreaser such as toluene, chlorinated solvents, xylene, acetates or the like. Subsequent to cleaning, the cleaned parts are heated to 750°F for about one-half hour to assure that all of the residue is driven off. The surface may be further cleaned and roughened by gritblasting or the like. Thereafter, the carbon monofluoride is coated by spreading a small quantity onto the surface and burnishing with a cotton cloth or wire brushing it into the metal. In most instances, the molds and other parts are formed of cast iron, which has a high degree of porosity. By repeated application on a roughened surface, a fairly good coating can be embedded and built on the surface. The parts which are lubricated in this fashion are then heated to a temperature of between 600° to 800°F and thereafter installed in the glass molding machine. No further lubrication is necessary and, should any decomposition occur, there is absolutely no smoke.

In one other method of lubrication, the carbon monofluoride was mixed in an aqueous solution with a few drops of tetrahydrofuran added to enhance the dispersion. The material was then sprayed into the hot molds during the molding operation, with the application of air pressure. Agitation of the aqueous solution was required in order to keep the carbon monofluoride uniformly distributed in the solution. This was accomplished with carbon monofluoride having a ratio of 1.12/1.00.

As is evident, in most instances, the dry lubricant approach will be preferred since, once applied, it will require no more of the machine operator's time. Also, in the dry form, the material may be applied with a greater degree of uniformity.

Handling of Glass Sheets During Shaping and Cooling

R.G. Frank; U.S. Patent 4,187,095; February 5, 1980; assigned to PPG Industries, Inc. describes a process for handling glass sheets during shaping and cooling. Each glass sheet in turn is heated to its deformation temperature and conveyed into a glass shaping station. The glass is lifted into engagement with an upper vacuum shaping mold while hot and held against the vacuum shaping mold by vacuum as the glass lifting mechanism retracts downward.

A transfer and tempering ring-like member having stop means upstream of an outline supporting surface conforming to and slightly inside the periphery of the bent glass sheet is brought into a position where its stop means are slightly misaligned upstream of the trailing edge of the glass sheet. The vacuum on the upper mold is released to deposit the shaped glass sheet on the ring-like member. The latter starts suddenly to displace the bent glass sheet into proper alignment with the ring-like member for movement into a cooling station where the shaped glass sheet is quenched as rapidly as possible to harden the glass surface and then transferred to an unloading conveyor at a transfer station. The ring-like member starts its rapid return from the transfer station to the shaping station before the shaped glass sheet is transferred to a position completely downstream of the ring-like member.

Apparatus for Protecting Tong-Suspended Glass Sheets from Buffeting

A.R. Febbo; U.S. Patent 4,185,983; January 29, 1980; assigned to PPG Industries, Inc. describes means to minimize swaying of the glass sheet along the first path of glass sheet travel after the glass sheet leaves a fluid imparting station (which may be a cooling station of a tempering apparatus or a coating application station of a coating apparatus) and an unloading station where the glass sheet is removed from tong-gripping support and transferred to a position between adjacent sets of pegs of a peg conveyor which provides a second path of glass sheet travel to an inspection and packaging station.

A preferred embodiment of the process comprises a tunnel-type furnace **10**, the exit portion of which is shown. A forward conveyor comprising stub rolls **12** extends in a generally horizontal path from its upstream portion to the left of Figure 2.6a to the right-hand end thereof, and extends through the furnace, an exit door **14** for the furnace, a glass sheet shaping station **16**, a first gate **18**, a fluid application station **20**, a second gate **22** beyond the fluid application station, a parking station **24**, and a glass sheet transfer station **25** immediately above one transverse end of a peg conveyor **26**. The latter is provided with sets of pegs **27** arranged in closely spaced sets along the extent of the peg conveyor.

Figure 2.6: Glass Sheet Shaping and Tempering Apparatus

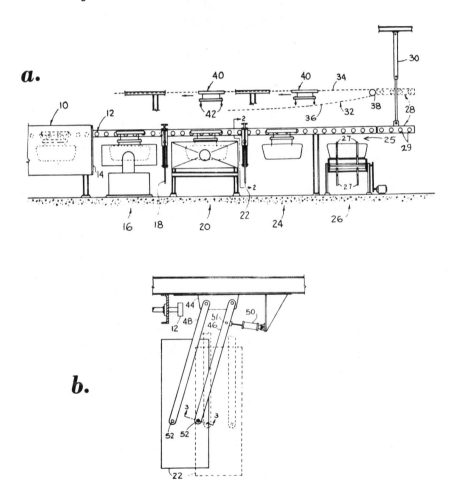

(continued)

Figure 2.6: (continued)

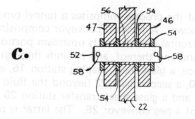

C.

(a) Longitudinal view
(b) Sectional view along line **2–2** of Figure 2.6a showing movable gate
(c) Sectional view along line **3–3** of Figure 2.6b showing actuating means
of movable gate

Source: U.S. Patent 4,185,983

A transfer conveyor section **28** containing a plurality of stub rolls **29** (similar to stub rolls **12**) is operatively connected to lifting and lowering means **30** for movement between a lowered position depicted in solid lines in alignment be-yond the downstream end of the forward conveyor and an upper position de-picted in phantom immediately above the position depicted in solid lines in Figure 2.6a.

A return conveyor drive chain **32** is located above the forward conveyor. The drive chain **32** has an upper run **34** in horizontal alignment with the upper posi-tion occupied by the transfer conveyor section depicted in phantom. The return conveyor drive chain also includes a return run **36**. The return conveyor drive chain is driven by a drive sprocket **38** at the downstream end of the chain. For the purpose of this description, the terms "upstream" and "downstream" are recited with reference to a first path of glass sheet travel defined by the stub rolls **12** of the forward conveyor.

The chain is supported by a chain guide housing for the upper run which en-sures that the upper run is supported along a horizontal path. The chain is pro-vided with longitudinally spaced lugs (not shown) adapted to engage the super-structure of tong support carriages **40** in a manner well known in the art. A carriage is provided for each lug. In addition, several carriages are spaced along the length of the forward conveyor when the apparatus is operating.

The return conveyor drive chain is arranged so that its upper run moves from right to left in the view depicted in 2.6a. An idler sprocket (not shown) similar to drive sprocket **38** is supported upstream of the entrance to the furnace **10** and elevator mechanism similar to the lifting and lowering means and a transfer conveyor section similar to the transfer conveyor section **28** is disposed upstream of the entrance to the furnace so that the additional transfer conveyor section moves into an upper position immediately upstream in alignment with the up-stream end of the upper run and a lower position immediately upstream of the upstream end of the forward conveyor comprising the stub rolls 12. Each of the transfer conveyor sections 28 is identical to the other.

The shaping station 16 of the illustrative apparatus embodiment comprises a pair of shaping molds having facing shaping surfaces of complementary curvature adapted to move between recessed positions at opposite sides of the first path of glass sheet travel and glass engaging positions to impress the shape of their shaping surfaces onto a hot glass sheet positioned therebetween. The fluid application station 20 of the illustrative embodiment comprises opposed nozzle boxes movable in unison relative to the opposite major surfaces of a hot glass sheet located therebetween to impose a temper in the glass sheet.

The conveyor is divided into sections, each operated individually to provide a program of glass sheet movement along the first path of glass sheet travel in a manner well known in the art. See U.S. Patent 3,178,045 for details.

The means for actuating the gate 22 are disclosed in detail in Figure 2.6b and Figure 2.6c. An overhead support plate 44 is provided to support a pair of double links 46 and 47, which are pivoted at their upper ends to the overhead support plate. In addition, a pair of idler links 48 (only one of which is seen) is also pivotally supported at their upper ends to overhead support plate 44. A pivotally supported piston 50 is pivotally connected at the end of its piston rod to a pivot pin 51 that also interconnects the links 46, 47 near their upper ends.

The links 46, 47 and 48 are pivotally connected at their lower ends to the gate and, in order to facilitate pivoting of the gate, washers 54 are provided against the opposite major surfaces of the links and sleeve bearings 56 are provided around each of the pivot pins 52. To ensure that the links remain properly connected in pivotal relation to the gate, cotter pins 58 are provided through diametrically extending apertures near the opposite longitudinal extremities of each of the pivot pins.

It will be seen that the pivotally supported piston is in position to actuate movement of the gate into the solid line position across the first path of glass sheet travel, as depicted in Figure 2.6b, wherein it shields a glass sheet gripped by tongs at the parking station 24 from blasts of tempering medium while a subsequent glass sheet is being cooled by blasts of tempering medium applied at the fluid application station. Thus, when the cooling step of tempering is completed and the glass sheet in the fluid application station is ready for transfer into the parking station, it is a simple matter to actuate the piston to retract the gate from its operative position to an inoperative position where it provides clearance to enable the glass sheet to move from the fluid application station to the parking station (as depicted in the phantom lines of Figure 2.6b) while the glass sheet depicted as being supported in the parking station simultaneously moves into position at the glass sheet transfer station 25 over the peg conveyor 26.

The latter is indexed whenever a glass sheet is deposited thereon so that the sheet previous to the one now occupying the parking station has been moved down one position along the peg conveyor to provide a clear space for the present glass sheet to be dropped onto the peg conveyor.

A tong support carriage 40, provided with glass-gripping tongs 42 and means adapted for engagement by the longitudinally spaced lugs along drive chain **32**, is loaded with one or more glass sheets at the entrance end of the furnace **10** by gripping the glass sheet or sheets with two or more glass-gripping tongs **42**. Rotation of the stub rolls **12** in a forward direction causes the carriage **40** to move in a downstream direction through the furnace. When a glass sheet in a series of glass sheets reaches a position immediately upstream of the exit door **14**, the latter is opened and the carriage with its tongs gripping the glass sheet (or sheets) moves into position at the shaping station **16** between a pair of complementary press bending molds located on opposite sides of the first path of glass sheet travel defined by the forward conveyor comprising stub rolls **12**.

After the press bending molds engage the glass sheet to impress their complementary shapes on the glass, the shaped glass sheet is moved into the fluid application station **20**. Gate **18** located between the shaping station and the fluid application station is retracted to permit the carriage with its tong-gripped glass sheet to move into the fluid application station. At the latter station, the glass sheet, still gripped by the tongs, is swept with opposing moving blasts of cold air that impinge against the opposite major surfaces of the glass sheet. The gate is closed immediately after the trailing edge of the glass sheet moves past the position occupied by the gate.

Gate **22** located beyond the fluid application station on the downstream end thereof, is also closed while the glass sheet is being subjected to the opposing blasts of tempering medium. A previous glass sheet shown in solid lines is disposed at the parking station **24** while a subsequent glass sheet is being shaped and cooled. When the glass sheet at the fluid application station is ready to be removed, the immediately preceding glass sheet at the parking station is transferred into position at the glass sheet transfer station **25** over a space between two adjacent sets of pegs at the upstream end of the peg conveyor **26**. The gate **22** between the fluid application station and the parking station is opened to permit passage of the glass sheet that has just been cooled while the preceding glass sheet moves from the parking station into a position at the glass sheet transfer station over the peg conveyor. The gate **22** is closed once more when the trailing edge of the cooled glass sheet clears the position occupied by the gate **22** in its operative position.

The process makes it possible for a glass sheet to remain with its major surfaces exposed in the vicinity of a fluid application station while a subsequent glass sheet is being processed at the fluid application station without introducing undue buffeting of the glass sheet. By this process, the glass sheet can be cooled by radiant cooling for an additional period of time without introducing swaying that may cause the glass sheet to become misaligned with a space between adjacent sets of pegs of the peg conveyor when the glass sheet is transferred from gripping engagement by tongs to lower edge support by the peg conveyor at the glass sheet transfer station.

Shaping and Tempering Process Employing Pivotal Transfer Apparatus

J.N. Eggert, R.A. Herrington and W.W. Oelke; U.S. Patent 4,138,237; Feb. 6, 1979; assigned to Libbey-Owens-Ford Company describe the production of bent, tempered sheets of glass and, more particularly, to a transfer apparatus for reorienting a glass sheet from a horizontal plane to a substantially vertical plane for optimum processing.

The mode of operation of the apparatus in bending and tempering one sheet of glass is as follows referring to Figures 27a and 27b:

Figure 2.7: Glass Sheet Heat-Treating Apparatus

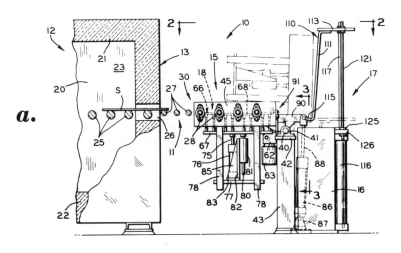

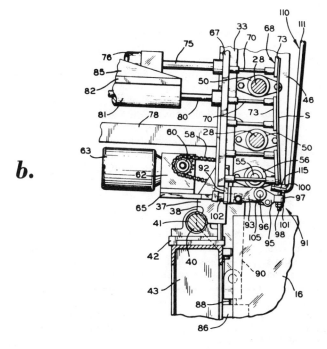

(a) Side elevational view
(b) Transfer apparatus in glass releasing position

Source: U.S. Patent 4,138,237

A flat glass sheet S is loaded onto the conveyor rolls 25 at the entrance end (not shown) of the furnace 13 for movement through the heating chamber 20 wherein the sheet is heated to substantially its softening point or bending temperature. This heated sheet passes through the opening 26 and is successively transferred onto conveyor rolls 27 and then conveyor rolls 28 of the pivotal conveyor section of the transfer apparatus 15. The sheet is accurately located above the mold 66 when the leading edge thereof engages the stop members 100 located in the path of the horizontally moving sheet.

A photoelectric cell (not shown), or other suitable detection device, senses the entry of the glass sheet into the bending area for initiating a bending cycle by energizing timing mechanisms which control the actions of the mold actuating cylinder 76 and the transfer apparatus actuating cylinder 86. The timing mechanism is so designed that the mold begins its ascent by actuation of cylinder 76 when a sheet is accurately positioned by means of engagement with the stop members. The mold moves rapidly upwardly causing the shaping rail 68 to engage the marginal edge of the sheet and thereby lift the latter off the rolls 28 quickly.

The combined effects of inertial and gravitational forces acting on the sheet as it is raised by the shaping rail above the rolls 28 causes the glass sheet to sag into conformity with the shaping surface of the molding shaping rail. The stop members, which are mounted on the movable platen 67, remain engaged with the leading edge of the sheet.

At the same time that the mold is raised to effect bending, the cylinder 86 is actuated to retract piston rod 88 for swinging the frame 30, together with all the components mounted thereon, about the pivot shaft 40 into a vertical attitude, as shown in Figure 2.7b and in phantom in Figure 2.7a. Bending of the sheet is completed quickly during the early swinging movement of the frame. The bent sheet, supported against the shaping rail and along its leading edge by the stop members, continues to be swung along with the mold into a generally upright position in close proximity to the legs 111 of receiving element 110 with the leading edge of the now vertically disposed sheet spaced just slightly above feet 115 of element 110. The momentum imparted to the sheet during its pivotal movement on the frame causes the sheet to continue in its arcuate path upon stoppage of the frame so that it passes a true vertical position and comes to rest against the legs of the receiving element.

During this arcuate movement past true vertical, the sheet pivots about the leading or lower edge of the sheet which is supported on the stop members. The cylinders 102 are then actuated to retract their respective piston rods for pivoting brackets 95 and lowering stop members 100 to deposit the glass sheet onto the feet of the receiving element. The sheet is not dropped onto the feet. Rather, the retracting stops 100 operate to lower the glass sheet gently onto the feet so that the lower edge of the sheet gingerly engages the feet for a smooth and easy transition. Cylinder 116 is actuated substantially simultaneously with the release of the bent sheet to retract piston rod 117 and lower the sheet into the quench tank 16 for immersion into the bath of coolant contained therein. The sheet is immersed within the bath for a time sufficient to properly temper the sheet and is then withdrawn from the bath by raising the receiving element 110, via cylinder 116, into its upper position shown in Figure 2.7a.

It should be appreciated that the sequence of operations in bending the sheet, pivoting it into an upright position and then transferring the same onto the element 110 upon which it is lowered into the quench bath is performed rapidly to retain the requisite heat necessary for proper tempering.

As a result of this process, a transfer apparatus is provided for reorienting a heated glass sheet, whether flat or bent, from a horizontal plane into a vertical plane. This feature permits combining a horizontal glass processing system with a vertically oriented quench system to successfully heat treat thin glass in a manner retaining the advantages of both systems. Thus, a sheet heated and, if desired, bent in a horizontal production line is transposed from its horizontal disposition into a vertical disposition for immediate vertical immersion in the coolant medium of a quench tank. Accordingly, very thin glass, having thicknesses on the order of 0.090 inch and less, can now be satisfactorily processed on the major portion of a horizontal production line by virtue of this transfer apparatus.

STRENGTHENING

Application of Potassium Fluoride and Metal Acetate

It is known that the surface of glass articles can be strengthened by producing a compressive stress layer at the surface of the article. One common method of achieving this is known as chemical tempering which creates a compressive stress at the surface primarily by a chemical alteration of the surface region of the article.

One method of chemical tempering involves the exchange of ions at the surface of the glass. An ion is made available in a substantial concentration at the surface of the article and diffuses into the glass article by a series of ion exchange reactions. The most common form of ion exchange presently in use involves the exchange of larger ions, such as potassium ions, for the sodium ions present in the glass.

L. Levene and R.W. McClung; U.S. Patent 4,134,746; January 16, 1979; assigned to Domglas Inc., Canada describe a process for strengthening the surface of an article of a soda-lime glass in which a solution of potassium fluoride containing a sufficient quantity of a metal acetate which prevents etching of glass by the fluoride ions is applied to the surface of the article at a temperature in the range of from 200° to 900°F; the article is maintained at an elevated temperature below the strain point of the glass to provide an exchange of potassium ions for the sodium ions in the glass to a depth sufficient to produce a substantial compressive surface layer; the article is cooled and washed to remove any residue from the surface.

Example: A solution was prepared by dissolving potassium fluoride dihydrate (61.9 g, 0.66 mol) in water (20 g). A second solution was prepared by dissolving nickel acetate tetrahydrate (1.09 g, 0.0044 mol) in water (16.3 g). The two solutions were mixed producing a clear solution having a potassium to nickel molar ratio of 150:1.

Glass rods were heated to 750°F and sprayed with the clear solution. The coated rods were subjected to a heat treatment at 750°F for 45 minutes. The surfaces of the rods were not etched. Compression layers of from 3 to 5 μm were obtained with an average strength of 12 kg/mm^2 (quartz wedge).

Lithium-Containing Glass Suitable for Ion-Exchange Strengthening

D.W. Rinehart; U.S. Patent 4,156,755; May 29, 1979; assigned to PPG Industries, Inc. has found that ion exchange glasses consisting essentially of silica, soda, lithia, alumina, and zirconia in specific concentration ranges have the desirable combination of low melting temperature (less than about 2850°F), wide working ranges (at least 50°F), the ability to attain high strength and deep compression layers with reasonably short ion exchange treatment times, good durability, and resistance to devitrification after reheating. The glass compositions are as follows (in percent by weight on the oxide basis):

Component	Range (%)	Preferred Range (%)
SiO_2	59-63	60-63
Na_2O	10-13	10-12
Li_2O	4-5.5	4-5.5
Al_2O_3	15-23	17-19
ZrO_2	2-5	3.5-5
$(Al_2O_3 + ZrO_2)$	19-25	21.5-24

Minor quantities (up to about 5% by weight total) of other glass-forming materials and glass modifiers may be included, such as MgO, MnO, TiO_2, Sb_2O_3, As_2O_3, K_2O, PbO, colorants, and mixtures thereof.

Addition of Alumina and/or Zirconia to Soda-Lime-Silica Glass

D.W. Rinehart; U.S. Patent 4,192,689; March 11, 1980; assigned to PPG Industries, Inc. has found that soda-lime-silica glass compositions having essentially the same physical properties as conventional flat glass compositions have their ion exchange properties enhanced, and in some cases also have their melting temperatures lowered by the inclusion of relatively small amounts of Al_2O_3 and/or ZrO_2. The alumina and zirconia contents are greater than that normally found in soda-lime-silica glass compositions but less than that in glass compositions specifically formulated for ion exchange. More specifically, alumina is included in a weight percent range of about 2.5 to 4.5% and zirconia in a weight percent range of about 2 to 4.5%.

It has been found that when soda-lime-silica glass compositions are thus modified by the inclusions of alumina and/or zirconia their strain point temperatures are advantageously raised, and in the most preferred embodiment, their melting temperatures are reduced as well. The lowered melting temperature means that less energy is consumed in producing a given amount of glass, and the increased strain point temperature advantageously enables ion exchange strengthening treatments to be carried out at higher temperatures and thus greater speed without the danger of thermally relaxing the induced stresses.

In order to accommodate the alumina and/or zirconia additions to the soda-lime-silica glass compositions, the silica content may be reduced slightly, and it was also found to be advantageous to reduce the CaO concentration in order to avoid devitrification problems.

The glass compositions consist essentially of the following constituent ranges in percent by weight on the oxide basis:

Composition	Percent
SiO_2	67–74
Na_2O	12–17
K_2O	0–8
$(Na_2O + K_2O)$	12–17
CaO	6–9
Al_2O_3 or	2.5–4.5
ZrO_2 or	2–4.5
$(Al_2O_3 + ZrO_2)$	3–9

Additionally, up to about 5% by weight total of conventional melting and fining aids and colorants may be included in the above composition.

Strengthening Glass Articles with Mixed Potassium Salts

M. Watanabe; U.S. Patent 4,206,253; June 3, 1980; assigned to Yamamura Glass KK, Japan describes a method of strengthening chemically a glass container, which comprises applying to the outer surface and inner surface of the glass container, an aqueous solution containing a small amount of surfactant and a high concentration of mixed potassium salts consisting of potassium nitrate having a relatively low melting point and at least one potassium salt having a relatively high melting point selected from the group consisting of potassium chloride and potassium sulfate in such a proportion that 10 to 80% by weight of the potassium salt be present in solid phase, when heated at an ion exchange temperature, to prevent the fused potassium salt from flowing off from the surfaces of the glass container.

The temperature of the glass container is lower than that of the aqueous solution, thereby depositing the potassium salt on the outer surface and inner surface of the container through the temperature difference to form a potassium salt adhered layer. The process then comprises drying, holding the container at an elevated temperature below the strain point of the glass but as near the strain point as possible for a period of time sufficient to form a compressive stress layer on the outer surface and inner surface of the glass container, cooling the glass container to room temperature and removing the residual potassium salts.

Example 1: A glass container having a strain point of 510°C was formed from an ordinary soda-lime-silica glass consisting of 72% SiO_2, 2% Al_2O_3, 14% Na_2O, 1% K_2O, 10% CaO, 0.4% MgO and trace impurities and minor constituents. The formed glass container was passed through a lehr and cooled to room temperature.

Potassium chloride and potassium nitrate were mixed in a proportion of 2:1 by weight and dissolved in warm water at 75°C to prepare a high concentration aqueous solution (saturated aqueous solution) of the potassium salts, to which a small amount (0.2 to 1.2% by volume) of an anionic surfactant polyoxyethylene sodium alkylsulfate was then added. This aqueous solution was sprayed uniformly onto the inner surface of the glass container at a relatively low temperature (45° to 60°C) and then the outer surface thereof was sprayed therewith.

The glass container was preheated at 130°C for 30 minutes in a dryer and the outer surface thereof was optionally oversprayed. The thus preheated glass container was subjected to a heat treatment at 505°C for 60 minutes, cooled and washed. Thin fragments of 300 μ were cut out of this glass container and then

Glass Technology

subjected to measurement of the compressive stress layer and stress value using a polarization microscope, thus obtaining results as shown in the table below.

Examples 2 through 7: The procedure of Example 1 was repeated with varying ratios of potassium salts. The results are also shown in the table.

Ex. No.	Aqueous Salt Solution		Stress Layer (μ)	Stress Value (kg/cm^2)
1	KCl:KNO$_3$ (2:1)	Outer surface	13	993
		Inner surface	13	890
2	KCl:KNO$_3$ (1:1)	Outer surface	15	1,100
		Inner surface	15	990
3	KCl:KNO$_3$ (3:1)	Outer surface	13	1,064
		Inner surface	13	960
4	KCl:KNO$_3$ (5:1)	Outer surface	14.5	1,061
		Inner surface	14	950
5	K$_2$SO$_4$:KNO$_3$ (1:1)	Outer surface	15.5	879
		Inner surface	15	790
6	K$_2$SO$_4$:KNO$_3$ (1:3)	Outer surface	15.5	705
		Inner surface	15.5	705
7	KCl:KNO$_3$:K$_2$SO$_4$ (3:1:1)	Outer surface	16	1,182
		Inner surface	16	1,070
1*	KCl only	Outer surface	8	850
		Inner surface	8	760
2*	KNO$_3$ only	Outer surface	10	700
		Inner surface	10	630
3*	K$_2$SO$_4$ only	Outer surface	12	930
		Inner surface	12	837
4*	KCl:KNO$_3$ (1:3)	Outer surface	11	780
		Inner surface	11	702

*Control.

Extractable-Alkali-Decreasing Treatment

In a modification of the previous process, M. Watanabe; U.S. Patent 4,164,402; August 14, 1979; assigned to Yamamura Glass Co., Ltd., Japan has found that when the internal surfaces of the glass containers are exposed to the action of an agent capable of decreasing the amount of extractable alkali from the surface at an elevated temperature immediately after the blowing process but before the annealing process and then the resulting containers are treated to form a compressive stress layer by exchanging sodium ions with potassium ions in both the external and internal surface portions thereof, the strength, particularly impact strength, of the resulting containers is synergistically greatly improved more effectively than with either the extractable-alkali-decreasing treatment or the ion-exchange treatment alone.

Preferably, the glass container may be provided with a coating layer of a metal oxide on the external surface prior to the extractable-alkali-decreasing treatment to increase abrasion resistance.

The process comprises the steps of:

(1) exposing the internal surface of a sodium-containing glass container to the action of an agent capable of

      decreasing the amount of extractable alkali from
      the annealing process;

(2)   passing the container through the annealing station; and

(3)   subjecting the external and internal surfaces of the glass
      container to a treatment for exchanging sodium ions in
      the surface portion with potassium ions to form a com-
      pressive stress layer in the surface portion.

The amount of extractable alkali is greatly decreased by treatment with ammonium chloride. Similarly, other extractable alkali-decreasing agents such as ammonium sulfate, aluminum chloride, elemental sulfur or sulfur dioxide may be
used for this purpose.

Example: Thin-walled, light glass bottles (wide mouth, coffee bottle, weighing
273 g, 2.5 mm thick) were produced by the press and blow process. The glass
was a conventional soda-lima glass having a strain point of 505°C and the following compositions: 72% SiO_2, 2% Al_2O_3, 14% Na_2O, 0.1% K_2O, 10% CaO, 0.4%
MgO and trace impurities.

The bottles were contacted on their external surfaces with vapor of tin tetrachloride immediately after the formation of these bottles and thus, prior to
their annealing. The temperature of these bottles at this time was between
about 500° and 600°C. A coating of stannic oxide was thus formed on the external surface. Then about 0.2 to about 0.5 g of an extractable-alkali-decreasing
agent shown in the table below was placed into the interior of the bottles and
the bottles were annealed. After passing through the annealing lehr the bottles
were washed with water.

69.0 g of potassium chloride, 34.7 g of potassium nitrate, 8.5 g of potassium
sulfate and 0.5 ml of an anionic surfactant were added in 200 ml of water and
the mixture was heated to 70°C to make a solution. The bottles were treated
with the solution by injection of the solution into the interior of the bottles
and then drying to form a layer of deposition of the mixed potassium salts on
the internal surface. The same layer of deposition of the mixed potassium salts
was formed on the external surface of the bottles by dipping the bottles into
the same solution and then drying. The bottles thus treated were heated at
500°C for 1 hour, cooled to room temperature and washed with water.

The impact strength of the resulting bottles was determined in accordance with
JIS S-2302 (striking point was 80 mm above the bottom). For comparison, the
strength of corresponding bottles which had been subjected to only the extractable-alkali-decreasing treatment or the ion-exchange treatment alone, and of bottles which had not been treated, were determined. The data are shown below.

Impact Strength of Untreated and Treated Bottles

Extractable-Alkali-Decreasing Agent		Untreated	Decrease in Extractable Alkali Alone	Ion-Exchange Alone	Decrease in Extractable Alkali + Ion Exchange
Sulfur	x	7.6	9.0	10.55	13.0
powder	min	2	4	5	6
	%	100	118	139	171

(continued)

Extractable Alkali Decreasing Agent		Untreated	Decrease in Extractable Alkali Alone	Ion-Exchange Alone	Decrease in Extractable Alkali + Ion Exchange
SO$_2$ gas	x	7.6	9.3	10.55	14.4
	min	2	4	5	7
	%	100	122	139	189
NH$_4$Cl	x	7.6	9.6	10.55	15.2
	min	2	3	5	6
	%	100	126	139	200
(NH$_4$)$_2$SO$_4$	x	7.6	9.2	10.55	13.5
	min	2	3	5	5
	%	100	121	139	178
AlCl$_3$	x	7.6	8.8	10.55	12.0
	min	2	3	5	5
	%	100	116	139	158
(NH$_4$)$_2$SO$_4$ + AlCl$_3$ (10:1)	x	7.6	8.8	10.55	13.0
	min	2	4	5	5
	%	100	116	139	171
NH$_4$Cl + AlCl$_3$ (10:1)	x	7.6	9.6	10.55	12.9
	min	2	5	5	7
	%	100	126	139	170
(NH$_4$)$_2$SO$_4$ + NH$_4$Cl (1:1)	x	7.6	9.8	10.55	13.2
	min	2	5	5	6
	%	100	129	139	174
Teflon	x	7.6	8.8	10.55	12.6
	min	2	3	5	7
	%	100	116	139	166
Water mist	x	7.6	9.1	10.55	12.7
	min	2	4	5	6
	%	100	120	139	167

x = mean value of 20 bottles in kg-cm; and min = minimum value in kg-cm.

Improving Durability of Spontaneous NaF Opal Glassware

A spontaneous sodium fluoride opal glass is a glass which spontaneously transforms from a clear state to an opalized state upon cooling from a melt, wherein sodium fluoride crystals are deemed to constitute the principal opacifying phase. Glasses of this type have recently been considered for tableware and related applications because of the dense, white opacity exhibited thereby.

The relatively poor durability exhibited by spontaneous NaF opal glasses gives rise to several problems in use, among which are weathering and surface chalking or crazing.

K.-E. Lu and W.H. Tarcza; U.S. Patent 4,187,094; February 5, 1980; assigned to Corning Glass Works describe a simple method for improving the durability of spontaneous sodium fluoride opal glassware, particularly against surface attack by water. That method broadly comprises, first, briefly exposing an opal glass article to a weak Na$_2$O-extracting medium, such as hot water, for a time sufficient to form a Na$_2$O-depleted surface layer thereon. Thereafter, the glass article is heated to a temperature in the range of about 200° to 500°C for a time sufficient to consolidate the Na$_2$O-depleted layer.

Opal glassware suitably treated in accordance with this method includes glassware formed of spontaneous opal glasses wherein NaF constitutes the principal opacifying phase. Such glasses are known in the art, typically comprising, in weight percent, about 8 to 13% Na_2O, 5 to 9% Al_2O_3, 71 to 78% SiO_2 and 3 to 6% F. In addition, the glass may include optional constituents such as K_2O, Li_2O, CaO, BaO, SrO, B_2O_3 and the like to modify the melting, forming, opalizing and other properties of the glass, as desired. However, the total of all such additions will normally be held to a level not exceeding about 10% by weight.

The effectiveness of the method in improving the durability of sodium fluoride opal glassware depends upon controlling the extent of soda removal from the surface of the treated glass article; limited-duration treatments and weak Na_2O-extracting media are used in order to avoid the complete removal of all extractable Na_2O from the surfaces of the article. If too much soda is removed, microcracking of the surface of the glass article will occur during the heating step of the durability-enhancing procedure. On the other hand, insufficient removal of soda renders the article subject to surface attack in later use.

Example: Two spontaneous sodium fluoride opal glass articles, consisting of pressed 6¾" bread and butter plates composed of a glass having a composition, in parts by weight as calculated from the batch, of about 73.6 parts SiO_2, 6.9 parts Al_2O_3, 9.0 parts Na_2O, 1.4 parts K_2O, 3.8 parts SrO, 0.8 part CaO, 1.6 parts B_2O_3 and 5.0 parts F are provided for treatment. Pressed glass plates of this configuration and composition exhibit dense white opacity and excellent surface gloss. However, when one of these plates is subjected to a 6-hour exposure to boiling water, simulating repeated washing of the plate, followed by a 15-minute heat treatment at 300°C to simulate oven baking, microscopically visible surface microcracking and some loss of surface gloss are observed to result.

In order to obtain improvements in surface durability against attack by water, the other plate is subjected to a durability improvement treatment comprising an initial immersion in boiling water for 5 minutes to achieve limited extraction of Na_2O from the surface layer of the plate. Thereafter, the plate is placed in a lehr operating at 300°C for approximately 10 minutes to consolidate the Na_2O-depleted surface layer thereon.

The treated plate is thereafter tested for surface durability in the same manner as the first plate, by immersion in boiling water for 6 hours followed by heating in a lehr at 300°C for 15 minutes. The article is then removed from the lehr and examined, and found to retain good surface gloss. Microscopic examination of the surfaces of the article at 30X magnification reveals no significant surface microcracking thereon.

Toughening of Glass in a Fluidized Bed of Particulate Material

P. Ward and G.M. Ballard; U.S. Patent 4,205,976; June 3, 1980; assigned to Pilkington Brothers Limited, England describe a method of thermally toughening a glass sheet in which the glass sheet is heated to a temperature near to its softening point and then lowered into a gas-fluidized bed of particulate material. A void region is established beneath the lower edge of the glass sheet as it is initially immersed in the fluidized bed.

As shown in Figure 2.8a of the drawings, a glass sheet 1 is suspended from a tong bar 2 by tongs 3. The tong bar is carried from hoist cables 4 by means of which the tong bar can be lowered to immerse the glass sheet 1 in a bed of gas-fluidized particulate material. The bed 5 is held in a container 6 and fluidizing gas is fed into the bed from a plenum chamber 7 through a membrane 8 forming the base of the container, which membrane comprises a number of layers of low permeability paper across which there is a relatively high pressure drop.

Figure 2.8: Apparatus for Toughening Glass Sheet in a Gas-Fluidized Bed

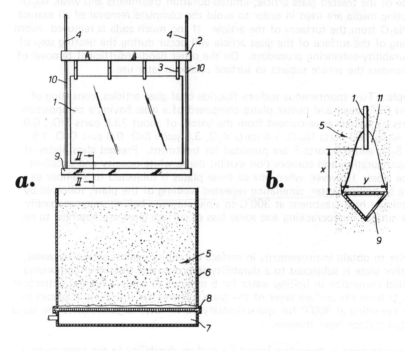

(a) Front elevational view
(b) Cross section along line II–II of Figure 2.8a

Source: U.S. Patent 4,205,976

The gas-fluidized bed of particulate material may be maintained in a quiescent uniformly expanded state of particulate fluidization by regulation of the velocity of fluidizing gas through the bed between the gas velocity which produces incipient fluidization and that which produces maximum expansion of the bed. An elongated tubular shielding member 9 of triangular cross section is suspended by rods 10 from the tong bar. The apex of the triangular cross section points downwardly away from the lower edge of the glass sheet. The maximum width of the member, which is the width of the base of the cross section, is designated **y** and the base is spaced at a distance x below the lower edge of the glass sheet. The shielding member extends beneath and parallel to the full length of the glass sheet.

In use of the apparatus of Figure 2.8a for toughening of the glass sheet 1, the glass sheet is heated to a temperature near to its softening point, for example, a temperature in the range of 620° to 680°C for soda-lime-silica glass, and is then lowered into the bed 5 of gas-fluidized particulate material, which is maintained at a temperature in the range of 30° to 150°C.

As shown in Figure 2.8b of the drawings, as the lower edge of the glass sheet enters the bed the shielding member 9 modifies the flow of fluidizing gas in the bed so as to establish a void region 11 which extends above the shielding member and envelopes the lower edge of the glass sheet. The void region is in the form of a bubble of the fluidizing gas which is substantially free of entrained particulate material. The rate of cooling of the lower edge of the glass sheet is considerably reduced by the presence of the void region, which thereby avoids the setting up of tensile stresses in the lower edge of the glass sheet such as would normally occur due to the greater rate of cooling of the lower edge of the glass sheet and which could otherwise cause fracture of the glass sheet upon entry into the fluidized bed.

The void region disperses subsequent to entry of the lower edge of the glass sheet into the fluidized bed so that all of the major surfaces of the glass sheet are then subjected to cooling by the particulate material of the fluidized bed to bring about a uniform overall toughening of the glass sheet including the region of the glass sheet adjacent its lower edge.

As an example, the fluidized bed is of porous γ-alumina having a particle density of 2.2 g/cm^3 and a particle size in the range 20 to 120 μ, the average particle size being 64 μ, and fluidizing gas was fed into the plenum chamber 7 of the container 6 so as to establish an upward flow of fluidizing gas through the membrane 8. The temperature of the fluidized bed was in the range 50° to 80°C. Glass sheets 3 mm thick had their lower edges edge-finished by being fully ground using a 400 diamond grit wheel. The sheets were heated and then lowered into the fluidized bed at a speed of 0.3 m/s. The width y of the shielding member is 51 mm; the included angle at the apex of the member is 60°; and the fluidizing gas velocity is 11 mm/s. To provide comparative figures, a number of glass sheets were first processed without the use of the shielding member.

Then a number of glass sheets were heated in a furnace while suspended from the tong bar 2 with the shielding member in position at various spacings x below the lower edge of the sheet. When the base of the shielding member was spaced only about 12 mm from the lower edge of the glass it was found that the shielding member prevented adequate heating of the lower edge of the glass sheet in the furnace so that a poor yield of only 14% was obtained because of the lower temperature of the bottom edge of the glass sheet. This difficulty was avoided by placing reflective silver foil on the upper surface of the shielding member to reflect additional heat towards the lower edge of the glass sheet. This difficulty can also be avoided by heating the glass sheet prior to positioning the shielding member below the glass sheet.

The results obtained were as follows: no shielding member, 33% yield; when x = 12 mm, 100% yield; when x = 25 mm, 80% yield; and when x = 50 mm, 100% yield. The yield figures are the number of unfractured toughened sheets obtained as a percentage of the total number of sheets processed. The temperature of the lower edge of glass sheet was 623° to 633°C.

Differentially Toughened Safety Glass from Localized Gas Flow

G. Greenhalgh; U.S. Patents 4,182,619; January 8, 1980; and 4,178,414; Dec. 11, 1979; assigned to Triplex Safety Glass Company, Limited, United Kingdom describes a method of toughening a glass sheet comprising advancing the glass sheet through a quenching station where the sheet is subjected to at least one localized gas flow and pulsing that gas flow at a repetition frequency related to the speed of advance of the glass through the quenching station to induce in the glass a distribution of regions of more highly toughened glass interspersed with regions of lesser toughened glass.

Figure 2.9 illustrates the fracture pattern of a toughened glass sheet suitable for use as the side window or rear window of a motor vehicle produced by the process. The glass sheet has a distribution, in rectangular array, of localized areas **1** of more highly toughened glass in the glass sheet interspersed with areas **2** of lesser toughened glass. Areas **3** of the glass have a medium toughened stress and in each of the areas **3** the principal stresses are unequal with the major principal stress acting in the direction indicated by the arrows **4**.

Figure 2.9: Differentially Toughened Glass Sheet

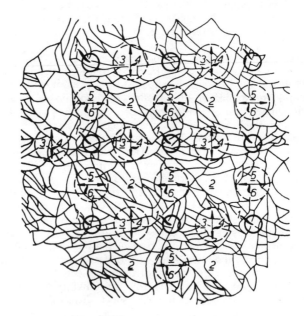

Source: U.S. Patent 4,182,619

Areas **5** of the glass also have a medium toughening stress and have unequal principal stresses with the major principal stress acting in the direction indicated by the arrows **6**. The major principal stress **6** in each area **5** acts in a direction substantially perpendicular to the direction of the major principal stress **4** in each of the areas **3**.

The size of the particles produced in a fractured glass sheet depends on the degree of toughening of the glass and in general the fineness of the particles increases with the degree of toughening. Hence the particles of relatively small size are produced in the more highly toughened areas 1, in the lesser toughened area 2 larger particles are produced, and in the areas 3 and 5 having a medium toughening stress particles of medium size are produced. This distribution of small, larger and medium sized particles is produced over the whole surface of the fractured glass sheet, and there are no splines in the fracture.

In a related process, *G. Greenhalgh; U.S. Patent 4,198,463; April 15, 1980; assigned to Triplex Safety Glass Company, Limited, England* describes a method of producing a glass sheet of thickness in the range 2.5 mm to 4.0 mm for use as a rear or side window for a motor vehicle.

The method comprises advancing the glass sheet between flows of quenching gas to produce in the sheet an average central tensile stress in the range from a maximum of 62 MN/m^2 for all glass thicknesses from 2.5 cm to 4.0 mm to a minimum of 56.5 MN/m^2 for 2.5 mm thick glass varying inversely with thickness down to a minimum of 44.0 MN/m^2 for 4.0 mm thick glass, and directing at least one gas jet at the advancing glass to produce at least one strip-shaped region in the glass sheet of more highly toughened glass such that the central tensile stress in the strip-shaped region is in the range from 2 MN/m^2 to 5 MN/m^2 greater than the average central tensile stress in the sheet, and such that there are major and minor principal stresses in the strip-shaped region acting in the plane of the glass sheet.

The difference between the major and minor principal stresses in the region should be in the range 5 MN/m^2 to 25 MN/m^2. The process may be applied to a glass sheet which is being advanced horizontally either on a roller conveyor or on a gaseous support.

For producing a plurality of strip-shaped regions of more highly toughened glass, the process further provides a method comprising advancing the glass sheet horizontally through a quenching station where the glass is quenched by flows of chilling air over both faces of the glass, directing towards at least one face of the glass gas jets which are spaced apart in at least one row transversely to the direction of advance of the glass, and regulating the speed of advance of the glass so that one face of the glass is subjected to localized gas flows to produce a distribution of parallel regions of more highly toughened glass interspersed with regions of lesser toughened glass.

Differentially Toughened Glass from a Fluidized Bed

D.C. Wright and B. Marsh; U.S. Patent 4,198,226; April 15, 1980; and D.C. Wright, B. Marsh and W. Wiechers; U.S. Patent 4,194,898; March 25, 1980; assigned to Pilkington Brothers Ltd., England describe a method of thermally toughening a glass sheet in a gas-fluidized bed of particulate material, comprising maintaining the gas-fluidized bed at glass quenching temperature, heating a glass sheet to a temperature above its strain point, extracting gas from a series of vertical, horizontally-spaced regions of the bed to maintain the particulate material in each of those regions in an unfluidized static condition, and lowering the hot glass sheet vertically into the bed so that the parts of the glass sheet to receive a lesser degree of toughening contact the regions of unfluidized static particulate material and the parts of the glass sheet between the regions are toughened to a higher degree by contact with fluidized particulate material, thereby producing a differentially toughened glass sheet having bands of lesser toughened glass alternating with bands of more highly toughened glass.

In one apparatus according to the process the gas-extraction means may comprise two elongated gas-extraction ducts arranged face-to-face and spaced apart in the container to define a path for an article therebetween, which extraction ducts are arranged to extract gas from the region of the fluidized bed between the ducts. Preferably the ducts are mounted horizontally in an upper part of the container. The two ducts may be parallel to each other.

Further the gas-extraction means may comprise a plurality of vertical gas-extraction ducts which are spaced apart from each other and are positioned in the container to extract gas from a plurality of localized vertical regions in the path of the article.

The gas-extraction ducts may be first and second banks of vertical parallel gas-extraction ducts mounted in the container, with the ducts in the two banks arranged facing one another and spaced apart to permit vertical entry of an article between the two banks. Each of the vertical gas-extraction ducts may be divided vertically into compartments with an individual gas-extraction pipe connected to each compartment.

Disposal of Shattered Glass During Tempering

This process broadly relates to the tempering of glass sheets and more particularly to apparatus for tempering glass sheets by the continuous horizontal process.

According to such process, the glass sheets are conveyed horizontally through a furnace in which they are heated to substantially the softening point of the glass and, upon exiting from the furnace, are conveyed upon horizontally spaced conveyor rolls between upper and lower blastheads and rapidly cooled thereby to establish a permanent stress pattern in the sheets.

It has been found that the stress conditions created in the glass sheets during tempering frequently cause the sheets to shatter during such operations. This presents a serious problem in that the broken glass becomes lodged between the conveyor rolls causing a back-up of succeeding sheets resulting in additional breakage as well as damage to the equipment. Close monitoring of the tempering operations must therefore be maintained to see that the broken glass is removed as quickly as possible. With present methods this is a time consuming and laborious procedure.

H.E. McKelvey; U.S. Patent 4,138,241; February 6, 1979; assigned to Shatterproof Glass Corporation describes a process whereby the removal of glass sheets that are shattered during the tempering operations can be easily and quickly effected in an efficient manner and with the expenditure of a minimum amount of time and labor.

Referring to Figure 2.10, the tempering apparatus includes generally an elongated furnace A in which the glass sheets B are heated and a quenching section C including the upper and lower blastheads D and E respectively for rapidly chilling the heated glass sheets as they issue from the furnace to establish a permanent stress pattern therein.

Figure 2.10: Side Elevation of Tempering Apparatus

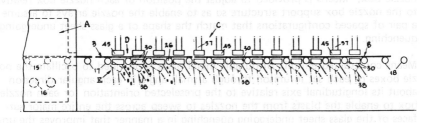

Source: U.S. Patent 4,138,241

The glass sheets are transported horizontally through the furnace by any suitable conveyor means such as the rolls **15**. The furnace can also be heated in any conventional manner such as by the gas fired radiant heating tubes **16**.

Upon issuing from the furnace the heated glass sheets move into the quenching section **C** where they are received and carried forwardly by a series of spaced conveyor rolls **17** located between the upper and lower blastheads **D** and **E**. The tempered glass sheets are discharged from the quenching section onto a series of carry-out conveyor rolls **18**.

The lower blasthead **E** embodies a plurality of elongated, parallel tubular members **30** which extend transverse to the direction of travel of the glass sheets. The tubular members are substantially rectangular in cross section and are arranged in groups of three positioned between adjacent conveyor rolls **17**, although one or any desired number may be used depending upon the distance between the conveyor rolls.

The upper blasthead **D** is similar to the lower blasthead **C** in that it also includes transverse, parallel tubular members **45** arranged in groups of three which are positioned directly above like groups of lower tubular members **30**.

The process embodies means for mounting the upper and lower tubular members **30** and **45** in such a way that they can be moved bodily away from the line of travel of the glass sheets to cause the broken glass to fall freely between the conveyor rolls and away from the apparatus. More specifically, the upper and lower tubular members are pivotally mounted at one end so that they can be swung upwardly and downwardly respectively relative to the conveyor rolls.

Tempering of Flat or Curved Sheets Supported Vertically

V.R. Imler; U.S. Patent 4,157,910; June 12, 1979; assigned to PPG Industries, Inc. describes an apparatus for tempering flat or curved glass sheets supported vertically. The apparatus comprises a first set of nozzle boxes providing a first array of nozzles disposed to one side of a position to be occupied by a glass sheet during tempering, a second set of nozzle boxes providing a second array of nozzles disposed to the other side of the position and facing the first array of nozzles. Each of the nozzle boxes is provided with means for adjusting the orientation of the nozzles that extend in parallel relation to one another from a common wall of the nozzle box relative to the longitudinal axis of the associated nozzle box. Means is provided to adjust the position of each nozzle box relative to the nozzle box support structure so as to enable the nozzle arrays to assume a pair of spaced configurations that match the shape of a glass sheet undergoing quenching.

Means constructed entirely of durable metal parts is provided to oscillate the nozzle boxes of each set in unison through a small angle in each angular direction about its longitudinal axis relative to the preselected orientation for each nozzle box to enable the blasts from the nozzles to sweep across the entire major surfaces of the glass sheet undergoing quenching in a manner that improves the uniformity of temper developed in the glass sheet and to provide a more efficient tempering operation than when the nozzles move in orbital paths in spaced flat planes or in curved paths while retaining a constant normal angle of impingement against a curved glass sheet surface during relative movement with respect to the curved major surfaces. The nozzle boxes preferably extend parallel to the axis of glass sheet bending.

It is thus seen that, instead of requiring movement of the entire tempering apparatus, a relatively small amount of the mass of the tempering apparatus is displaced to provide the angular oscillation for each of the nozzle boxes. The need for flexible couplings in the individual nozzle structures used to provide nozzle movement for quenching flat glass sheets in the prior art is avoided by this process.

V.R. Imler; U.S. Patent 4,140,511; February 20, 1979; assigned to PPG Industries, Inc. describes an improvement over the previous process by limiting the angular deviation of the angular oscillation of the nozzle boxes in unison to a maximum deviation of 20°, and preferably about 15°, from the preselected orientation at which blasts of tempering medium are directed through the nozzles in a direction approximately normal to the adjacent portion of glass sheet surface they oppose.

Such limited angular oscillation improves the heat transfer rate for flat or curved glass sheets, improves the uniformity of temper developed in the glass and provides more efficient tempering than when nozzles move in orbital paths either in spaced flat planes relative to flat or curved sheets or in curved paths while retaining a constant normal angle of impingement against a curved glass sheet surface or move with a cycle of pivoting involving angular displacements that exceed 20° from normal during the pivoting cycle during relative movement with respect to its curved major surfaces.

Restraining of Glass During Tempering

V.R. Imler; U.S. Patent 4,150,963; April 24, 1979; assigned to PPG Industries, Inc. also describes a process whereby buffeting of glass sheets hung vertically from tongs in the quenching section of a glass tempering operation is minimized by providing the quenching zone with a plurality of lightweight discs spaced apart on wires extending along one side of the position taken by a glass sheet in the quench. The pressure with which tempering medium is applied to the opposite sides of the glass sheets is controlled so as to force each glass sheet to one side into contact with the discs. Each disc provides minimal interference with the flow of the tempering medium and minimal heat absorption, and thus a large number of the discs may be used, thereby distributing the restraining force over a wide area.

Because the discs are carried on wires which have resiliency in the transverse direction, the initial impact between the glass and the discs may be cushioned, and the wires tend to be forced to follow the curvature of the bent glass, thereby bringing more discs into contact with the glass surface and dividing the load among a greater number of contact points.

Another advantage attained by the wire support arrangement of this process is that the ends of the wires can be supported outside the quench area itself, where they are readily accessible for making adjustments without the necessity of halting production. Additionally, this arrangement permits the support for the wires to be independent from the support means for the tempering nozzles. This permits a preferred, advantageously stable method of operation whereby the discs and the glass remain stationary with respect to one another while the nozzles are oscillated or reciprocated in any direction or mode.

COATING

Metal Coating of a Glass Ribbon

R. Van Laethem, R. Leclercq, P. Capouillet and A. Van Cauter; U.S. Patent 4,188,199; February 12, 1980; assigned to BFG Glassgroup, France describe a process for forming a metal or metal compound coating on a face of a continuously longitudinally moving glass ribbon.

The process includes the steps of contacting such face while it is at an elevated temperature, at a zone along the ribbon path, with a fluid medium consisting of or containing one or more substances which undergo chemical reaction or decomposition to form the metal or metal compound on the face and discharging at least part of the fluid medium against the face as a stream or streams. The stream, or at least one of the streams, has a a velocity component in the direction of movement of the ribbon and is inclined to the face so that the acute or mean acute angle of incidence of such stream on the face, measured in a plane normal to the face and parallel with the direction of the ribbon displacement, is not more than 60°.

Advantageously, the speed of movement of the glass ribbon and the volume rate of supply of the substance to the coating zone are such that the growth rate of the coating thickness is at least 1000 Å of coating thickness per second. Very good results have been achieved in processes where the thickness growth rate was 1200 Å/sec and even 1500 Å/sec.

In certain embodiments the substance undergoing the chemical reaction or decomposition is delivered to the face in gaseous phase. Preferably the stream discharged against the glass ribbon face is gaseous and is discharged so that such stream contacts the ribbon face simultaneously at all positions across the whole or substantially the whole of the ribbon width. As thus performed, the process enables coatings to be built up rapidly, e.g., at a rate of at least 700 Å of coating thickness per second. Such embodiments promise to be of special importance when coating glass ribbons in the course of continuous production at high speeds, e.g., speeds in excess of 2 m/min and even in excess of 10 m/min, such as are often attained by the float process.

It is very suitable to form metal oxide coatings from a substance or substances supplied in the gaseous phase. Coatings of other metal compounds can, however, also be formed from the gaseous phase, e.g., a coating of metallic boride, sulfide, nitride, carbide or arsenide can be formed by reaction of a corresponding organo-metallic compound with a halogenated boron compound, H_2S, NH_3, CH_4 or an arsenic-containing compound, in the absence of oxygen. Metal coatings can also be formed from the gaseous phase. For example, a coating of nickel can be formed by decomposing nickel carbonyl under the action of heat provided by the heated ribbon, in a reducing atmosphere or at least in the absence of oxygen.

Instead of delivering in gaseous form the substance or substances undergoing chemical reaction or decomposition, the required coating metal or metal compound can be formed from a compound in liquid phase, wherein the fluid medium discharged against the face of the glass ribbon comprises at least one stream of droplets comprising a metal compound which by pyrolysis forms the coating metal or metal compound on such face.

Example: An apparatus was used for coating a ribbon of glass in the course of its manufacture in a Libbey-Owens type drawing machine. The coating apparatus was disposed so as to spray the glass ribbon at a position where the glass temperature was of the order of 600°C. The spray gun, which was of conventional type and was reciprocated transversely of the ribbon path, was operated under a pressure of a few kg/cm². The gun was directed so that the angle between the glass ribbon and the axis of the spray was 30° and the spray cone angle was 20°.

The spray gun was supplied with an aqueous or organic solution of tin chloride, e.g., a solution obtained by dissolving in water 400 g/ℓ of hydrated tin chloride ($SnCl_2 \cdot 2H_2O$) and adding NH_4HF_2 in an amount of 65 g/ℓ. On contact with the hot glass ribbon, tin oxide doped with fluorine ions was formed as a coating on the glass. The discharge from the spray gun was regulated to a value such that a coating having a thickness of 8400 Å formed.

The coating had a neutral tint viewed by reflected light. The coating had a high transmissivity in respect of visible light and had a very high reflective power in respect of radiation in the far infrared wavelength band. The coating possessed good electrical conductivity. Microscopic examination of the coating showed that it was of a homogenous structure, comprising a regular arrangement of crystals, in contact with the glass.

In another process, the foregoing procedure was followed but using an aqueous solution of $TiCl_4$ as the feed to the spray gun so as to form a coating of TiO_2 on the glass ribbon, the coating having a thickness of 800 Å.

Silicon-Containing Coating

M. Landau; U.S. Patent 4,188,444; February 12, 1980; assigned to Pilkington Brothers Ltd., England describes a method of coating glass with a silicon-containing coating, in which the glass is moved past a coating station while the temperature of the glass is at least 400°C. Silane-containing gas is released close to the glass surface at a substantially constant pressure across the glass surface and under nonoxidizing conditions so that the silane pyrolyzes to deposit a silicon-containing coating on the glass surface. For imparting a predetermined alkali resistance to the silicon-containing coating, the silane-containing gas includes a proportion of a gaseous electron-donating compound which imparts the alkali resistance.

The process is useful for the treatment of many commercially produced glasses in different forms which can be moved past a coating station, for example, window glass, optical glass and glass fibers. Such glasses generally contain oxides of at least two elements and are usually lead-silicate glasses, alkali metal silicate glasses and alkaline earth metal silicate glasses, especially soda-lime-silica glasses. In some cases, depending on the alkali resistance of the glass and the proportion of the electron-donating compound used, the alkali resistance of the coated surface may be greater than that of the glass substrate.

Preferably the silane-containing gas comprises nitrogen as carrier gas and up to 6% by volume of a gaseous electron-donating compound. Particularly suitable electron-donating compounds are olefins, especially ethylene. The silane-containing gas may comprise monosilane in nitrogen as carrier gas and up to 6% by volume of a gaseous olefin.

The ratio of electron-donating compound to silane in the gas may be in the range 0.1 to 2.0. Preferably the ratio is in the range 0.2 to 0.5. Proportions outside these ranges may be advantageous in some circumstances.

Further, the silane-containing gas may comprise 1% to 7% by volume of monosilane, 0.5% to 6% by volume of ethylene, and optionally a proportion of hydrogen, the remainder being nitrogen.

The silane-containing gas may also comprise 0.3% to 7% by volume of monosilane, 0.2% to 6% by volume of gaseous electron-donating compound, and optionally a proportion of hydrogen, the remainder being nitrogen. When hydrogen is present, the silane-containing gas may comprise up to 10% by volume of hydrogen. A larger proportion of hydrogen may be used.

The electron-donating compound may be an acetylenic hydrocarbon, for example, acetylene. Further the electron-donating compound may be an aromatic hydrocarbon, for example, benzene, toluene or xylene. The electron-donating compound may also be ammonia.

The glass coated with a reflecting silicon coating is characterized in that the coating has a refractive index in the range 2.5 to 3.5, and an alkali-resistance such that the coating shows no sign of damage to the unaided eye after immersion in 1 N sodium hydroxide at 90°C for at least 60 seconds. Preferably the coating shows no sign of damage for at least 5 minutes.

Preferably the coating is such that 30% of the light from a CIE Illuminant C source incident on the coated side of the glass is reflected. The refractive index of the coating may be in the range 2.8 to 3.4.

Enamel Coating for Opal Glassware

Opal glasses are produced by forming, or maintaining, in a glass a second phase that has a sufficiently different refractive index from that of the glass to effectively scatter transmitted light and thereby create an opaque appearance.

In the absence of glass colorants or pigments, an opal glass generally appears white, and the density of the opal depends, among other things, on the relative difference in refractive indexes and the amount of opal phase present in the glass.

Decorative enamels have been employed extensively in the past to impart color and/or gloss to opal glass articles. Such enamels customarily are composed of a clear glass base, referred to as the flux, and a mill-type addition, referred to as the pigment, the latter supplying the coloring effect in the enamel. The enamel, consisting of intimately mixed flux and pigment, is applied to the surface of the glass in frit form and fired to form an adherent, continuous coating on the glass.

D.R. Wexell; U.S. Patent 4,158,080; June 12, 1979; assigned to Corning Glass Works describes a family of enamel fluxes that forms the basis for enamels used in decorating opal glasses where the glasses have softening points in excess of 760°C and coefficients of thermal expansion (25° to 300°C) of 70-85 x 10^{-7}/°C. The enamel fluxes have softening points in the range of 550° to 575°C, toxic metal release values below FDA prescribed standards, coefficients of thermal expansion in the range of 65-75 x 10^{-7}/°C and preferably at least 5 units below the coefficient of the opal glass, and fire to a high gloss in a firing cycle of less than 10 minutes having a maximum temperature below 720°C.

These enamel fluxes are composed essentially, in weight percent on an oxide basis, of 28 to 35% SiO_2, 45 to 55% PbO, 5 to 8% ZrO_2, 4 to 6% B_2O_3, 1 to 3% TiO_2, 1 to 2% Li_2O, and 0.5 to 2.0% $M_2O(Na_2O + K_2O)$, the total Li_2O plus M_2O being not over 3%. In order to minimize cadmium release when this metal is present in the enamel pigment, the Li_2O and M_2O contents, respectively, should not exceed 1.5% and 1.0%.

Enamel Coating for Borosilicate Glass

In spite of the long and widespread popularity such glassware has enjoyed, no satisfactory enamel has been available for firing on borosilicate glass surfaces. Accordingly, borosilicate baking ware has been marketed as a clear glass product, that is undecorated, over the years. Where markings became absolutely essential on borosilicate glassware, for example, on measures or volumetric ware, efforts have been made to develop ion exchange stains as a color medium.

D.R. Wexell; U.S. Patent 4,158,081; June 12, 1979; assigned to Corning Glass Works describes a glass or glass-ceramic body with a coefficient of thermal expansion below 40 x 10^{-7}/°C and having an adherent, high gloss enamel fired on at least a portion of the external surface of the body. The enamel flux is composed, in percent by weight on a calculated oxide basis, of 25 to 40% SiO_2, 40 to 60% PbO, at least 75% PbO + SiO_2, 3 to 10% ZrO_2, 4 to 12% B_2O_3, and 0.2 to 2.0% Li_2O, being free of TiO_2, optionally containing up to 2% $Na_2O + K_2O$ and up to 4% CdO, having a coefficient of thermal expansion of 48-70 x 10^{-7}/°C, a softening point below 660°C and low lead and cadmium release values under either acid or alkaline conditions.

Metal Oxide Film to Control Solar Energy

H.E. Donley; U.S. Patent 4,170,460; October 9, 1979; assigned to PPG Industries, Inc. describes a process for coloring glass articles whereby the surface of a glass substrate is modified by dissolving at least one metal, such as gold, silver, copper, nickel, platinum or palladium, into a surface of the glass substrate. This modification is preferably accomplished by contacting the glass surface with a pure metal or metal alloy under reducing conditions and at a temperature sufficient to permit migration of the metal into the glass surface where it may be present in metallic form or incorporated into the oxide matrix of the glass. The modified surface, preferably while still at an elevated temperature and preferably after brief exposure to an oxidizing atmosphere, is contacted with a metal-containing coating composition under such conditions as to cause pyrolization of the coating composition and deposition of a metal oxide coating on the surface of the substrate.

The resultant articles exhibit increased solar energy control capabilities, improved durability, and flexibility in the selection of reflected and transmitted colors when compared with articles made by the prior techniques.

Employing the method of this process produces certain uniform colors, previously obtainable only by interference techniques employing multiple-layer films, by using pyrolytic techniques. A single coating provides a variety of color effects heretofore obtainable only with multiple-layer coatings, thereby eliminating significant reheating costs. This single coating can be applied in a continuous run procedure using pyrolysis for application of a metal oxide film after dissolving a color imparting metal into the immediate surface of the glass substrate.

In order to obtain articles having particular reflectance and transmittance characteristics and exhibiting desired color characteristics when viewed either in reflectance or transmittance, metals or combinations thereof having a particular index of refraction are selected and the thicknesses of the modified glass-metal layer and of the metal oxide film are controlled.

A pure metal or metal alloy is dissolved into the surface of the glass matrix by maintaining molten, solid, or vaporized metal in contact with the glass surface at an elevated temperature, preferably above the softening point of the glass and under reducing conditions. In preferred embodiments, the desired metal is maintained on the glass surface as a molten pool.

The dissolution of pure metal or metal alloy may be permitted to proceed by diffusion with thermal energy alone to enhance the rate of metal dissolution or the rate may be accelerated by electrochemical techniques such as by providing an electric potential across the contacting metal and the glass. When an electric potential is used as the driving force, the metal or alloy is maintained on the surface of a glass substrate which is to be modified while the opposite surface is contacted with an electroconductive material.

The metal confined on the surface of the glass may be an alloy of tin, lead or bismuth with an element selected from the group of elements consisting of gold, silver, platinum, palladium, nickel or copper or combinations thereof. The metal or metal alloy migrates into the glass matrix establishing a metal content in the glass which is greatest near the contacted surface and decreases toward the interior of the glass. The metal-enriched portion of the glass substrate has a higher refractive index than the interior glass matrix.

The surface of the glass which has been modified by the metal is then contacted, preferably while still at an elevated temperature, with a coating composition which pyrolyzes or otherwise reacts to form a metal oxide coating. The coating step may occur after a brief or a long exposure of the metal-modified glass to oxidizing conditions or may occur without interim oxidation.

The coating composition principally comprises a metal coating reactant which will pyrolyze or otherwise react to form a metal oxide coating upon contact with the glass. It may also include a solvent and/or a carrier gas for the coating reactants employed. The coating composition may be dispensed toward the glass as a liquid or vapor.

Typical of the coating reactants employed in the process are organometallic salts known in the coating art such as acetates, hexoates and the like. While many such organometallic salts are suitable to pyrolyze on contact with hot glass to form a metal oxide coating, superior films result from various metal acetylacetonates in an organic vehicle. Preferred is a family of compositions containing one or more of the acetylacetonates of cobalt, iron and chromium.

Electroless Deposition of Cuprous Oxide

J.S. Breininger and C.B. Greenberg; U.S. Patent 4,170,461; October 9, 1979; assigned to PPG Industries, Inc. describes a wet chemical method for the direct deposition of cuprous oxide onto a transparent nonmetallic substrate such as glass wherein the electrolessly deposited cuprous oxide film is heated to effect a change in the color of transmitted light.

A detailed description of the process is as follows: Large sheets of glass, preferably soda-lime-silica glass about $\frac{7}{32}$" thick, are cleaned by conventional procedures, preferably a blocking operation carried out with rotating felt blocks which gently abrade the glass surface with an aqueous slurry of a commercial cleaning compound, preferably cerium oxide.

The surface to be coated is contacted with a dilute aqueous solution of a sensitizing agent, preferably 0.01 to 1.0 g/ℓ of stannous chloride. The sensitized surface is preferably activated by depositing on it a thin catalytic silver film preferably by contacting the sensitized surface with an alkaline solution of ammoniacal silver nitrate and a solution of a reducing agent, preferably dextrose. The thickness of the silver film preferably reduces the luminous transmittance of the sheet to about 40 to 80%.

It is important that the silver film be substantially free from silver oxide as the presence of the oxide appears to favor deposition of copper in the subsequent coating step. Therefore, if it is likely that the silver film has undergone significant oxidation prior to the cuprous oxide coating step, it is preferred to rinse the activated surface with a dilute solution of an oxide inhibitor such as sodium borohydride, ammonium polysulfide, formaldehyde, or preferably sodium thiosulfate.

The silver activated surface is contacted with an electroless plating bath comprising a copper salt, a complexing agent, a reducing agent and sufficient alkali to raise the pH above about 12.9. The preferred copper salt is copper sulfate and the preferred complexing agent is Rochelle salt, sodium potassium tartrate, although other complexing agents such as gluconic, citric, malic and lactic acids and their alkali metal salts may be used. Formaldehyde, particularly a 37% aqueous solution is a preferred reducing agent, although other common reducing agents such as dextrose or hydrazine sulfate can be used. The preferred alkali is sodium hydroxide.

In a most preferred embodiment, a silver activated glass substrate is contacted for several minutes at ambient temperature with an aqueous solution comprising per liter about 2 to 5 g copper sulfate; 15 to 40 ml formaldehyde solution; and 1 to 5 g Rochelle salt and sufficient sodium hydroxide to maintain the pH of the solution above about 12.9.

Contacting a silver-activated glass substrate having a luminous transmittance of about 50% with such a solution for sufficient time to deposit a cuprous oxide film of sufficient thickness to lower the luminous transmittance to about 10 to 40% results in a cuprous oxide coated article which appears blue-green at the film surface, yellow-green at the glass surface and greenish by transmission. The most preferred coated glass sheets have, in addition to a luminous transmittance of about 10 to 40%, a luminous reflectance of about 10 to 40% from the film surface and a luminous reflectance of about 30 to 60% from the glass surface. The coated sheets are then heated, preferably at a temperature of at least about 300°F, and more preferably, above about 600°F, for sufficient time to effect a change in the color of transmitted light and a significant reduction in the reflectance from the uncoated glass.

Example: Flat sheets of clear soda-lime-silica glass are cleaned and sensitized with an aqueous solution of 0.5 g/ℓ stannous chloride. The sensitized surface is contacted for 25 seconds at ambient temperature with an alkaline aqueous solution containing 2.1 g/ℓ silver nitrate, 10 ml/ℓ ammonium hydroxide (28% aqueous solution), and 0.32 g/ℓ sodium hydroxide, and a solution of 2.6 g/ℓ dextrose to deposit a thin catalytic silver film which lowers the luminous transmittance of the glass sheet to about 50%.

The silver-activated surface is rinsed with a solution of 0.1 g/ℓ sodium thiosulfate and contacted for 3½ minutes at ambient temperature with a solution containing 3.8 g/ℓ copper sulfate, 29 ml/ℓ formaldehyde (37% aqueous solution), 3 g/ℓ Rochelle salt and 25 g/ℓ sodium hydroxide to deposit a cuprous oxide film.

The cuprous oxide coated sheet appears blue-green at the film surface with a reflectance of about 30%, yellow-green at the glass surface with a reflectance of about 44%, and greenish by transmission with a luminous transmittance of about 24%.

The cuprous oxide coated sheet is heated to 900°F for 15 minutes in order to simulate the temperature conditions of a glass edged multiple glazed unit fabrication method. After such heat treatment, the reflectance from the uncoated glass surface is decreased to 23% and the article appears brown by transmission.

Magnetizable Surface Layer

R.F. Reade; U.S. Patents 4,198,467; April 15, 1980; and 4,198,466; April 15, 1980; both assigned to Corning Glass Works describes a process to provide a glass article having an integral surface layer containing very fine-grained crystallites of $NiFe_2O_4$, $CoFe_2O_4$, or $(Co,Ni)Fe_2O_4$ such that the article manifests ferrimagnetic properties.

This can be accomplished with glasses having compositions within a narrowly-defined segment of the Li_2O- and/or Na_2O-FeO-Al_2O_3-SiO_2 system which additionally contain NiO and/or CoO. The composite articles consist of an interior glassy portion and an integral surface layer of not more than several thousand angstrom thickness consisting essentially of very fine-grained crystallites of $NiFe_2O_4$, $CoFe_2O_4$, or $(Co,Ni)Fe_2O_4$ dispersed within a glassy matrix.

The operable glass compositions consist essentially, in weight percent on the oxide basis, of about 1 to 17% R_2O, wherein R_2O consists of Li_2O and/or Na_2O in the following indicated proportions when either is present alone of 2.5 to 5.5% Li_2O and 1 to 16% Na_2O, 1 to 3.75% FeO, 0.75 to 5% RO, wherein RO consists of NiO and/or CoO in the following indicated proportions when either is present alone of 0.75 to 3% NiO and 0.75 to 4% CoO, 20 to 32% Al_2O_3 and 50 to 72% SiO_2. Such glass formers as P_2O_5 and B_2O_3 may desirably be present in amounts of up to 10% P_2O_5 and up to 3% B_2O_3. Fluoride may also be included in amounts up to about 2% as a fluxing agent.

The presence of B_2O_3 and/or F appears to have the further beneficial effect of promoting the growth of ferrite crystals. As_2O_3 is frequently added to perform its customary function as a fining agent. In general, the preferred compositions will consist essentially solely of Li_2O- and/or Na_2O, FeO, Al_2O_3, SiO_2, NiO, or CoO with, optionally, P_2O_5, B_2O_3, and/or F. Furthermore, and very importantly, nucleating agents, e.g., TiO_2, SnO_2, and ZrO_2, will be substantially absent from the compositions such that ferrite surface development will be enhanced and internal crystallization inhibited.

The method for producing such composite articles involves the following three general steps:

(1) A glass-forming batch to provide the above-cited composition ranges is melted;

(2) The melt is simultaneously cooled below the transformation range thereof (optionally to room temperature) and a glass article of a desired configuration shaped therefrom; and

(3) The glass article is subjected in an oxidizing environment to a temperature between about 725° to 875°C for a sufficient length of time to cause the growth in situ of nickel ferrite, cobalt ferrite, or cobalt nickel ferrite crystallites in a thin surface layer.

Inasmuch as crystal growth in situ is well recognized to comprehend a time-temperature relationship, extensive crystallization will require long exposure periods, e.g., up to 24 hours and longer, at temperatures within the cooler extreme of the crystallization range, whereas a period of only 1 to 2 hours may be necessary at the higher end of the crystallization range.

WELDING

Welding of Plate Glasses by Electrical Current

Y. Ebata, T. Ueno, N. Kataoka and A. Akao; U.S. Patent 4,173,460; November 6,

1979; assigned to Agency of Industrial Science and Technology; Central Glass Company, Limited, Japan describe a process and apparatus for welding the adjacent edges of plate glasses, in which the edges of the plate glasses are welded to each other by heating the edges to a high temperature by passing electric current to the edges of the plate glasses.

Referring to Figure 2.11, there is shown a preferred embodiment of an apparatus for welding plate glasses. The apparatus is composed of a table or stand **12** made of a heat resistant material. In this instance, a first plate glass **14** is mounted on the table **12** through supporting members **16** each of which is made of ceramics or porcelain to electrically insulate the table from the plate glass.

Figure 2.11: Plate Welding Apparatus

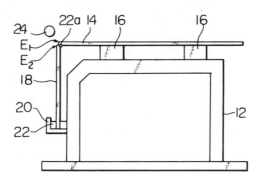

Source: U.S. Patent 4,173,460

The first plate glass **14** has, in this instance, the dimensions 42 cm length, 23 cm width and 5 mm thickness. As shown, an edge E_1 of the first plate glass **14** is disposed to be contacted to an edge E_2 of a second plate glass **18** which is supported by a supporting member **20** of an insulating material, which supporting member **20** is secured to the table **12**. The table is electrically insulated from the second plate glass **18** by the insulating effect of mica **22**.

The reference numeral **24** denotes a heating unit which is positioned in close proximity to the edge portions of the first and second plate glasses by a known supporting means (not shown), each edge portion including each edge E_1 or E_2. This heating unit **24** functions to locally heat the edge portions which are contacted with each other.

With the thus arranged apparatus, the whole of the contacted plate glasses **14** and **18** located as shown in Figure 2.11 is heated to a predetermined temperature such as about 350°C in an electric oven. Then, the edge portions of the first and second plate glasses **14** and **18** are heated to a predetermined temperature (such as 500°C) which is higher than that in the other portions of the plate glasses and lower than the softening temperatures of the first and second plate glasses **14** and **18**. Accordingly, the edge portions of the first and second plate glasses **14** and **18** are kept at about 500°C, while the other portions of the plate glasses are kept at about 350°C. By thus heating the whole of the plate glasses, the plate glasses can be prevented from breaking due to local heating of the edge portions.

A voltage of 5,000 V is impressed on the electrodes **22a** and **22b** (not shown) so that electric current of about 0.7 to 0.9 A is passed through the electrodes to the edge portions of the first and second plate glasses **14** and **18**. It is to be noted that an electric circuit having the least electric resistance can be completed through the edge portions since the locally heated edge portions are decreased in electrical resistance as compared with the other portions of the plate glasses **14** and **18**.

Consequently, the temperature in the edge portions reaches to a level (such as 1200°C) at which the edges E_1 and E_2 can be welded to each other. Then, the edges E_1 and E_2 of the first and second plate glasses **14** and **18** are welded to each other and the welded portion is smoothly rounded by virtue of surface tension of the slightly molten glasses.

Thereafter, electric current supply to the heating unit **24** and electrodes is stopped and electric current supply to the heating wire (not shown) is controlled to slowly cool the whole of the first and second plate glasses **14** and **18** to room temperature to anneal the welded first and second plate glasses. This provides a stable welded glass without internal stresses. Otherwise, supply of electric current may be continued after the welding is finished to further uniformly heat the whole of the welded glasses to, for example, 630°C and thereafter the welded glasses may be rapidly chilled or tempered to obtain a strengthened tempered glass.

Manufacture of Double-Glazed Window Units

W.G. Jeffries; U.S. Patent 4,132,539; January 2, 1979; assigned to PPG Industries, Inc. describes a method of heating marginal edge portions of a glass sheet to a fusing temperature which includes the steps of sequentially applying an electrical potential to marginal edge portions of the sheet followed by alternately applying an electrical potential to opposed corners of the sheet.

This process also relates to a method of fusing glass sheets together to form a multiple glazed unit wherein the method includes the steps of applying a conductive stripe to marginal edge portions of at least one sheet through which a current is passed to heat the marginal edge portions of the sheet to a fusing temperature. Thereafter, the heated marginal edge portions of the sheets are welded together to form a multiple glazed unit.

The improvement includes sequentially passing electrical current through marginal edge portions of the sheet having the conductive stripe followed by alternately passing a current through opposed corners of the sheet. The apparatus is of the type having an electrode mounted in spaced relation to each corner of the glass sheet and a source of electrical potential.

Electrical connecting and disconnecting facilities, e.g., contactors, are provided to electrically connect or disconnect the electrodes to the source. Facilities, e.g., relays, are provided for acting on the contactors sequentially connecting a pair of adjacent electrodes to the source while electrically disconnecting the remaining electrodes from the source and for alternately connecting an opposed pair of electrodes to the source while electrically disconnecting the remaining two electrodes from the source.

Aqueous Colloidal Graphite Electroconductive Stripe

H. Franz; U.S. Patent 4,205,974; June 3, 1980; assigned to PPG Industries, Inc. describes a related process whereby double-glazed window units are made by electrically heating the margins of two glass sheets to form a continuous peripheral weld wherein an electroconductive stripe is deposited on the margins of one of the glass sheets from an aqueous composition comprising colloidal graphite, a perfluoroalkyl surfactant and optionally a water-soluble thixotropic agent.

Preferred aqueous colloidal graphite compositions are prepared by diluting with water concentrated graphite suspensions such as Electrodag 137 (Acheson Colloids Company). The aqueous colloidal graphite compositions preferably further comprise a nonfoaming slightly carboxylated perfluoroalkyl surfactant to achieve proper wetting of a metal-containing glass surface or a glass surface coated with a metal-containing film.

Preferred surfactant compositions include perfluoroalkyl ethoxylates and perfluoroalkyl ammonium compounds. Two particularly useful compositions are Zonyl FSJ and Zonyl FSN surfactants. Enough surfactant is added to achieve sufficient wetting so that a uniform stripe is formed. Typically, surfactant concentrations of about 0.001 to 5% are useful while concentrations of about 0.01 to 1.0% are preferred.

In a particularly preferred embodiment, the aqueous colloidal graphite composition containing a nonfoaming slightly carboxylated perfluoroalkyl surfactant further comprises a water-soluble thixotropic agent. The thixotropic agent thickens the striping composition so that it does not run on the glass surface or drip from the striping apparatus, but does not interfere with the wetting properties achieved by the surfactant.

Preferred thixotropic agents, in addition to being compatible with the colloidal graphite and surfactant in the aqueous medium, should be capable of completely burning off so as to leave no residue on the glass surface. Carbohydrates are useful in this regard. A preferred thixotropic agent is a high molecular weight polysaccharide such as Kelzan (Kelco Company). Being a carbohydrate, the thixotrope leaves no residue on the glass surface. The concentration of thixotrope may vary considerably. For the Kelzan xanthan gum product, with a molecular weight greater than 1,000,000, concentrations in the range of 0.1 to 1% are most useful.

The aqueous colloidal graphite composition containing a nonfoaming, slightly carboxylated perfluoroalkyl surfactant is preferably applied to the periphery of a glass sheet in accordance with U.S. Patent 2,597,106; that is, with a resiliently supported roller moving in the opposite direction from a glass sheet being conveyed through the striping apparatus.

After a uniform electroconductive stripe has been applied to the periphery of the glass sheet, a second glass sheet is assembled with the first, generally in such a manner that the electroconductive stripe is on the upper surface of the upper glass sheet.

Preferably the glass sheets are preheated. Electrodes are provided, preferably at the corners of a rectangular window assembly, to direct an electric current through the electroconductive stripe causing the stripe to be heated. The adjacent portions of glass are likewise heated by the electric current flowing through the stripe.

As electrical heating continues, the marginal portions of glass reach a temperature at which the stripe of electroconductive material burns off. At such sufficiently high temperatures, however, the glass itself becomes electrically conductive so that the electrical heating current continues to flow through the marginal portions of the glass, ultimately heating the glass periphery to its softening point whereupon the edges of the two glass sheets are fused to form a continuous weld about the periphery of the assembled sheets, which are then separated to form an insulating space.

Example: A composition for the deposition of an electroconductive stripe on a glass sheet to facilitate glass welding to form a double-glazed unit is prepared by adding 1% of a slightly carboxylated perfluoroalkyl surfactant to a typical aqueous colloidal graphite composition. In this example, 1 drop of a nonionic perfluoroalkyl ethoxylate, Zonyl FSN surfactant, is added per milliliter of a solution comprising 40 ml of Electrodag 137 colloidal graphite from Acheson and 300 ml of water.

The composition is applied about the periphery of a float glass sheet. A continuous uniform stripe is formed. Proper wetting is evidenced by the fact that no narrowing or beading of the stripe is observed. The glass sheet is then welded to a second glass sheet. A good weld is formed, more uniform and with less corner crazing than if no surfactant is employed in the composition.

Weld Bead Containing Metallic Elements

R.M.R.G. Louis; U.S. Patent 4,142,881; March 6, 1979 describes a process for welding two glass members together to form a weld bead through which connecting wires extend. The glass members are first brought into intimate contact with metal parts which can be heated by a high frequency current. The glass members are held close to one another with the wires disposed in the space between them, and the glass is then melted until it welds by heating the metal parts with the aid of high frequency heating coil windings surrounding the metal parts.

Figure 2.12 shows two glass rings **1** and **2** respectively which are to be welded together without direct heating by means of a flame. The glass rings **1** and **2** are first brought into intimate contact with the end of a respective metal tube **3, 4**. The two metal tubes **3** and **4** are of a metal compatible with the glass.

The intimate contact between each glass ring **1, 2** and the respective metal tube **3, 4** can be achieved in two different ways, either the end faces **3'** and **4'** of the tubes **3** and **4** can be pressed against the corresponding end faces **1'** and **2'** of the rings **1** and **2**, or each glass ring can be welded onto the end portion of the respective metal tube, illustrated in Figure 2.12. This welding operation is effected in in a well known way which is not further described. When this work is completed, each metal tube **3, 4** has an end portion completely engaged in the corresponding ring **1, 2**. The tube **3, 4** penetrates longitudinally into the glass ring **1, 2** over about half the length of the latter.

Figure 2.12: Weld Bead Apparatus for Incorporating Metallic Elements

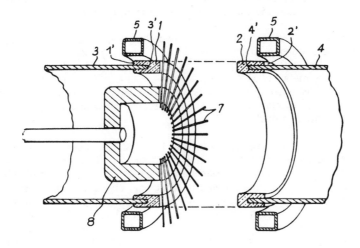

Source: U.S. Patent 4,142,881

In Figure 2.12, the rings **1** and **2** are shown to have equal or only slightly different diameters, and they are shown axially aligned with one another. The spacing
between the rings is exaggerated in order to clearly show that a large number of
individual, spaced, radially extending rods **7** are disposed in the joint plane be-
ween the adjacent end faces of the glass rings **1** and **2**. Before the welding operation is effected, these rods **7** are clamped between two jaws **8** (only one of which
is shown in Figure 2.12 for the sake of clarity) so that the rods **7** are held by the
jaws **8** in the illustrated position.

Each of the jaws **8** extends longitudinally within a respective metal tube **3, 4**
and a respective glass ring **1, 2**. The jaws are pushed against one another in the
axial direction. The jaws may be made of ceramic material, quartz, rubber, or
any other insulating material, or may be made of metal if the rods **7** are of small
diameter. The jaws have a large diameter compared to the diameter of the rods.

One or both of the jaws may have a series of radially extending grooves in its
end face to receive the rods and ensure that the rods are correctly spaced. For
example, if the glass rings **1, 2** each has an internal diameter of about 36 mm
it is possible to dispose 100 spaced radial rods **7** in the joint between the two
rings. In this case the rods should have an external diameter of the order of
0.5 to 0.6 mm in order to ensure that each rod is separated from the adjacent
rod by a distance substantially equal to the external diameter of the rods. The
rods **7** are made of a metal which is compatible with the glass and is also a conductor of electricity.

In the process illustrated in Figure 2.12 the two glass rings **1** and **2** are to be
welded by way of two high frequency heating coil windings **5**, one winding being
disposed on each side of the joint plane. This ensures that the joint plane is free
of all obstructions and that there is the necessary space for the rods **7**.

When each high-frequency heating coil winding **5**, which is disposed externally around a respective metal tube **3, 4**, is energized, the tubes **3, 4** are heated by induced current, and the tubes in turn heat the glass rings by conduction.

To perform the welding operation, the two glass rings **1, 2** are pressed axially towards one another against the arrangement of radial rods **7** and current is fed to the high-frequency heating coil windings **5**. The glass of the rings **1, 2** is heated by the metal tubes **3, 4** and is locally melted. The molten glass flows around the rods whereby the glass is welded to the rods and the glass rings are welded directly together in the intervening spaces between the rods. A weld bead is therefore formed.

Example: A weld was made as shown in Figure 2.12 between two glass rings **1, 2**, each having an external diameter of 40 mm and a wall thickness of 2 mm. Each metal tube **3, 4** was of Kovar metal and had a wall thickness of 0.5 mm. Each winding **5** was a closed loop extending over about 360°, made of copper, and having an internal diameter of 44 mm. Each winding **5** was hollow and a flow of cooling water was maintained through it. The high-frequency generator used supplied an industrial high frequency of 1 Mc; it had a power of 2 kW. Each winding **5** should surround the pieces of glass as closely as possible. In this respect, the localized swelling produced by the welding operation must be taken into account.

In the figure, the spacing of the parts is exaggerated for the sake of clarity. In practice, the windings **5** must be very close and must also be brought as near as possible to the glass rings **1, 2**; moreover, they should extend over the end portion of the metal tubes **3, 4** embedded in the glass rings.

COLORING

Glass Incorporating Both Transparent and Opaque Portions

J.E. Pierson and S.D. Stookey; U.S. Patent 4,134,747; January 16, 1979; assigned to Corning Glass Works describe certain glass compositions which produce "reverse" opals, i.e., the glasses develop a white or single color opacity in those portions thereof that are not exposed to high energy or actinic radiation and remain transparent and, optionally, can be colored in those portions so exposed.

Where a simple reverse opal glass is desired, i.e., a glass wherein the opacified area is white or of a single color and the transparent portion is essentially uncolored or of a single color, the composition thereof will contain Na_2O, SiO_2, F, a compound containing at least one thermally reducible ion selected from the group of copper, gold, silver, palladium, and platinum, a compound containing at least one photoreducible ion selected from the group of copper, gold, and silver, and a thermoreducing agent which will preferably be SnO and/or Sb_2O_3. Where ultraviolet radiation having wavelengths between about 2800 to 3500 Å comprises the actinic radiation, CeO_2 will be included in the composition.

As is apparent, copper, silver, and gold have the dual capability of being subject to thermal reduction and photoreduction so, therefore, their utility in satisfying both functions of the reducing mechanism renders their use the preferred embodiment.

One method for preparing the simple reversible opals contemplates four basic steps:

(1) A batch of the proper composition is melted and shaped into a transparent glass article of a desired configuration having submicroscopic particles of a metal selected from the group of Ag, Au, Cu, Pd, and Pt dispersed therein;

(2) A portion of the glass article is exposed to high energy or actinic radiation at ambient temperature for a sufficient length of time to develop a latent image therein;

(3) The glass article is heated to a temperature between about the transformation range of the glass and the softening point thereof for a sufficient length of time to (a) cause the growth of NaF crystals on the submicroscopic particles of at least one metal selected from the growth of Ag, Au, Cu, Pd, and Pt in the unexposed portion of the article to a sufficient size to scatter visible light, and (b) to cause nucleation and growth of microcrystals of NaF in the exposed portion of the article of insufficient size to scatter visible light; and

(4) The article is cooled to ambient or room temperature.

Where a relatively large amount of a silver-containing compound is employed for thermal reduction, the opal portion of the article may take on a yellow hue; copper can produce a red tint; and gold can lead to a red or blue color.

In contrast, heat treatment of the exposed portion of the glass article can result in total transparency. Hence, by governing the amount of exposure to which the glass is subjected, the subsequent heat treatment leaves a portion which is transparent. In general, the longer and/or more intense the exposure, the more transparent the exposed portion becomes after heat treatment.

Where reverse opal glass is desired wherein the transparent portion thereof can be variously colored, the base composition must include a silver-containing compound and will also contain a halide selected from the group of chloride, bromide, and iodide. Several additional steps must be added to the method described above for the simple reverse opals: the exposed portion of the article is again exposed to high energy or actinic radiation and at least the exposed portion of the article is heated to a temperature between about the transformation range of the glass and the softening point thereof for a sufficient length of time to cause, in the exposed portion, metallic silver to be deposited as discrete colloidal particles less than about 200 Å in the smallest dimension.

The metallic silver is either deposited within the microcrystals, the metal-containing part of the microcrystal being less than about 200 Å in the smallest dimension, or deposited on the surface of the microcrystals, the portion of the microcrystal coated with metal being less than about 200 Å in the smallest dimension. The microcrystals have a concentration of at least 0.005%, but less than about 0.1%, by volume, and the size thereof does not exceed about 0.1 μ in diameter, thereby being too small to scatter visible light. Where the entire article is heated, further growth of the NaF crystals in the unexposed portion of the article may occur. Then the article is cooled to ambient or room temperature.

In general, the glass compositions will consist essentially, in weight percent on the oxide basis, of about 10 to 20% Na_2O, 1.5 to 3% F, preferably 1.8 to 2.6%, 0.001 to 0.03% total of at least one metal selected from the group of Ag, Au, Cu, Pd, and Pt, 0.1 to 1% Sb_2O_3 and/or 0.01 to 1% SnO, the total Sb_2O_3 + SnO not exceeding about 1%, and the remainder SiO_2. Where ultraviolet radiation constitutes the actinic radiation, about 0.01 to 0.2% CeO_2 will be included in the base composition.

Variegated Glass in a Continuous Sheet

The term, variegated glass, as used herein, refers to ornamental glass made by incompletely mixing two or more differently colored glass components, while they are in a molten state. While still molten, the glass is shaped to a desired form, and annealed, before the individual color components diffuse and blend into a single, homogeneous color tone. Typically, in the finished glass, the starting colors appear as distinct, readily identifiable color domains, in the form of striae, individual streaks, swirls and waves.

J.D. Rhodes and R.B. Ek; U.S. Patent 4,133,666; January 9, 1979; assigned to Spectrum Glass Company, Inc. describe a method of producing variegated glass in sheet form whereby two or more components of glass, each of a different preselected color, are heated to a molten state, and are received by a forehearth, while in an unmixed condition. The unmixed components are caused to flow as a "plug" of molten glass toward a discharge end of the forehearth, at which counterrotating, forming rollers receive such "plug flow" and press it into a continuous, elongate sheet which then passes on to an annealing lehr.

Upstream of the forehearth's discharge end, the plug of molten glass forms an unmixed confluent of the differently colored glass components, which appear as distinct color streams, each stream constituting a discrete color domain and having an abrupt color transition at the interface with another such domain. The plug flow is established by providing a forehearth having a narrow flow channel of width substantially less than that of the forming rollers and the ultimate sheet of glass produced thereby, and by sloping such narrow flow channel a sufficient incline to cause a substantially uniform velocity profile of glass across the width of the channel.

As such confluent flows toward the discharge end of the forehearth, it is subjected to controlled stirring which partially breaks up the initially large homogeneous streams into smaller, but still discrete color domains, which are then dispersed throughout the flow.

Adjacent the discharge end of the forehearth, the abovementioned narrow channel diverges to match the greater width of the forming rollers and the stirred confluent is partially dammed by a weir so as to form a reservoir of molten glass at a location on the upstream side of the forming rollers, and the volumetric flow rate of the plug flow is adjusted at the input end of the narrow channel section of the forehearth so that the amount of molten glass reaching the discharge end is just sufficient to supply the forming rollers. Such rollers are operated at a selected feed rate within a predetermined, limited range of rates, to produce a continuous sheet of glass of a given, substantially uniform width. This volumetric flow rate is at a minimum, just sufficient to meet the feed requirement of the forming rollers, without causing a buildup of excessive molten glass at the feed side of the rollers.

Additionally, the molten glass that flows naturally, without external assistance, over the weir and to the forming rollers, is frequently supplemented by scooping supplementary quantities of the stirred molten glass that exists within the reservoir behind the weir, and shoving the scooped glass over the weir to the feed side of the rollers to alleviate transient shortfalls of molten glass there. Such transient shortfalls occur as the result of unsteady flow of the molten glass caused by the step of adjusting the volumetric rate of flow to a minimal level, i.e., just sufficient to meet the required feed rate of the forming rollers.

By damming the partially mixed confluent at a location spaced upstream from the forming rollers, and adjusting the volumetric flow rate of the moving plug of molten glass as described above, it has been found that overmixing of the glass components and diffusion of colors due to prolonged "residency time" which occur when excessive flow of molten glass is allowed to back up as a reservoir immediately behind the forming rollers, are substantially eliminated.

Also, since the plug of glass is drawn into the forming rollers after passing over the weir, a more consistent mix of the color components is achieved in the resulting sheet because the weir obstructs the direct flow into the rollers of the bottommost layer of glass which is slow moving, and thus usually overblended. The presence of the weir forces this overblended glass to be dispersed in unnoticeable amounts into the larger volumes of unblended glass in the upper regions of the flow, rather than allowing it to appear in concentrated amounts at the bottom surface of the rolled sheet.

Preferably, the flowing plug of molten glass is stirred by a combination of apparatus and method steps as follows:

(1) A means and a step for imparting a coarse, transversely oriented stirring motion to the molten glass at the location in the narrow section forehearth channel where the different color components are combined;

(2) A means and a step for injecting gas bubbles into the confluent of molten glass from a bottom surface of the forehearth at a location upstream of the weir whereby the injected bubbles rise to the surface of the molten glass and thereby further breakup and disperse the discrete color domains;

(3) A means and a step for imparting a fine mixing action to the molten glass at a location adjacent the discharge end of the forehearth at the region of divergence of the channel width, and for moving supplementary quantities of the molten glass from the reservoir behind the weir over to the feed side of the forming rollers as needed to supplement the abovementioned transient shortfalls in the unaided flow of the glass to the rollers.

EDGE TREATMENT

Heat Treatment of Plate Glass Edges

It is well known in the art that the corners of the edge of a plate glass are rounded off by being ground with a grinding wheel carrying diamond abrasive or a circular grinding belt carrying abrasive. This grinding operation is, in general, carried out by rotating the grinding wheel or the grinding belt and by moving it along the edge of the plate glass.

However, such grinding operations encounter the following problems: the corners of the edge of the plate glass are considerably liable to chip off at an initial period of the grinding operation, and the surface rounded off is liable to have fine stripe markings along the length of the edge of the plate glass. Such defects are conspicuous particularly when using the grinding wheel, and those defects greatly contribute to lowering the values of the final products. Additionally, by the abovementioned grinding operations, it is liable to occur that there remains fine streaks on the rounded edge surface of the plate glass, which streaks may lead to the breakdown of the plate glass.

Furthermore, in case of preparing window glasses, it has been necessary to further grind the rounded edge of the plate glass with a fine grinding wheel after the abovementioned step of rounding off the edge, in order to improve the smoothness and to polish the surface of the rounded edge. During this further grinding, it is necessary to apply a liquid such as water to cool and wash down the portion being ground. This is very troublesome and contributes to deterioration of the working environment.

Y. Ebata, T. Ueno, N. Kataoka and A. Akao; U.S. Patent 4,165,228; August 21, 1979; assigned to Agency of Industrial Science and Technology and Central Glass Company Limited, Japan describe a process and apparatus for rounding off an edge of a plate glass, in which the edge of the plate glass is locally heated to a temperature higher than that of other portions of the plate glass and lower than the softening temperature of the plate glass after the whole of the plate glass is heated, and electric current is thereafter applied to the edge through a pair of electrodes which are in contact with the both side surfaces of the edge of the plate glass, respectively.

Grinding the Edges of Cup-Shaped Glasses

E. Ilk; U.S. Patent 4,185,419; January 29, 1980 describes a working method for the automatic machining or processing of the edges of cup-shaped glasses, which method allows absolute uniformity of the edge of each glass regardless of the wall thickness of the glass, the method further providing for improved quality as compared to the purely manual manufacture, whereby not only the edge should be shaped with absolute uniformity (consistency), but also the height of the finished glasses should be absolutely identical among themselves.

These objects are solved by a working method comprising the following method steps:

    (1)   Marking or scribing of the edge or rim;

    (2)   Heating and bursting-off of the edge;

    (3)   Pregrinding the edge in three steps;

    (4)   Internal and external trimming of the edge;

    (5)   Precision grinding or finishing of the edge; and

    (6)   Polishing of the edge in two steps.

For additional labor saving, following the edge finishing operation the glasses are automatically washed and dried, such that the glasses may be placed into the shipping packing in a manner to prevent breakage of glasses from occurring.

As far as the apparatus is concerned, the object of the process is solved by a processing rotary platen comprising a rotatable turret for conveying the glasses between the stations of a stationary table, with the cup-shaped glasses being suspended with their feet or bases in the peripheral portion of the turret with the cups thereof directed downwards, the stationary table carrying the machining or processing tools in the peripheral portion thereof and being adapted to be raised and lowered; a pair of identical processing stations each, disposed in side-by-side relation and serving to process a pair of likewise side-by-side positioned glasses of the same processing stage; and a Geneva wheel-type transmission for advancing the glass-suspending turret by the pitch of a pair of processing positions each.

In order to provide for continuous operation of the diamond points which are rapidly worn in the marking process, the marking stations may be provided with a plurality of diamond pins or styluses combined into a rotatable unit, which pins are adapted to be successively swung against the periphery of the glasses to be marked in the fashion of a machining turret.

For the absolutely uniform or consistent machining of the already ground edge, the automatic machine preferably includes edge trimming stations provided with diamond grinding wheels engaging the edge of the cup, and a conical or cylindrical hold-up element, the components being disposed on opposite sides of the edge or having the edge of the glass positioned between them, respectively.

In order to obviate the necessity for the extremely rapidly wearing grinding or polishing belts for polishing the edges of the glasses, the automatic machine is preferably further characterized by polishing stations comprising a pair of contrarotating grinding wheels formed of sandstone and adapted to be vertically pressed against the edge.

For the final processing of the glasses after the grinding operation, the automatic machine further includes a subsequently positioned washing and drying station for the glasses, with the drying station comprising a rotating endless belt of a highly absorbent material upon which the glasses are placed with the top edge directed downwards and which is continuously dried by being squeezed out between a pair of rollers.

The method allows a quality not only equal to, but even better than the quality of manually ground glasses. Since breakage of glasses is avoided and the services of highly trained operators are no longer required, this solution provides a substantial technical advance.

HYDRATION

Hydration of Silicate Glass in Water-Containing Atmosphere

J.E. Pierson and W.H. Tarcza; U.S. Patent 4,201,561; May 6, 1980; assigned to Corning Glass Works describe a method for hydrating such fine-dimensioned bodies as beads, granules, powders, ribbon, etc., of alkali metal-containing silicate glasses wherein the water content absorbed therein can be carefully controlled and the amount of such water will be effective to impart thermoplastic properties thereto.

This objective can be achieved in a single-step hydration procedure. The glass compositions, operable in this process, consist essentially, in mol % on the oxide basis, of 3 to 25% Na_2O and/or K_2O and 50 to 95% SiO_2, the sum of those constituents comprising at least 55 mol % of the total composition. Additions advantageously included to improve melting and forming of the glass and/or to modify the chemical and physical properties thereof include such metal oxides as Al_2O_3, BaO, CdO, B_2O_3, CaO, MgO, PbO, ZrO_2, WO_3, MoO_3, TiO_2, SrO and ZnO. With the exception of PbO, CaO, ZnO and B_2O_3 which can demonstrate utility up to about 25%, MgO which is operable up to about 35%, and BaO and Al_2O_3 which can advantageously be present in amounts up to about 15%, individual additions of other optional metal oxides will preferably be held below about 10%.

The presence of CaO will frequently result in an opaque body which, obviously, would limit its utility to those applications where transparency is not required. Li_2O appears to inhibit hydration so it should not be included, if at all, in amounts greater than 5%. The well-recognized glass colorants such as CdS-Se, Co_2O_3, Cr_2O_3, CuO, Fe_2O_3 and NiO may be incorporated into the glass composition in the customary amounts up to a few percent. It should be recognized that these latter ingredients can be tolerated in amounts up to about 10% where their function is not limited to their effect as a colorant. Finally, where necessary, conventional fining agents can be included in customary amounts.

The single step procedure contemplates contacting fine-dimensioned bodies of such compositions at temperatures in excess of 225°C with a H_2O-containing gaseous environment having a relative humidity less than 50%.

Hydration of Silicate Glass in Alcohol-Water Solutions

R.F. Bartholomew, W.L. Haynes and L.M. Sanford; U.S. Patent 4,133,665; January 9, 1979; assigned to Corning Glass Works also describe a method of hydrating anhydrous glass bodies having compositions consisting essentially, in mol % on the oxide basis, of about 3 to 25% Na_2O and/or K_2O and 50 to 95% SiO_2, the sum of those components constituting at least 55 mol % of the total composition.

An especially useful group of glasses exhibiting good chemical durability and excellent transparency after hydration has anhydrous compositions consisting essentially, in mol % on the oxide basis, of about 72 to 82% SiO_2, 10 to 17% Na_2O and/or K_2O, and 5 to 20% PbO and/or ZnO. Up to 5% Al_2O_3 and up to 3% B_2O_3 and/or BaO and/or MgO may also be included.

The method consists of the following general steps:

(1) A batch for a particular glass composition is compounded, melted, and the melt formed into desired shapes;

(2) The anhydrous glass shapes are contacted with a gaseous atmosphere or a liquid solution containing a mixture of H_2O and a short chain aliphatic alcohol at a temperature greater

than 100°C and at a pressure in excess of atmospheric pressure for a period of time sufficient to hydrate at least a surface portion of the glass shapes; and

(3) The hydrated glass shapes are cooled to room temperature.

GLASS FIBERS

FIBER PRODUCTION

There are four techniques currently in use by which glass fibers can be made. They include the following.

(1) Longitudinal Blowing: Longitudinal blowing is a glass fiber manufacturing process according to which melted glass flows from the forehearth of a furnace through orifices in one or two rows of tips protruding downwardly from a bushing, the glass being thereby formed into multiple glass streams which flow down into an attenuating zone where the streams pass between downwardly converging gaseous blasts. The blast emitting means are located in close proximity to the streams so that the converging blasts travel in a downward direction substantially parallel to the direction of travel of the glass streams. Generally the glass streams bisect the angle between the converging blasts. The blasts are typically high pressure steam.

There are two longitudinal blowing techniques. In the first technique the attenuating blasts engage already drawn fibers and the product resulting is typically a mat, commonly known as steam bonded mat, suitable for reinforcement. In the second longitudinal blowing technique the attenuating blasts strike directly on larger streams of molten glass and the product resulting is typically an insulation wool commonly known as steam blown wool.

In a variation of the first longitudinal blowing technique, the entire bushing structure and associated furnace are enclosed within a pressure chamber so that, as the streams of glass emerge from the pressure chamber through a slot positioned directly beneath the glass emitting tips of the bushing, this variation being commonly referred to as low pressure air blowing, and products being commonly known as low pressure air blow bonded mat and staple yarn.

(2) Strand: The strand glass fiber manufacturing process begins in the manner described above in connection with longitudinal blowing, that is, multiple glass streams are formed by flow through orifices in tips protruding downwardly from a bushing. However, the strand process does not make use of any blast for attenuation purposes but, on the contrary, uses mechanical pulling which is accomplished at high speed by means of a rotating drum onto which the fiber is wound or by means of rotating rollers between which the fiber passes.

(3) Aerocor: In the aerocor process for making glass fibers, the glass is fed into a high temperature and high velocity blast while in the form of a solid rod, rather than flowing in a liquid stream as in the longitudinal blowing and strand processes discussed above. The rod, or sometimes a coarse filament, of glass is fed from a side, usually substantially perpendicularly, into a hot gaseous blast. The end of the rod is heated and softened by the blast so that fiber can be attenuated therefrom by the force of the blast, the fiber being carried away entrained in the blast.

(4) Centrifuging: In the centrifuging glass fiber manufacturing process, molten glass is fed into the interior of a rapidly rotating centrifuge which has a plurality of orifices in the periphery. The glass flows through the orifices in the form of streams under the action of centrifugal force and the glass streams then come under the influence of a concentric and generally downwardly directed hot blast of flames or hot gas, and may also, at a location concentric with the first blast and further outboard from the centrifuge, come under the action of another high speed downward blast, which latter is generally high pressure air or steam. The glass streams are thereby attenuated into fine fibers which are cooled and discharged downwardly in the form of glass wool.

Centrifugal Fiberization of Hard Glass

In commonly employed centrifugal systems, it has been customary to employ so-called soft glasses, i.e., glass compositions which are specially formulated to have temperature/viscosity characteristics providing a viscosity which will pass freely through the orifices in the spinner wall at a temperature well within the limits of the temperature which the material of the spinner is capable of withstanding without excessive corrosion and deformation.

For the above purpose, the glass compositions employed have customarily incorporated appreciable quantities of one or more barium, boron, and fluorine compounds, which tend to lower the melting temperature, devitrification or liquidus temperature and the viscosity, and which have, therefore, been effective in avoiding the necessity for employment of molten glass at excessively high temperatures.

However, the use of compositions containing substantial amounts of boron or fluorine or even barium requires that certain precautions be taken, especially in the case of boron and fluorine because objectionable volatile constituents may be developed and carried through and out of the molten glass production system and, in this event, if this possibility of pollution is to be avoided, special treatment of the discharged gases would be necessary in order to separate and appropriately dispose of these constituents.

J.A. Battigelli, F. Bouquet, I. Fezenko and J.-J. Massol; U.S. Patent 4,203,746; J.A. Battigelli, F. Bouquet and J.-J. Massol; U.S. Patent 4,203,774; and I. Fezenko; U.S. Patent 4,203,747; all dated May 20, 1980; all assigned to Saint-Gobain Industries, France describe a process which increases the production of a given plant facility of the kind employing a centrifugal spinner delivering streams of glass into an annular attenuation blast surrounding the spinner, while at the same time, it substantially eliminates certain sources of pollution, making possible the use of glass compositions of lower cost, and provides a fiber product having improved temperature-resistant characteristics.

Turning first to the composition of the glass (examples being given hereinafter), while the method and the equipment including the spinner construction may be used with presently used compositions, it is contemplated that the glass composition be formulated to contain no fluorine and little if any barium and boron. Such glass compositions are hard glasses, having higher melting and devitrification temperatures.

Referring to the embodiment of Figure 3.1, a vertical spinner supporting shaft is indicated at **10**, this shaft carrying at its lower end a hub for mounting the spinner, the hub being indicated at **11**. The spinner itself is indicated in general at **12**. The spinner is made up of a peripheral wall **13** having a multiplicity of rows of spinner orifices, and the upper edge of the wall **13** is connected to the hub **11** by the central mounting portion or neck **14**. The orifices in the spinner wall are illustrated only in the sectioned portions of the spinner wall but it is to be understood that a multiplicity of orifices are provided in each of a plurality of vertically spaced rows of orifices. At its lower edge, the spinner is provided with an inwardly projecting flange **15** to which the upper edge of a cylindrical part or element **16** is connected, this cylindrical part serving a reinforcing or bracing function, as will further be explained.

Figure 3.1: Fiberization Equipment for Hard Glass

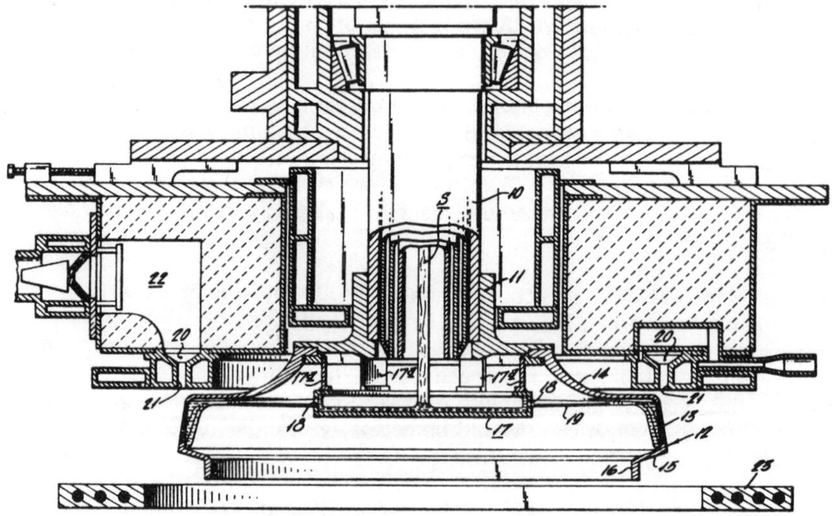

Source: U.S. Patent 4,203,746

Mounted within and rotating with the spinner is a distributing basket **17** having a single series of distributing orifices **18** which are located substantially in the plane of the uppermost row of orifices in the peripheral wall of the spinner. As shown, the basket **17** is mounted on the hub **11** by means of depending brackets **17a**. A stream of glass is delivered downwardly and centrally through the spinner mounting structure, as is indicated at **S**, being delivered to the inside of the bottom wall of the basket **17**, and spreading laterally on that bottom wall to the perforate peripheral wall of the basket, so that the glass builds up a layer on the inside of the basket wall, from which streams indicated at **19** are projected through the orifices radially outwardly to the inside surface of the peripheral wall of the spinner adjacent to the uppermost row of orifices from which zone the glass flows downwardly on the inside surface of the spinner wall.

This downward flow is unobstructed, there being no interior confining wall or chamber structure inside of the peripheral wall, and the flow has laminar characteristics, when viewed under stroboscopic light, in which there is the appearance of smooth waves. It is from this unobstructed or unconfined laminar flow layer that the glass enters the orifices in the peripheral wall of the spinner and is projected therefrom outwardly from all of the spinner orifices, in a multiplicity of streams or primaries which are subjected to attenuation by the annular gas blast.

In connection with the arrangement of the distributor basket, it is pointed out that most of the distributor baskets employed in prior art techniques are provided with several rows of orifices vertically spaced from each other in order to provide for distribution of the glass to the perforated peripheral wall of the spinner throughout much of the vertical dimension of the perforate spinner wall.

It was found, however, that in providing the multiplicity of orifices required to effect the vertical distribution of the glass in accordance with the common technique of the prior art, certain disadvantages and difficulties were encountered, especially in connection with spinners of relatively large size, both with respect to diameter and vertical height of the perforate peripheral wall.

One of the most important problems relates to heat loss from the streams of glass being delivered from the distributor basket to the inside of the peripheral wall of the spinner. Such heat loss is directly proportional to the total surface area of the delivered streams. With a large number of small streams, as in prior arrangements, the total surface area is much greater than with the arrangement herein disclosed in which the distributor basket is provided with only one row of orifices of larger size, thereby effecting delivery of the same quantity of glass with much smaller total surface area.

Indeed, in a typical case, the arrangement as herein disclosed provides for delivery of a given quantity of glass in streams having only about one-seventh of the surface area of prior arrangements.

The improved arrangement, therefore, eliminates the excessive heat loss from the glass being delivered from the distributor basket to the peripheral spinner wall, which was a major disadvantage of the prior art equipment. Moreover, with the smaller streams of glass used in prior techniques, the temperature loss in delivery from the distributor basket to the peripheral wall of the spinner is much less uniform as between different streams, than is the case where a smaller number of larger streams are provided, as in the arrangement of this method.

Another factor of importance is that the technique herein disclosed contemplates increasing the diameter of the spinner. With glass streams of small diameter delivered from the distributor basket, as in prior arrangements, the increase in the spinner diameter tends to result in fluttering of the streams thereby adversely influencing uniformity of operating conditions. The use of a smaller number of larger streams overcomes such fluttering.

Still further, with many smaller streams of glass delivered to the inside of the perforate peripheral spinner wall throughout most of the perforate area of that wall some of the streams arrive at the perforated wall in substantial alignment with individual perforations in the wall, whereas others arrive at the perforated wall in imperforate areas between the perforations of the spinner wall; and this has introduced nonuniform dynamic conditions tending to adversely affect the uniformity of the fibers being produced.

With the foregoing in mind, instead of employing a multiplicity of supply streams distributed vertically over the peripheral spinner wall, the improved arrangement provides for establishing and maintaining an unrestrained, unconfined and downwardly flowing layer of molten glass on the inside surface of the perforated peripheral wall, the feed of the glass being effected to the upper edge of that layer and the layer flowing downwardly in laminar fashion over all of the perforations of the spinner wall, so that the dynamic conditions for projecting the stream of glass through and from each perforation of the peripheral wall are substantially the same, thereby eliminating a source of nonuniformity of the fibers produced.

This development or establishment of the downwardly flowing unconfined layer is effected by the distributor basket arrangement described above in connection with Figure 3.1, i.e., by the employment of a basket or distribution system in which all of the glass to be fiberized is delivered to the spinner wall through a single series of orifices close to or in a plane located at or close to the level of the uppermost row of perforations in the spinner wall. This single series of orifices desirably comprises a total of only 75 to 200 orifices, which is about $\frac{1}{10}$ to $\frac{1}{3}$ of the number commonly used in multiple row distributor baskets.

For the purpose of the attenuation, the structure, as shown in Figure 3.1, includes an annular chamber **20** with an annular delivery orifice **21**, the chamber **20** being fed from one or more combustion chambers such as indicated at **22** supplied with appropriate means for burning fuel and thus producing the desired hot attenuating gases. This provides a downwardly directed annular stream of attenuating gas in the form of a curtain surrounding the spinner.

When employing hard glass compositions, instead of utilizing a temperature differential between the upper and lower edges of the spinner, the improved technique establishes approximately the same temperature at the upper and lower edges of the spinner, and this temperature is established at a level (for instance 1050°C) which is above and yet relatively close to the devitrification temperature.

For establishing the desired temperature at the lower edge portion of the spinner, it is contemplated to provide more intense heating of the lower edge of the spinner than has heretofore been utilized. Thus, the heater **23** in Figure 3.1 should have at least two to three times the power of heaters heretofore used. A heater of 60 kW capacity at 10,000 Hz is suitable.

It is preferred that conditions be maintained establishing a temperature of the glass in the region of both the top and bottom portions of the peripheral spinner wall at a level from about 10° to 20°C above the devitrification temperature of the glass being used.

For most purposes, it is also preferred that the lower edge portion of the peripheral wall of the spinner should be at least 1½ times the thickness of the upper edge portion of the peripheral wall of the spinner; and in some cases, it may be desirable to proportion the spinner wall so that the lower edge portion is as much as 2½ times the thickness of the upper edge portion. A spinner having a lower edge portion approximately twice the thickness of the upper edge portion is typical. For example, in such a typical spinner, the upper edge portion of the spinner may be 3 mm in thickness and the lower edge 6 mm in thickness.

The equipment and technique herein disclosed may be employed with quite a broad range of glass compositions, for instance, as indicated in the table below.

Constituents	A General Range	B Glass Containing Barium	C Manganese + Iron
SiO_2	59–65	59–65	60–64
Al_2O_3	4–8	4–8	5–6.5
Na_2O	12.5–18	12.5–18	14.5–18
K_2O	0–3	0–3	0–3
$R_2O = Na_2O + K_2O$	15–18	15–18	16–18
Al_2O_3/R_2O	0.25/0.4	0.25/0.4	0.25/0.4
CaO	4.5–9	4.5–8	5–9
MgO	0–4	0–4	0–4
MgO/CaO	0/0.75	0/0.75	0/0.75
MgO + CaO	7–9.5	7–9.5	8–9.5
MnO	0–4	1–3.5	1.5–4
BaO	0–5	2–3.5	trace
Fe_2O_3	0.1–5	0.1–1	0.8–3.5
$MnO + BaO + Fe_2O_3$	3.5–8	4–8	3.5–6.5
B_2O_3	0–2	0–2	trace
Miscellaneous	≤1	≤1	≤1
(amount SO_3)	≤0.6	≤0.6	≤0.6

Fiber Formation by Gas Blast Attenuation

M. Levecque, J.A. Battigelli and D. Plantard; U.S. Patents 4,159,199; June 26, 1979; and 4,194,897; March 25, 1980; both assigned to Saint-Gobain Industries, France describe a method of gas blast attenuation including the use of a pair of gaseous jets having axes lying in a common plane and directed to impinge upon each other preferably at an acute angle, the attenuable material being introduced into the influence of air induced by one of the jets in the region of impingement of the jets, and the combined jet flow of the two jets carrying the attenuable material into a zone of interaction between the jet flow and a larger gaseous blast.

Referring to Figure 3.2, between the two pairs of tornadoes in the region of impingement of the jets upon each other, a zone L of laminar flow associated with the tornadoes is developed, this zone having high intensity in-flow of induced air, and it is into this laminar flow zone at the side of the upper pair of tornadoes that the stream of glass is introduced. As clearly appears in Figure 3.2, the stream **S** of the glass is developed from the glass bulb, which bulb or cone is located in a position horizontally offset from the jet delivery device. However, because the glass of the bulb **G** is in attenuable or flowable condition as released from the delivery device and the stream **S** of the attenuable glass is deflected from the horizontally offset position of the bulb toward the laminar flow zone **L**, this deflection occurring as a result of the intense in-flow of induced air, this effect assures delivery of the stream of attenuable material into the laminar zone. Indeed, even with some misalignment of the glass delivery device with respect to the pairs of jets, the in-flow of induced air will automatically compensate for such misalignment and bring the glass stream into proper position.

Figure 3.2: Gas Blast Attenuation Apparatus

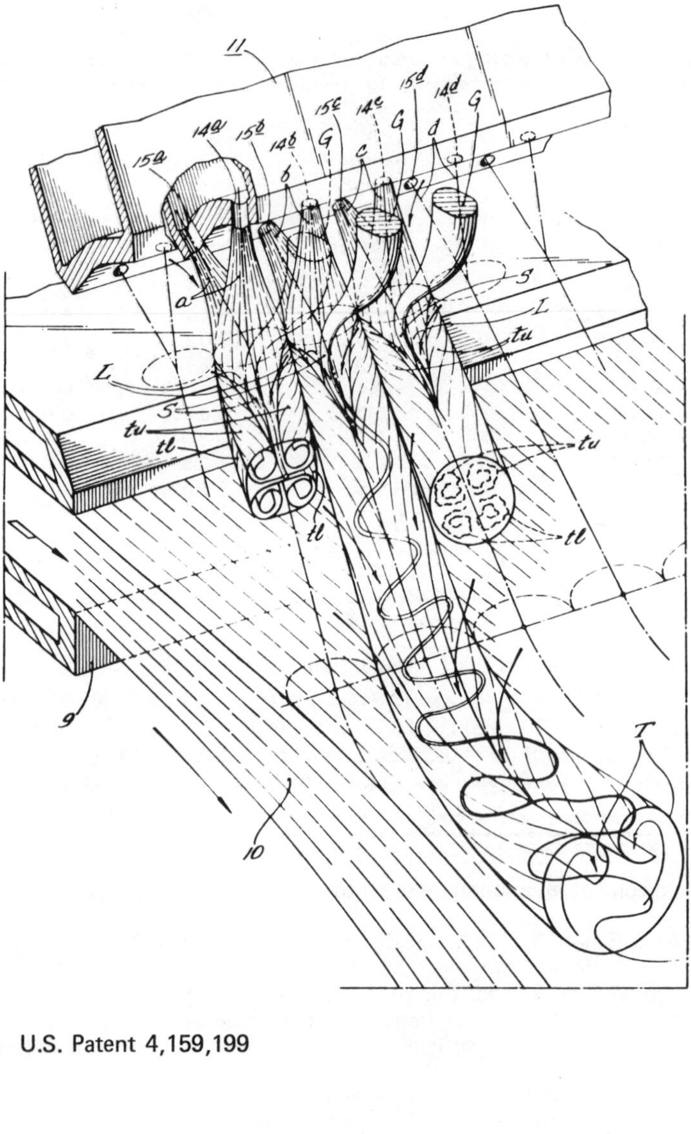

Source: U.S. Patent 4,159,199

From the above, it will be seen that by developing the pairs of tornadoes with the intervening zone of laminar flow at each fiberizing center, and by delivering the attenuable material in attenuable condition into the region near the zone, the induced air automatically carries the stream of attenuable material into the zone of laminar flow and automatically compensates for misalignment, thereby providing a highly stable introduction of the attenuable material into the system.

The arrangement as described above and the action of the induced air currents provides for stable introduction of the attenuable material into the system, even where the glass delivery devices are appreciably spaced away from the jet delivery devices, which is desirable in order to facilitate maintenance of appropriate temperature control for both the glass delivery devices and the jet delivery devices.

As seen in Figure 3.2, the pairs of tornadoes tu and tl tend to merge downstream of the laminar zone L, and as the flow progresses downstream the tornadoes tend to lose their identity, as is indicated (toward the right of Figure 3.2) by the sectional showing of the two pairs of tornadoes originating with the jets c–c. The merged jet flow of each pair of jets then proceeds downward to penetrate the blast 10 as is indicated in Figure 3.2 for the jet flow originating with the pair of jets b–b, and within the flow of the blast, the jet develops the zone of interaction characterized by an additional pair of tornadoes indicated at T, this interaction being identified as toration and fully explained in the U.S. Patents 3,874,886 and 3,885,940.

It will also be seen that the plane containing the axes of the jets cuts through the blast in a direction substantially parallel to the direction of the flow of the blast.

Also, as shown in Figure 3.2, each stream S of glass is subjected to a preliminary attenuation in the jet flow between the zone of laminar flow or point of introduction of the glass, and the point of penetration of the jet into the blast, and the partially attenuated stream is subjected to further attenuation in the zone of interaction of the jet flow with the blast. As indicated in the drawings, these two stages of attenuation are effected without fragmentation of the glass stream, so that each stream produces a single fiber.

The action of the jets at each fiberizing center, particularly in the development of the pairs of tornadoes with the intervening zone of laminar flow, is achieved by employment of a pair of jets each of which is preferably of substantially the same kinetic energy per unit of volume; preferably also, the jets of each pair are of approximately the same cross-sectional area and cross-sectional shape, but some leeway is permissible with respect to the relation between the cross-sectional areas of the two jets of a pair, particularly if the kinetic energy per unit of volume of each jet is substantially the same.

Energy-Efficient Fiber-Producing Apparatus

N. Ohsato, K. Tanaka and E. Mizushima; U.S. Patents 4,135,903; January 23, 1979; and 4,185,981; January 29, 1980; both assigned to Nippon Sheet Glass Co., Ltd., Japan describe a method and an apparatus for producing high quality fibers from a heat-softening material at a considerably higher energy efficiency than in the known conventional methods.

Basically, this method comprises:

   (1)   heating a heat-softening material to form a viscous melt and flowing it continuously; and

   (2)   jetting out a specified high-speed gas stream to the melt that has been flowed out.

Melting and flowing of the heat-softening material are accomplished, for example, as shown in Figure 3.3, by heating the heat-softening material in a melting crucible **2** by a suitable known method to change it into a viscous melt **4**, and then directly and continuously flowing the melt **4** through a nozzle **8** provided in the crucible by the weight of the melt and/or by the pressure inside the crucible.

Figure 3.3: Fiber-Producing Apparatus

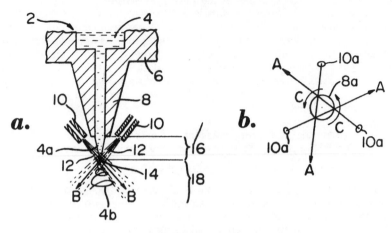

(a) Sectional view.
(b) Bottom view of apparatus shown in Figure 3.3a.

Source: U.S. Patent 4,185,981

Preferably, the flow-out nozzle **8** is provided downwardly at a bottom wall **6** of the crucible, as shown. It may be provided in an optional direction on the bottom wall or side wall, etc. of the crucible so as to flow out the melt in any desired direction. It is preferred that the cross-sectional shape of the flow-out nozzle be circular and thus the melt **4** be flowed out from the crucible **2** with its cross-sectional shape being rendered substantially circular.

Then, a high speed gas stream is jetted out onto the melt which has been flowed continuously from the jet-out nozzle **8**. At least three straight-line high speed gas streams **12** which are jetted out from at least three nozzles **10** disposed at intervals in the peripheral direction around the flow-out nozzle **8** (Figure 3.3b shows only a hole **8a** of the nozzle **8** and a hole **10a** of the nozzle **10** in order to simplify the drawing), or an annular high speed gas stream jetted out from an annular high speed gas jetting nozzle disposed around the nozzle **8** is jetted against the melt. It is important that the gas stream to be jetted against the melt should contain a component in the tangential direction (the direction shown by arrow **A** in Figure 3.3b) along the cross section of the melt flowing from the nozzle **8**, and a component (shown by arrow **B** in Figure 3.3a) which gradually approaches the central axial line of the melt toward the flowing direction of the melt and then gradually departs from the central axial line.

By the action of the high speed gas stream described above, the melt is rotated about its central axial line, and formed into a substantially conical shape **4a** whose section gradually decreases toward the flowing direction in a first zone **16** which ranges from the outlet port of the flow-out nozzle **8** to a part at which the gaseous stream most closely approaches the central axial line of the melt (point **14**, to be referred to as a focal point).

Where the central axial lines of the straight-line gas streams most closely approach the central axial line of the melt at different points on the central axial line of the melt, this focal point is the one at which an average of the distances between the central axial lines of the gaseous stream and the central axial line of the melt becomes minimum.

In the first zone **16**, at least three straight-line high speed gaseous streams **12** or the annular high speed gaseous stream which act on the surface of the melt that has been flowed out have a component in the tangential direction (the direction shown by arrow **A** in Figure 3.3b). Hence, a rotating moment in the direction shown by arrow **C** in Figure 3.3b acts on the melt, and causes the melt to rotate in the direction of arrow **C** around the central axial line of the melt.

Furthermore, since in the first zone **16**, the high speed gaseous stream also has a component in a direction (shown by arrow **B** in Figure 3.3a) gradually approaching the central axial line of the melt toward the flowing direction of the melt, the melt is restrained so that its cross section gradually decreases toward its flowing direction, whereby it is formed into a conical shape **4a** which is tapered toward the flowing direction. It will be readily understood that the rotating angular speed of the melt being rotated gradually increases towards the flowing direction of the melt.

Preferably, the conical shape of the melt formed in the first zone **16** is such that its diameter at its smallest end is about 0.1 to 1 mm. The distance between the outlet of the nozzle **8** and the smallest end of the cone is generally required to be larger than 20% of the inside diameter of the nozzle **8**.

In a second zone **18** subsequent to the first zone **16**, the melt is continuously flown as a fiber from the tip of the cone in the flowing direction and outwardly in the radial direction in a vortex form **4b** which is spiral or helical or both. As a result of the flying of the melt from the tip of the cone **4a** as one fiber, a pulling force in the flowing direction of the melt acts on the cone **4a**, and is transmitted to the melt within the crucible **2** to promote the flowing of the melt through the nozzle **8**.

Thus, the amount of the melt lost from the cone as a result of the flying out of one fiber from its tip is made up for by a fresh flow of the melt. Consequently, the cone **4a** is maintained stably and continuously without the flowing of the melt from the nozzle **8** becoming discontinuous. On the other hand, the melt

4b flown in the form of fiber from the tip of the cone **4a** undergoes a pulling force by the force of inertia of its rotation and the high speed gas stream acting on it, and is thereby drawn and attenuated.

Method of Forming and Collecting Fiber Particles

A.P. Symborski, R.M. Fulmer and D.W. Thomas; U.S. Patents 4,194,896; March 25, 1980; and 4,163,653; August 7, 1979; both assigned to Owens-Corning Fiberglas Corporation describe a method and apparatus for forming glass filaments provided by:

(a) a plurality of spaced apart forming sections wherein each section is comprised of:

   (1) a feeder adapted to supply a plurality of streams of glass to be attenuated into continuous filaments, a zone being defined by the paths of the free-falling streams,

   (2) applicator means laterally spaced from the zone adapted to apply a coating to the advancing filaments,

   (3) first guide means laterally spaced from the zone adapted to gather the filaments into a strand, the guide means and applicator being positioned such that the coating is applied to the filaments at a region external to the zone and intermediate the feeder and the first guide means along the path of advancement of the filaments, and

   (4) secondary attenuation means laterally spaced from the zone adapted to advance the filaments as waste;

(b) a primary attenuation means adapted to simultaneously attenuate the streams from each section into filaments; and

(c) second guide means positioned to advance the strands from each section external to all of the zones when the strands are being advanced by the primary attenuation means, the feeders, applicator means, first guide means and second guide means being positioned such that any of the strands are advanced along paths external to all of the zones and in the absence of contact with any other of the strands when any of the waste strands are advanced by the secondary attenuation means.

Also, described is a method and apparatus for forming and collecting discrete fibers comprising: supplying a plurality of streams of molten glass; attenuating the streams into filaments and cutting the continuous filaments into discrete fibers by a forming means; positioning a delivery means having a first collection zone and a second collection zone spaced from the first zone to receive the discrete fibers; controlling the forming means to operate at a first speed or a second speed slower than the first speed; and directing the discrete fibers to the first zone when the forming means is operated at the first speed and to the second zone when the forming means is operated at the second speed.

As shown in Figures 3.4a and 3.4b, the forming system is comprised of spaced apart continuous filament forming sections **4, 5** and **6** which are associated with a primary attenuation means or chopped strand forming means **40**. Thus, a single attenuation means is adapted to attenuate and advance all of the filaments from all of the forming sections. As such, the primary attenuation means **40** can be a winder, pull wheel, or a chopper for forming discrete fibers or chopped strands. As shown, each forming section **4, 5** and **6** is comprised of a stream feeder or bushing **10**, size or coating applicator **18**, gathering shoes or guide means **22**, an idler roll or guide means **30** and secondary attenuation means **50**, all of which can be of the type well known in the glass fiber forming art.

In operation, each of the feeders **10** supply a plurality of streams of molten glass to be attenuated into filaments by the action of the primary attenuation means **40** or secondary attenuation means **50**. In any glass fiber forming operation, occasionally, some of or all of the filaments from any bushing will break and the attenuation process will be halted. During restart, bead down can occur wherein the streams of molten material leave the feeder as beads of molten glass. If left unattended in a free-fall condition, the beads or streams will move along a path defining a vertically extending zone beneath feeder **10**. To establish a continuous and efficient operation, it is imperative that such forming equipment as the applicators **18**, gathering shoes **22**, idler rolls **30**, and scrap pull rolls or secondary attenuation means **50** be positioned external to such zone to keep such equipment out of the free-fall path of the heads or streams.

As shown in Figure 3.4b, applicator **18**, gathering shoe **22**, idler roll **30**, and secondary attenuation means **50** are laterally spaced to the same side of zone **54**. Generally, it is preferred that such equipment be on the side of the zone **54** opposite the side of the zone where the operator is normally positioned. With such an orientation, the operator has easy access to the bottom region of the feeder while permitting easy access to coating applicator **18**, gathering shoe **22**, idler roll **30** and scrap pull roll **50**.

During production, feeders **10** of forming sections **4, 5** and **6** supply a plurality of streams of molten glass to be attenuated into continuous filaments **14, 15** and **16**. Each of such filaments are advanced along a path external to the zone **54** of each of the plurality of forming sections. Each of the applicators **18** is fixed external to the zone and laterally spaced therefrom such that the applicator surface, which can be a rotatable roll wet with a suitable liquid size and/or binder or coating, is located laterally spaced from the zone. The liquid is applied to the filaments at a region laterally spaced from the zone. Gathering shoes **22** are laterally spaced further from the zone in the same direction as applicators **18** to bring the filaments into contact with the applicator roll. Gathering shoes **22** further serve to gather the filaments into a substrand and guide the strand along a path external to the zones **54**.

From gathering shoes **22** the substrands **26a, 26b, 27a, 27b, 28a** and **28b** are advanced to idler rolls **30** associated therewith to form strands **26, 27** and **28**, respectively. Strands **26, 27** and **28** are advanced to second guide means **35** which is laterally spaced from the same side of zones of all of the forming sections **4, 5** and **6**. Gathering shoes **22** can be considered a first guide means or the combination of gathering shoes **22** and the idler rolls **30** can work in conjunction as a first guide means. The second guide means **35** is positioned in a horizontal plane below the horizontal plane or zone containing the first guide means such that the strands

Figure 3.4:   Apparatus for Forming and Collecting Discrete Fibers

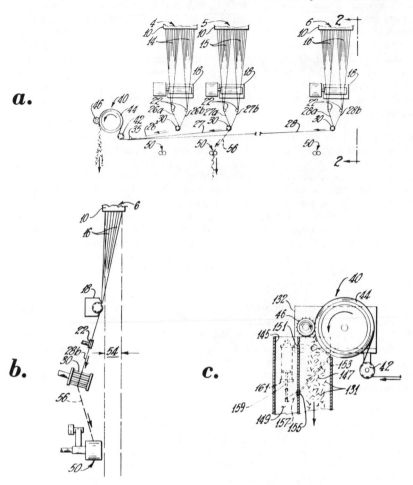

(a) Front elevational view.
(b) Side view along line **2–2** of
     Figure 3.4a.
(c) Detailed view of primary attenua-
     tion means.

Source: U.S. Patent 4,194,896

26, 27 and **28** are maintained in a spaced apart relationship until such strands substantially reach second guide means **35**. As shown in Figure 3.4a, second guide means **35** is located in a horizontal plane beneath the horizontal plane containing idler rolls **30** of sections **4, 5** and **6**. With the strands **26, 27** and **28** remaining in a spaced apart relationship until approximately reaching guide means **35** it is easier to restart individual strands than if the strands were confined prior to that point.

Strands **26**, **27**, and **28** are then advanced as a larger strand to primary attenuation means **40**. As shown in Figure 3.4a, primary attenuation means **40** is comprised of an idler roll **42**, a cot roll **44**, and a cutter roll **46** having a plurality of radially projecting cutter edges as found in choppers well known in the art.

Cot roll **44** acts as a pull wheel and idler roll **42** is positioned to bring the strands into engagement with the surface of cot roll **44** in a nonslipping engagement such that all of the streams from all of the bushings **10** are simultaneously attenuated into filaments. As the strand contacting the surface of cot roll **44** passes through the region where cutter roll **46** pierces the elastomeric surface of cot roll **44** the continuous glass filaments or strands are cut into discrete segments.

As shown, second guide means **35** is positioned such that strands **26**, **27**, and **28** are all advanced along paths external to each of the zones **54** of section **4**, **5** and **6**. Thus, if one of the sections should be disrupted at any time, the streams or beads thereof will not contact the remaining advancing strands and disrupt the entire operation.

During start-up or restart, the filaments can be normally in contact with the applicator **18**, gathering shoe **22** and idler roll **30**, with the strand being advanced as waste material since the secondary attenuation means **50** is normally of the type that the filaments are pulled at a lower speed than by the primary attenuation means.

As shown in Figure 3.4c, first rotatable means or cot roll **44** is journaled in frame **132**, and cutter roll or second rotatable member **46** is also journaled in frame **132**. Chopper or primary attenuation means **40** serves to attenuate the streams of molten material into filaments through the action of the cot roll **44** which acts as a pull wheel and cut or chop the continuous filaments into discrete segments.

Drive means, such as a conventional electrical motor, is adapted to rotate cot roll **44** and cutter roll **46**, and receptacle or delivery means **145** is positioned to receive the chopper strand **131** as the chopped strand leaves the forming means **40**.

Receptacle **145** is comprised of a first zone **147** spaced from a second zone **149** by means of wall **151**. Movable element **153** is pivoted at one end of wall **151** by means of rod **155** joined thereto and journaled in the side walls of the receptacle **145**. Attached at one end of rod **155** is arm **157** which is attached at the opposite end to shaft **159** of motive means or air cylinder **161**.

As shown in Figure 3.4c, movable element **153** is adapted to direct the chopped strand **131** to the second zone **149** of receptacle **145** when the shaft **159** is retracted. When the air cylinder **161** is energized to extend shaft **159**, movable element **153** retracts into the normal operating position to allow the chop strand **131** to move to and through first zone **147**. Generally, first zone **147** is adapted to direct the chop strand a suitable shipping container, while second zone **149** is in communication with a suitable scrap collection system.

During the operation of such multifeeder wet chop systems, it is desirable to maintain production from the feeders **10** in spite of process interruptions, such as break-outs and the like, from one of the other feeders. With high speed operations, it may be difficult and undesirable to rethread the disrupted feeder with the system operating at full speed. But the forming means or chopper **40** cannot be stopped, otherwise the other feeders will be interrupted. Therefore, the speed of the chopper, the cot roll and/or cutter roll must be reduced to a second speed, slower than the first speed, to allow the operator to thread-up the strand to be restarted and yet maintain the other bushings in the operational mode.

When the system is operating at the second or slower speed, the filaments and/or chopped strand formed may be of the type or quality unacceptable for inclusion with the desired product. To prevent the secondary chop strand or waste from contaminating the desired product, delivery means or receptacle **145** is adapted to direct the chopped strand to the first zone **147** when the forming means **40** is operating at the first speed and to direct the chopped strand **131** into the second zone **149** when the attenuation/forming means **40** is operating at the second or slower speed. This is accomplished by means of a control system co-ordinating the drive means (not shown) and the delivery means **145**.

During operation when the operator is about to rethread a strand from a dis-rupted bushing, the operator activates the control system which reduces the speed of the forming system **40** and energizes air cylinder **161** to shift movable element **153** to direct the waste into the second zone **149**.

To prevent the system from running at the second speed for too great a length of time which may set up thermal imbalances in the other normally operating feeders, the control system is adapted to automatically shift back into the nor-mal or high speed mode after a predetermined length of time. That is, a system is provided wherein normal production resumes after the strand has been re-started in the absence of further actions by the operator.

Method of Introducing Glass Strand onto Feed Roller

Japanese Patent 50-27089 discloses a method of manufacturing chopped strands of glass filaments directly from strands of glass filaments which are formed such that a multiplicity of glass filaments extruded through orifices at the bottom of a spinning furnace and applied with a sizing agent are corrected to form the strand. More particularly, the strand of glass filaments is taken up by a single feed roller to contact the circumferential surface thereof through a predeter-mined angle so that glass filaments of the strand are drawn and attenuated by frictional force due to the contact with the surface of the feed roller.

A cutter roller is disposed to contact under pressure the circumferential surface of the feed roller, thereby cutting the strand into chopped strands of a prede-termined length. The position of the feed roller where the cutter roller contacts the circumferential surface of the feed roller is selected so that the frictional force imparted to the strand is greater than a drawing force by which normal attenuation of glass filaments of the strand is attained.

In general, in the operation of the above mentioned system, the strands are formed of glass filaments extruded approximately at a rate of 30 to 50 m/min from the respective spinning furnaces. The strands are taken up for attenuation by the feed roller which rotates at high circumferential speed, approximately

1,500 to 3,000 m/min. Therefore, it is very dangerous to introduce a broken strand onto the feed roller while the system continues operation. As a consequence, upon introducing the broken strand onto the feed roller, the feed roller must be slowed down or completely stopped for ensuring a safe situation to operators. In other words, the introduction of the broken strand must be made at the sacrifice of the productivity of the system.

Furthermore, since glass is a thermo-softening material, streams of molten glass extruded through orifices in the bottom of the spinning furnace must be always drawn under a proper tension to allow the attenuation into filaments. If the drawing force is removed, the molten glass streams would conglomerate into a relatively large bead or beads on the undersurface of the spinning furnace. It is extremely cumbersome to remove such a glass bead. Therefore, more than two operators are required for ensuring proper extrusion of glass filaments from all of the spinning furnaces, which is one of the factors which hinder the optimization of the operation of the system.

Furthermore, in the beginning of the operation of the conventional system, the strands consisting of glass filaments in nonattenuation are first brought into engagement with the circumferential surface of an end portion of the feed roller, and thereafter the speed of the feed roller is gradually increased to a normal attenuation speed.

After having been completely attenuated, the strands are moved to a cutting portion of the feed roller and the cutting operation is started. However, during the cutting operation, the nonattenuated filaments taken up at the end portion of the feed roller tend to loosen gradually so that pieces like fluffs may fly out and drop and are mixed with the normal chopped strands, which leads to lowering the quality of a product. Additionally, the nonattenuated filaments are relatively thick and have high rigidity so that it is difficult to have them completely engage or wound on the circumferential surface of the feed roller.

Therefore, a portion of the strand length tends to move away from and wave about the feed roller with relatively large radii. This is very dangerous to the operators.

K. Nakazawa, T. Kikuchi and T. Fujita; U.S. Patent 4,175,939; November 27, 1979; assigned to Nitto Boseki Co., Ltd., Japan describe a method and apparatus for introducing a glass filament strand onto a feed roller which can substantially eliminate the above and other problems encountered in a system of the type in which a plurality of glass filament strands are drawn and attenuated by a single feed roller and chopped off by a single cutter roller.

According to this process, in the method of manufacturing chopped strands of glass filaments as described in Japanese Patent 50-27089, there is provided a method of and an apparatus for introducing a strand of glass filaments onto a feed roller in which an auxiliary feed roller is disposed adjacent one end of the main feed roller in axial alignment therewith and is driven independently of the main feed roller and an auxiliary cutter roller is arranged to cooperate with the auxiliary feed roller so as to cut or chop off the strand taken up by the auxiliary feed roller.

When one of the strands is accidentally broken during operation of the system, the cutting operation for the remaining unbroken strands may be continued without reduction of a high rotation speed of the main feed roller, while the auxiliary feed roller is maintained substantially stationary or is operated to rotate at a low speed of 30 to 50 m/min. The end portion of the broken strand is then wound on the auxiliary feed roller and thereafter the speed of the auxiliary feed roller is gradually increased to a normal attenuation speed of 1,500 to 3,000 m/min at which the main feed roller is rotated.

After complete attenuation of the broken strand is attained, the strand is transferred from the auxiliary feed roller to the main feed roller. Since the strand taken up by the auxiliary feed roller is continuously chopped off by the auxiliary cutting roller, the wave of the strand about the feed roller as abovementioned is avoided as well as the coming loose and flying out in the form of fluffs.

Furthermore, since the main feed roller is not needed to be slowed down during the introduction of the broken strand, the productivity of the system will not be reduced. Also, since the introducing operation may be conducted with safety and ease, the system may be attended by one operator who is not so skilled.

Fiber Mat Production Using Variable Speed Attenuator

W.W. Drummond; U.S. Patent 4,158,557; June 19, 1979; assigned to PPG Industries, Inc. describes a method and apparatus for forming a more uniform continuous strand mat, such as a glass fiber mat. The attenuator, which is traversed across the mat formation surface, may be either a wheel attenuator or a belt attenuator.

The attenuator is driven by a belt, chain or the like, which is driven by a remote motor which is not itself traversed across the mat formation surface. This substantially reduces the weight of the traversing attenuator and permits greater speeds of traverse which, in turn, permits a faster mat formation rate at a given width. It also allows for production of greater widths of mat at a given rate of mat length.

In addition, the attenuator is designed such that as it traverses across the mat formation surface, its speed of projecting the strands onto the mat formation surface is varied, due to the constant speed belt or chain driving it. Thus, when the attenuator is traversing away from the bushing or packages of previously formed strand, its strand projection speed is reduced from the set speed of the driving belt or chain in an amount equal to its rate of traverse.

When the attenuator is traversing towards the bushing or packages of previously formed strand, its speed of projection of the strands onto the mat formation surface is increased by its rate of traverse. Thus, the net speed of attenuation or advancement of the filaments and strands remains a constant.

For example, driving the attenuator of the process by a constant belt drive of 2,000 ft/min (605.90 m/min), when traversing away from the strand source at 100 ft/min (30.48 m/min), the speed of projection reduces to 1,900 ft/min (575.32 m/min), thus resulting in a net speed of attenuation of 2,000 ft/min (605.90 m/min).

When the attenuator is traversing towards the bushing or packages of previously formed strand at a speed of 100 ft/min (30.48 m/min), the attenuator projection increases to 2,100 ft/min (636.28 m/min), thus resulting in a net speed of attenuation of 2,000 ft/min (605.90 m/min). Thus, in either direction of traverse, the filaments and strand are attenuated or advanced at a constant force from the strand source, thus resulting in more consistent diameter filaments and thus resulting in a better quality mat.

Turning to Figure 3.5, a single forming position for producing continuous strand glass fiber mat is illustrated. It should be noted that a complete mat formation line will employ a plurality of these positions along the mat formation surface length.

Figure 3.5: Glass Fiber Mat Formation Apparatus

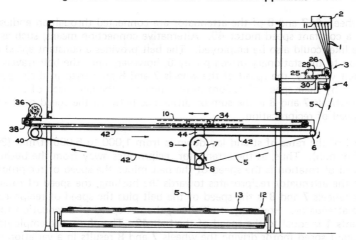

Source: U.S. Patent 4,158,557

Glass filaments 1 are attenuated through bushing tips 11 at the bottom of a heated bushing 2 which contains molten glass. The filaments are passed across an application surface 26 where they are coated with a binder and/or size from a sump 30. The application surface is shown as a roller connected by a belt 29 to a motor 25. It will be apparent to the skilled artisan that this applicator can be any conventional applicator such as a belt or pad applicator, a spray applicator or the like.

The filaments are then passed across the face of a gathering shoe 4, which is a grooved wheel or cylinder formed of a material such as graphite. The filaments are gathered by the gathering shoe into one or more strands 5. If a plurality of strands are formed, the gathering shoe may be replaced by a comb, as is known to those skilled in the art. The strands pass across driven pulley 6, around a first wheel 8 of an attenuator 9, which is illustrated in this figure as a wheel attenuator, and around the second wheel 7 of the attenuator.

Attenuation is accomplished through cohesive forces between the wet coated strands and the rotating wheels 8 and 7. The strands are projected downwardly from the wheel **7** onto a mat formation surface **12**, which is typically an endless belt or chain conveyor, where the continuous strands **5** are laid as a mat **13**. The wheels **8** and **7** are attached by a bracket **44** to a carriage **34**. The carriage reciprocates across and above the face of the mat formation surface such that the strand is laid continuously across the mat formation surface.

The carriage is, in turn, attached to a chain **38** which is driven by a belt connected to a reversing motor **36**. The carriage rides within a track **10** as it reciprocates across the strand formation surface **12**. Typically, the speed of reciprocation for the attenuator **9** across the mat formation surface **12** is in the range of 75 to 150 ft/min (25.9 to 45.7 m/min). Preferably, the attenuator traverses in a perpendicular direction to the mat formation surface. However, the attenuator may be oriented to lay down the mat at an oblique angle to the mat formation surface.

The wheels **6, 7** and **8** of the attenuator are connected through an endless belt **42** to a constant speed motor **40**. Alternative connection means, such as chains or the like, could also be employed. The belt provides a constant speed of rotation to the stationary driven pulley **6**, however, since the belt travels at a constant velocity, the speed of the wheels **7** and **8** are varied in their speed of rotation by their speed of reciprocation, such that the net speed of attenuation of the wheels **7** and **8** is the sum or difference between the speed of the belt and speed of reciprocation.

Typical of the rates of strand lay down is from 1,000 to 5,000 ft/min (304.8 to 1,524 m/min). Thus, as the attenuator reciprocates away from the bushing **2**, its speed of rotation is the speed of the belt minus the speed of reciprocation and as the attenuator reciprocates towards the bushing, the speed of rotation of the wheels **7** and **8** is the speed of the belt plus the speed of reciprocation of the attenuator. However, at all times, the net speed of attenuation for the filaments **1** is equal to the speed of the belt **42**. Therefore, the employment of a constant speed motor driving the wheels **7** and **8** results in a variation of wheel speed which compensates for the variation in reciprocation speed such that the net attenuation speed remains constant. This, of course, results in more consistent diameter filaments being formed throughout the reciprocation of the attenuator **9** and thus in a more uniform mat.

The slight difference in speed of strand lay down as the attenuator is traversed towards the strand source and away from the strand source is not significant to mat uniformity. As the mat is laid, each traverse normally overlaps the previous traverse, thus resulting in a uniform mat along its length. This is further aided by the fact that typically six or more layers are laid from attenuation positions along the length of the mat formation surface **12**.

Manufacture of Fiber Mats of Uniform Thickness

R.E. Loeffler; U.S. Patent 4,168,959; September 25, 1979; assigned to Johns-Manville Corporation describes a method and apparatus for producing a fiber glass mat of substantially uniform density and in one embodiment at substantially increased rates. According to this process, maneuverable forming tubes and binder nozzles are provided. In the operation of the apparatus, a battery of fiber generators discharges a gaseous stream of fibers into a corresponding group of stationary, elbow shaped conduits, which turn the fiber streams from a horizontal direction to be discharged downwardly towards a moving collection surface.

Forming tubes are telescopically mounted over the downwardly sloped portions of the conduits and each can be moved in a pendulum-like fashion to the left or right of the center line of the stationary conduit to be secured in one of various orientations. Whenever there is a mat of uneven thickness being produced, the forming tubes can be individually maneuvered to direct the deposition of fibers so as to fill the low regions on the mat profile and to lessen the deposition of fibers on the high regions, so that a satisfactorily level mat-profile is quickly achieved.

In the operation of a fiber mat machine according to the process, a record is made of the forming tube setting best for each of the various fiber glass products produced on the machine. Thus, on machine start-up of any given product or changeover from one product to another the forming tubes may be adjusted to the appropriate predetermined setting.

Another feature of this process lies in the fact that the telescopically mounted forming tubes, as a group, may be mounted with their discharge ends located at various selected distances above the collection surface.

This feature facilitates conversion of the machine to produce mats for differing end uses. For example, fiber glass insulation is generally produced at higher pull rates and burner pressures than fiber glass filter media and the attendant heat dissipation factor generally requires a shorter forming tube which readily can be provided by the apparatus.

In one embodiment the fiber generators are arranged in two horizontally spaced apart lines with the generators of the first line being laterally staggered with respect to the generators of the second line. Gaseous blasts from generators of the first line issue in the direction of the second line of generators, and vice versa. The forming tubes and the lower portions of the elbow shaped conduits associated with the first line of burners interlay the forming tubes and conduits associated with the second line of burners. Forming tubes and conduits are sufficiently narrow to provide for their accommodation in a single transversely extending row within the width of the collection surface and also allow for the pendulum-like and vertical movement described above.

This process also encompasses a binder application system having binder headers extending longitudinally along each side of the line of forming tubes and also positioned in the spaces between adjacent forming tubes. From each binder header extends a plurality of valve controlled means having nozzles disposed adjacent the discharge ports of the forming tubes for applying an atomized spray of binder from any angle or position necessary for efficiently and effectively coating the fibers.

The flexibility of the binder application means complements the adjustability of the forming tubes and permits binder distribution to be held uniform when forming tubes are rearranged. Further, the universality of nozzle positioning allows precise adjustments that ensure a minimum of binder waste and uniformity of binder distribution upon the mat.

Bushing for Glass Spinning Apparatus

According to a current technique of manufacturing glass fibers, a bushing is used to contain molten glass which is the material of the glass fiber. The bushing is provided at its bottom with an orifice plate having 400 to 2,000 orifices. The molten glass in the bushing is forced out through the orifices, by the static head and viscosity, to form a number of glass cones which are then mechanically drawn to be turned into fibers.

This method, however, has been found inconvenient in that the orifice plate is apt to be bent and deflected to cause the lower surface to become convex downwardly, as the time elapses, due to a combined effect of weight of the molten glass in the bushing, spinning tensile force, high spinning temperature which can reach 1100° to 1300°C and so on.

This deflection of orifice plate causes a difference in the rates of heat radiation from the glass cones suspended from the orifice plate, especially between the orifices located near the periphery of the orifice plate and the orifices located near the center of the orifice plate, resulting in an unstable spinning.

In addition, the deflection of the orifice plate badly affects the stability of the glass cones to incur the breakdown of the latter, due to an increased component of force pulling the glass cones laterally along the orifice plate. Moreover, the deflection of the orifice plate tends to cause so-called flooding in which the cones creep through capillary action along the underside of the orifice plate to join adjacent cones. Once the flooding takes place, it is quite difficult to resume the separate flows of glass cones.

H. Shono, S. Ishikawa, I. Wakasa and M. Adachi; U.S. Patent 4,178,162; December 11, 1979; assigned to Nitto Boseki Co., Ltd., Japan describe a bushing for a glass fiber spinning apparatus, having at its bottom an orifice plate provided with a great number of orifices consisting of plain holes arranged so densely that molten glass cones consisting of masses of molten glass having flowed out of the bushing through respective orifices tend to join with adjacent ones, so as to cause a flooding condition at the downside of the orifice plate, the bushing having a beam member or members disposed therein and connected at both ends to the opposing walls of the bushing, the beam member extending in parallel with and spaced from the orifice plate, and at least one rod member connecting the beam member to the orifice plate.

Referring to Figure 3.6, a bushing is provided at its bottom with an orifice plate 2 in which are formed a great number of orifices 1 in the form of extremely densely arranged plain holes. The bushing has opposing walls 3, 3' to which respective ends of beam members 4 are secured by means of, for example, welding. These beam members extend in parallel with and spaced from the orifice plate. The beam members are connected to the orifice plate through rod members 5 having a small diameter. The beam members and the rod members in combination constitute stiffening or reinforcing members for the orifice plate.

The beam members are disposed to extend at right angles to the longitudinal direction of the bushing when the latter has a rectangular profile, and the number of the beam members is optionally selected depending on the size of the bushing, while the number of rod members is suitably selected in accordance with the length of the beam members.

Needless to say, the beam members may be disposed to extend radially from the center of a circle or may be arranged in parallel with one another, when a cylindrical bushing is used, and any other pattern of the arrangement of the beam members can be adopted, e.g., in the form of a lattice or the like, as necessary.

Figure 3.6: Bushing for Glass Spinning Apparatus

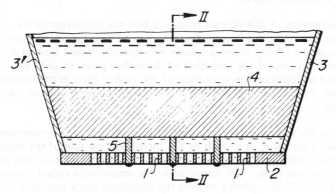

Source: U.S. Patent 4,178,162

The distance between the beam member and the orifice plate is preferably within a range of between 1.5 to 10 mm. A too small distance would adversely affect the supply of the molten glass to the orifices, while a too large distance would cause an excessively large elongation or strain of the rod members to deteriorate the stiffening effect on the orifice plate.

The beam member can have any desired shape such as flat plate, rod and the like, and is preferably made of platinum. Thus, in order to minimize the expense, the beam members are made small, insofar as they can ensure sufficient stiffening effect. Also, the rod members connecting the beam members to the orifice plate should be made as small as possible. For the same reason, the number of beam members employed in a bushing, as well as the number of the rod members for each beam member, should be made as small as possible, consistent with a satisfactory stiffening effect.

The pitches at which the beam members and the rod members are disposed, respectively, are typically 10 to 50 mm, and 5 to 20 mm, although they depend on the size of the orifice plate and the number of plain hole orifices formed in the latter.

Because of the structural features as stated above, the orifice plate of the bushing is rendered free from the force which would cause the downward deflection of the orifice plate, since the force is conveniently born by the reinforcing members consisting of the beam and the rod members.

In addition, the reinforcing members occupy only a very limited area on the orifice plate. More specifically, the number of the orifices rendered invalid or useless by the provision of the reinforcing members is very small for the total number of the orifices, so that the temperature differential between the glass cones from respective orifices is conveniently reduced.

Thus, the entire area of the orifice plate can be effectively used for densely and uniformly locating the orifices, in good contrast to the conventional arrangement in which the distance between the adjacent orifices is made large at portions where the reinforcing members are located, or the orifices are arranged in a plurality of separate groups.

Another advantage of this process resides in that the separation of the joining glass cones into respective independent cones can be performed efficiently. This separation of the joining cones is usually effected by upward flow of air jetted from air nozzles toward the underside of the orifice plate, so as to cool the orifice plate to facilitate the separation of the molten glass in the flooding condition from the orifice plate and thereby promote the separation of the joining glass cones.

However, as the separation is performed to some extent, the temperature of the orifice plate rises again, as a result of the increased flow rate of the molten glass through the orifices and the correspondingly increased amount of heat brought out by the molten glass. This temperature rise undesirably causes rejoining of the glass cones. This tendency renders the separation work considerably difficult, especially when the number of the orifices employed is large.

However, according to this process, the separation of the joining glass cones is commenced at first at portions where the spaces between adjacent orifices are larger, due to the provisions of the rod members, than other portions, to define a plurality of groups of joining glass cones. The separation then goes on in each group. It will be understood that such a procedure of separation is less likely to cause the rejoining of the glass cones, and enhances separation efficiency.

In fact, the time can be reduced to almost a half of that required by the conventional technique, in separating the joining glass cones.

Resistively Heated Silicon Carbide Bushing

J.W. Hinze; U.S. Patent 4,185,980; January 29, 1980; assigned to Owens-Corning Fiberglas Corporation describes an electrically heated glass feeding apparatus which comprises a bushing essentially comprising silicon carbide, the bushing being adapted to maintain molten glass therein and to issue a stream of glass therefrom and the bushing including a pair of opposed electrical terminals or tabs.

Means are provided for passing an electric current through the bushing between the terminals to resistively heat the bushing, the means comprising an electrical contact of an oxidizable, electrically conductive, carbon-containing material in direct electrical contact with one but not the other of the terminals and a metallic electrical connector in direct contact with that oxidizable material.

The oxidizable, electrically conductive, carbon-containing material will be so disposed in the electrical contact so as to substantially preclude direct physical contact of the terminal, or tab, of the bushing and the current supplying connections. Desirably that material will be graphitic and in a preferred embodiment the material will be a member, or spacer, in the form of a graphite tape. Suitably, the tape is that supplied as Grafoil (Union Carbide).

Generally, Figure 3.7a illustrates a glass feeder, or bushing **10** which is fabricated of silicon carbide. The manner of fabrication can be any of the conventional techniques used for fabricating refractory structures such as, for example, hot pressing or injection molding. Generally, bushing **10** in cross section will include a rectilinearly shaped wall **12** and disposed at the outer portion thereof is an outwardly extending flange **14** which is adapted to be secured to a supply of molten glass as, for example, a forehearth generally designated **16**.

Conventional means may be employed for such purpose. The lower portion of bushing **10** has a bottom wall which includes a plurality of orifices **18** from which streams of molten glass issue. These streams are attenuated into fibers **20** which are in turn passed through a suitable size applicator **22** and they then converge to a suitable gathering means **24** in the form of a strand **26** which is then collected on a suitable roll-up drum **28**.

Figure 3.7: Resistively Heated Bushing of Silicon Carbide

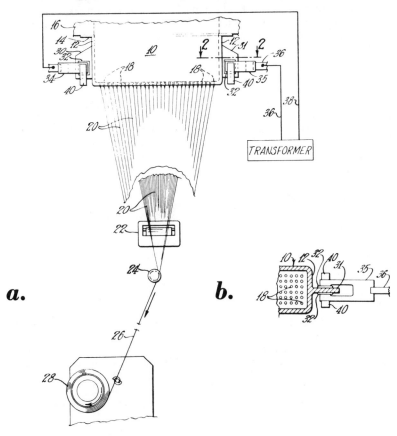

(a) Illustrates an embodiment of this process.
(b) View taken along line **2-2** of Figure 3.7a.

Source: U.S. Patent 4,185,980

Extending outwardly from opposed portions of wall **12** are a pair of electrical terminals, in the form of tabs **30** and **31** respectively, by which current is passed between those terminals and through bushing **10** so as to resistively heat it. In this way, the bushing maintains molten glass from forehearth **16** at a suitable forming temperature. Each of the respective electrical terminals **30** and **31** is in direct electrical contact with a separate, oxidizable, electrically conductive member **32** (best seen in Figure 3.7b).

In the preferred embodiment, each member **32** will be a graphite tape and the tape will be disposed on opposing faces of each terminal in order to provide for a better electrical contact. Current is supplied to bushing **10** by means of a conventional arrangement typically employed in directly electrically heating precious metal bushings. Thus, tabs **30** and **31**, each of which contains graphite tape **32**, will be respectively attached to metallic bus bar connectors, for example, copper members **34** and **35**, respectively in the form of a fork-shaped member, which members in turn are connected by suitable wiring **36** and **38** to a power source for maintaining a potential across the bushing **10**. Suitable clamping means **40**, for example, in the nature of a C-Clamp, will be employed to tighten the legs of the forked-shaped connectors **34** and **35** against the graphite tape so as to minimize the contact resistance of the electrical contact.

In fabricating the apparatus as indicated above, it will be found desirable to position the graphite tape such that no direct contact is made between connectors **34** or **35** with the respective tabs **30** and **31**. Thus, since the members **32** are electrically isolated by means of bushing **10**, all current will flow from a connector, for example, **35**, through its appurtenant graphitic member **32** and into the electrical terminal **30**, then through the bushing; in such manner, the bushing will be heated and it will be found that the formation of silica adjacent each of the electrical contacts will be substantially inhibited.

Bushing Block with Cylindrical Flow Passage

N.E. Greene, S. Srinivasan and L.A. Stenger; U.S. Patent 4,161,396; July 17, 1979; assigned to Owens-Corning Fiberglas Corporation describe a bushing block construction disposed between the glass flow channel in a forehearth and the stream feeder or bushing wherein a substantially vertical glass flow passage or passage means in the bushing block is configurated to provide a minimum surface defining the glass flow passage to thereby reduce the area of refractory exposed to erosion processes.

Figure 3.8 is a semischematic elevational view illustrating a melting furnace and a forehearth construction for supplying heat-softened refined glass to stream feeders or bushings associated with the forehearth. Glass batch is melted and refined in a furnace or tank **10**, a forehearth **12** being connected with the melting and refining furnace, the forehearth having a channel **13** containing heat-softened refined glass flowing from the furnace. The glass in the forehearth channel is maintained at the proper temperature by conventional heating means (not shown).

The glass is processed in the furnace and forehearth to render it suitable for attenuation to filaments or fibers. Associated with the forehearth are one or more stream feeders or bushings **16** of hollow rectangular configuration, there being two stream feeders shown in Figure 3.8. The forehearth floor adjacent each bushing is provided with a flow block **14** provided with a lengthwise elongated glass flow passage **15** of conventional construction. Disposed between each of the bushings **16** and the adjacent flow block **14** is a bushing block **19** having a glass flow passage **20**, the passages **15** and **20** accommodating flow of heat-softened glass from the forehearth channel **13** into a bushing.

Figure 3.8: Forehearth and Bushing Arrangement

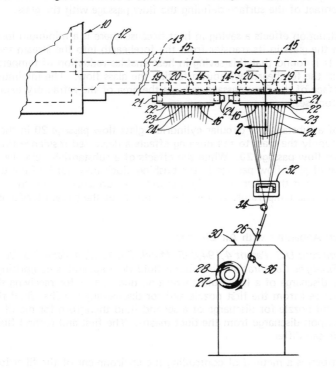

Source: U.S. Patent 4,161,396

The bushing block **19** functions to thermally isolate a bushing or stream feeder from the forehearth. The bushing block **19** is fashioned of suitable high temperature resistant refractory and extends substantially the length of a stream feeder or bushing **16**.

The central region of the bushing block **19** is provided with a substantially cylindrical passage or flow passage **20** which is defined by a substantially circular cylindrical surface.

To function successfully, the flow capacity of the passage **20** in the bushing block should be greater than or equal to the discharge capacity of the bushing so that the bushing and not the flow passage controls the throughput of glass. The upper portion of the bushing block **19** may be fashioned with angularly disposed or ramp surfaces to facilitate flow of glass from the forehearth to the passage 20 with a minimum of resistance. It is to be understood that the ramp surfaces may be eliminated if desired.

Several advantages are attained in utilizing a generally circular cylindrical passage 20 in a bushing block for transferring glass from a forehearth channel into a bushing. The circular cylindrical flow passage provides for a maximum volume of flow of glass from a forehearth channel into a stream feeder with a minimum of area contact of the surface defining the flow passage with the glass.

This construction effects a saving in heat because there is a minimum loss of heat from the glass in its transfer from the forehearth into the stream feeder or bushing. It is found that this saving in heat enables reduction of temperature of the glass in the forehearth without affecting the glass flow. The minimum area of the surface of the flow passage reduces the area of the refractory exposed to erosion processes.

The use of a substantially circular cylindrical glass flow passage 20 in the bushing block to supply the glass to the bushing effects a decreased residence time of the glass in the flow passage 20. While the effects of a substantially circular cylindrical type of glass flow passage in the bushing block may not be fully understood, it is found that there is vastly improved performance of the stream feeder or bushing. The temperature of the glass throughout the stream feeder is more stable.

Fluid Flow Apparatus with Air Blowers

T.K. Thompson; U.S. Patent 4,194,895; March 25, 1980; assigned to Owens-Corning Fiberglas Corporation describes a fluid flow apparatus comprising a first nozzle for discharge of a first fluid therefrom, duct means for receiving the first fluid discharged from the first nozzle and for discharging the first fluid therefrom, and a second nozzle for discharge of a second fluid therefrom for modifying the first fluid upon discharge from the duct means. The first and second fluids can be alike or can differ.

Also described is a method of controlling the environment of the fiber-forming region of a glass-fiber-forming bushing comprising establishing a first flow of fluid in which the fluid in the center region of the first flow has a flow velocity at least as high as the flow velocity of fluid in all other regions of the first flow, directing a second flow of fluid into contact with the first flow to establish a combined flow in which the fluid in the center region of the combined flow has a flow velocity lower than the flow velocity of fluid in other regions in the combined flow and contacting the fiber-forming region with the combined flow.

Figure 3.9 shows bushing 10 connected to a forehearth (not shown) of a furnace for melting glass or glass-forming materials. The bushing is provided with a plurality of orifices at which cones 12 of molten glass material are produced for the attenuation of glass filaments 14 therefrom for collection on winding apparatus 20. The filaments are passed over sizing applicator 15 and also over gathering pulley 16 which gathers the filaments into strand 18 for winding into package 26 on the winder apparatus.

Winder 20 has a winding collet 22 mounted for rotation about a horizontal axis for the collection of strand into packages. A collection tube (not shown) can be placed over the collet for collection of the wound package thereon. A variable speed drive (not shown) within the housing of the winder rotates the collet. Conventional winder speed controls (not shown) modify the rotational speed of the collet during formation of packages.

Figure 3.9: Fluid Flow Fiber-Forming Apparatus

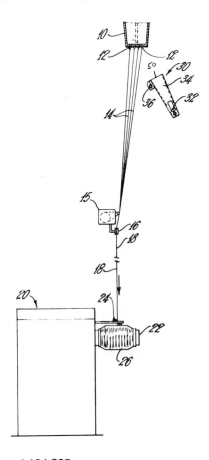

Source: U.S. Patent 4,194,895

Strand traversing apparatus **24**, such as a spiral wire traverse, is provided for distributing the strand along the length of the collet during strand collection.

Bushing **10** in Figure 3.9 is shown as a tipless bushing. However, the bushing can have a plurality of orificed projections, or tips, through which the molten glass is supplied for attenuation into fibers.

The glass-fiber-forming process shown in Figure 3.9 is provided with a fluid flow or blower assembly for controlling the fiber-forming environment. Control of the fiber-forming environment adjacent the glass cones at the regions of the orifices is particularly important. As shown, blower assembly **30** comprises a first flow nozzle **32**, a conduit or duct means **34** and a second flow nozzle **36**.

In the embodiment shown, the first flow nozzle is positioned to discharge into channel or duct means **34** and the air discharged from nozzle forms a first fluid flow which passes through the duct means. Duct means **34** can be of any configuration. As shown, the end walls are parallel, and front and back wall diverge upwardly and outwardly from the point of discharge of the first fluid into the duct means. As shown, nozzle **32** and the duct means form an induced air blower apparatus such that ambient air is induced into the duct means at the nozzle. The induced air is combined in the duct means for discharge with the air from the nozzle as the first fluid flow.

The duct means, or channel, directs the fluid passing therethrough and establishes a fluid flow stream such that an established fluid flow stream is discharged from the duct. Establishment of a flow stream is particularly important with an elongated blower, such as rectangular blower, so as to establish a flow exiting therefrom which has a generally uniform velocity profile along its length.

Second flow nozzle **36** modifies the fluid flow being discharged from the discharge region **50** of the duct means.

Motor Speed Control Device

J.D. Phillips; U.S. Patent 4,145,201; March 20, 1979; assigned to Owens-Corning Fiberglas Corporation describe a process to produce uniform filaments from a source of plastic materials such as molten glass by starting the molten glass through a bushing, and engaging the filament with the means for pulling glass filaments at a first rate and then altering the pulling rate at which the fiber is pulled from the source of plastic material a predetermined time after the fiber pulling is initiated.

Further, this process provides for changing the pulling rate when the temperature of the bushing changes, to maintain the filaments in a uniform diameter.

Referring to Figure 3.10, a means is shown for drawing the filament **3** from the source of thermoplastic material such as molten glass shown as **5**, through an orifice **7** in bushing **9**. The supply of molten glass **5** is shown in partial form.

The filament is engaged with the winder collet **11** by being wrapped around the collet after being initially drawn through the orifice **7**. The collet **11** is then rotated by motor collet driver **13** through a suitable drive connection shown as **15**. The power to motor **13** is supplied by power line **17**. A current sensor **19** is attached to power line **17**. A second sensor **21** is shown attached to motor **13** and senses rotary motion of motor **13** and more particularly cessation of that rotary motion. The sensor **21** may also be attached to collet **11** as is known in the art or to the drive connection **15** between motor **13** and the collet **11**.

A means for pulling the filament at a first rate and for altering the rate subsequent to a predetermined time is shown generally as numeral **23** with the first sensor **19** connected to the power line and sensing an interruption in the power to the motor, and with the second sensor **21** connected to the motor **13** and sensing a cessation of motion in the motor. Sensor **21** may also be provided to the drive connection **15** or to collet **11** for the same purpose.

Figure 3.10: Fiber-Producing Apparatus Utilizing Speed Control

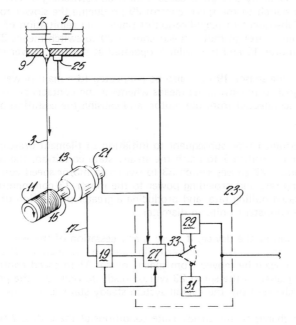

Source: U.S. Patent 4,145,201

Further shown is a temperature sensor 25 also within the means for altering the pulling rate. As is shown, temperature sensor 25, rotary motion sensor 21, and power sensor 19 are all connected to timer and switching mechanism 27 which provides power from either motor speed control 29 or motor speed control 31, through switch arm 33.

Power is shown applied either through a first speed control 29 for pulling the fiber at first rate, or to a second speed control 31 for pulling the fiber at the second pulling rate. The timer and switching mechanism 27 may select either the first speed control 29, causing the motor to run at the lower speed, or the second speed control 31 causing the motor to run at the higher speed. In addition, a timer 27 may cause the switch 33 to be delayed in switching from one speed control to the other in response to a signal from sensor 19, 21, or 25.

The operation of the device is described by referring to the figure. At the initiation of the filament-producing process, the material is drawn through the orifice 7 and attached to the pulling means shown as the winder collet 11. To compensate for the difference in conditions at the orifice 7 at initiation from that occurring after the steady state in the filament processes is reached, the pulling rate must be altered. These steady state conditions, for example, could be a higher orifice temperature. If the orifice 7 temperature at the initiation of the fiber-drawing process is cooler, a small diameter filament, and with more yards per pound of drawn filament, would be produced if the initial pulling rate was the steady state pulling rate for the higher steady state orifice temperature.

Accordingly, when an interruption occurs, either by the cessation of power through line **17** or by any other cause, timing and switching mechanism **27** causes **33** to switch to the speed control **29** producing the slower rotation of the collet **11**. Subsequent to reapplication of power and reinitiation of the process, power is applied through motor speed control **29** to switch means **33**, power cable **17** to motor **13** and the collet is operated at the first slower rate.

Additionally, the sensor **19** may also be connected to a main power switch to provide a signal to the switching means whenever the contacts are open indicating power disconnected from the motor, and causing the cessation of the winder process.

At a predetermined time subsequent to initiation to filament drawing process, necessary for the orifice **7** to reach the steady state condition, the switching and timer mechanism **27** causes switch **33** to switch to motor speed control **31** altering the pulling rate and providing power to the motor **13** commensurate with a second increased pulling rate, and producing a greater production of filaments but with the consistent uniform diameter.

In the operation of the device, if there is any cessation of the winding process, by blockage or by removal of power deliberately or through a malfunction, switch **33** will again be moved from speed control **31** to speed control **29** and the means for altering the speed **23** will be ready to reinitiate the process at the lower speed change from the overall system steady state conditions.

Further, any change to the steady state conditions at the orifice **7** will be sensed by temperature sensor **25** and relayed to switching means **23**. Accordingly, a decrease in temperature will cause the switching and time mechanism **27** to switch the motor from the first speed control **31** to the speed control **29** decreasing the pulling rate accordingly.

The switch and time mechanism as well as the motor speed control units **29** and **31** can be constructed from known units and are elaborated here.

In the case of altering the drawing rate responsive to a change in the temperature, it would ordinarily not be necessary to use the timing mechanism as in the case of altering the pulling rate subsequent to a reinitiation of the filament pulling process.

The sensor **19** may be a relay which opens and responds to a cessation of power · **17** and provides a low signal to switching and timing mechanism **27**, which, in turn, may use relays to switch **33**. Additionally, a removal of power by opening of a switch may also produce either a high signal, being the voltage across the open switch, to sensor **19** which may cause a subsequent opening of a relay contact, and low signal to switch the timer **27** causing movement of the switch **33** from speed control **31** to speed control **29**. The timing mechanism may be any suitable timing mechanism, which responds to a high signal, after initiation of power as relayed to speed control **27** by sensor **19** to move switch **33** from speed control **29** to speed control **31** subsequent to reinitiation of the winding process. Additionally, cessation of rotary motion may be sensed by sensor **21** causing a similarly low signal to sensor **27** which produces a signal which moves the switch accordingly. Suitable and well-known temperature sensors may also be used to control the switching mechanism as similarly described in regard to the responsive switching mechanism **27** to the signals of sensors **19** and **21**.

As is well known, solid state devices may be used in place of relays. The change in speed described above may be an increase or a decrease dependent upon the expected conditions at initial start-up, initiation after an interruption, and the expected changed conditions during the speed change and nonoperating time.

Additionally, a proper subsequent change in pulling rate responsive to temperature changes may be an increase or decrease, dependent upon the extent and direction from steady state conditions. For example, a cooling of the material after a short interruption may call for a slower pulling rate at reinitiation and then a faster pulling rate when steady state conditions are reached. However, should the interruption result in the material cooling too far, it may be necessary to use a higher than steady state pulling rate to maintain the fibers at a uniform diameter. A similar situation can exist at start-up where the start-up conditions may deviate from the steady state conditions to the extent that a higher or lower initial pulling rate may be required, followed by either a respective lower or higher pulling rate.

Microcomputer-Controlled Winder

J.W. Lonberger; U.S. Patent 4,147,526; April 3, 1979; assigned to Owens-Corning Fiberglas Corporation describes an improved microcomputer controlled winder for attenuating a plurality of streams of molten glass into fibers and for collecting the fibers as a strand on a wound package.

According to the process, a winder speed is digitally controlled to maintain a predetermined attenuation and collection rate for a strand of glass fibers. A constant speed motor is connected through an electromagnetically actuated clutch for driving a winder collet. An integrated circuit microcomputer or microprocessor which receives feedback data on the actual winder collet speed, generates a digital output which is used for phase firing two SCRs. This, in turn, controls power to the magnetic clutch for controlling coupling between the constant speed motor and the winder collet.

The winder collet speed is controlled in accordance with a third order polynomial which provides a predetermined speed curve. The actual curve for each product is determined by the constants in the polynomial. The polynomial is programmed into a microcomputer or other digital controller for the winder. Preferably, the constants for the polynomial are stored in a separate memory which stores the constants for defining the speed curve for a number of different products. Merely by telling the microcomputer which product is to be manufactured, the appropriate constants will be read from the memory and used in solving the polynomial for any point in the speed curve from an initial starting time. This differs from prior art systems in which digitized analog data for a single speed curve is stored and used for generating an error signal which controls the winder speed.

Gas Streams to Reduce Boric Oxide Deposition

By reason of environmental restrictions pertaining to air pollution and contamination, glass compositions for forming textile fibers or filaments are being employed wherein the glass compositions contain boron but little or no fluorine. In employing such fluorine-free glass compositions for forming streams of glass for attenuation to fibers, the major chemical species in the high temperature environment at the stream flow region is boric oxide.

The vapor pressure of boric oxide B_2O_3 decreases very rapidly with temperature so that the boric oxide condenses on the fin shields resulting in a comparatively rapid buildup of solid boric oxide on the fin shields. This condition necessitates frequent cleaning of the fin shield assembly to remove the accumulated condensation products from the metal fin shields or members.

G.R. Machlan; U.S. Patent 4,140,506; February 20, 1979; assigned to Owens-Corning Fiberglas Corporation has found that a gas, such as water vapor or steam at a temperature of above 250°F or more, provides a gas environment above the fin shields or metal members and between rows of depending projections on the stream feeder floor which is effective to attain the chemical reactions with volatiles emitted from a glass having boron therein but little or no fluorine to greatly reduce or minimize the accumulation or buildup of solids or condensation products on the fin shields or metal members and to render the glass beads formed during start-up longer and thinner to reduce the tendency of flooding of the glass over the feeder floor or tip section.

Nozzle Plate Alloy Composition

I. Wakasa, T. Noji and S. Takahashi; U.S. Patent 4,159,198; June 26, 1979; assigned to Nitto Boseki Co., Ltd., Japan have found that an alloy containing between 82 and 92% by weight platinum, between 3 and 10% by weight gold, and between 3 and 12% by weight palladium forms a single complete solid solution, and has an excellent resistance to wetting by molten glass and a superior machinability, and a nozzle plate made of the above alloy has a superior durability.

Example: An alloy containing 90% by weight platinum, 5% by weight gold, and 5% by weight palladium was melted under vacuum in an alumina crucible to produce an ingot. A nozzle plate was made by rolling the ingot weighing 1,200 g into a nozzle plate blank of 2 mm thick, and machining 4,000 nozzle holes in the nozzle plate blank having a nozzle hole density of 34 nozzle holes per square centimeter.

No cracking occurred during the processing. Glass fibers were spun with the use of this nozzle plate, and it was found that this nozzle plate can be used continuously for a period of time of 3 months or more.

A publicly known alloy containing 85.5% by weight platinum, 9.5% by weight rhodium, and 5% by weight gold was melted under vacuum in an alumina crucible to produce an ingot. A nozzle plate blank was manufactured by rolling the ingot weighing 1,200 g to a thickness of 2 mm. However, small cracks were formed in the nozzle plate blank during the rolling operation, thus it was impossible to use the nozzle plate blank until it was repaired.

Of these cracks, the visible ones were repaired by welding all the surfaces of the nozzle plate. However, invisible cracks deep in the nozzle plate were propagated by the expansion and contraction of the nozzle plate due to temperature changes during the operation, resulting in a leakage of the molten glass. It was thus impossible to prepare a practical nozzle plate having 800 or more nozzle holes.

Cutting of Glass Strands with Lasers

J. Kallenborn; U.S. Patent 4,158,555; June 19, 1979; assigned to PPG Industries, Inc. describes a method of cutting glass fiber strands wherein the strands are driven across the beam of an interruptable laser. The laser beam melts the strand at each pulse of the beam cutting the strand into short lengths. The cut strands can be immediately collected and packaged or can be further subjected to a drying procedure prior to packaging. The strands formed by this method have fused and comparatively even ends as opposed to the ragged uneven ends of mechanically chopped glass strands.

Glass Melting Using Electric Furnace

C.R. MacPherson, R.E. Boyce, and A.G. Smith; U.S. Patent 4,146,375; March 27, 1979; assigned to Reichhold Chemicals, Inc. have found that E glass can be successfully melted for direct filamentation, utilizing the electric furnace principle of U.S. Patent 3,429,972, notwithstanding the fact that it rather violently, turbulently, and in an extremely localized zone, fuses glass at extreme temperatures approaching 4000°F, with practically no opportunity for refining or soak, as compared to the rather leisurely melting and refining permitted by the long and spacious glass tanks used heretofore.

In carrying out the steps of the method, as will be seen from Figure 3.11a, the chute **1**, connected to any conventional source of powdered raw batch glass material, continually supplies E glass, raw batch **2** to electrical resistance furnace **3**. In the preferred embodiment, the pot furnace **3** is essentially that disclosed in U.S. Patent 3,429,972.

If desired, by using any of several flexible duct devices, raw material chute **1** may be rotated about the inner periphery of resistance furnace **3** to thereby continuously and evenly distribute raw batch material around the inner periphery of furnace **3** according to a predetermined rate.

Alternatively, rotating skimmer means may be utilized for continuously maintaining the proper level of raw batch in the top portion of furnace **3**, which raw batch serves also as heat insulation means for reducing heat loss from the melting furnace.

Raw batch materials are continuously melted in furnace **3** at a temperature of between about 3500° and 4000°F, and a resultant steady stream of molten glass **4** is continuously discharged from the bottom of furnace **3**, the glass having a temperature of from about 2900° to 3100°F at the point of discharge.

In the embodiment shown, molten glass **4** is continuously discharged upon sloping, heat resistant guide means **5**, preferably of platinum, from which the glass flows by gravity, directly into refiner means **6**.

Refiner means **6** preferably is of rectangular construction, of suitable refractory material, and its temperature is maintained and stabilized at a temperature of from about 2300° to 2600°F by conventional fuel burners **7** which may be positioned in either the crown, or along the sides as shown.

From refining means **6**, the molten glass continuously passes beneath skimmer wall **8** into forehearth means **9** where, by conventional fuel burners **10** disposed either in the crown or sides, temperatures are maintained at the filamentation temperature of E glass, or about 2300° to 2450°F. Electrically heated bushing means **11**, having electrical end terminals **12**, continuously receive molten glass from the forehearth means.

Referring to Figure 3.11b, there is illustrated the pool of glass **4** in forehearth **9**, with bushing means **11** disposed therebeneath. Filaments **13** are pulled from a multiplicity of orifices **14** in the bottom of bushing means **11**, passed over roller size applicator means **15**, gathered at point **16** into a strand **17**, and collected on a collet **18** held in sleeve-like arrangement by high speed winding mandrel **19**. The strand **17** uniformly traversed by traversing means **20** as it is wound.

Generally, the preferred embodiment of the foregoing method contemplates that the temperature of the glass in refiner means **6** will always be stabilized and maintained approximately 100°F higher than the drawing temperature maintained in forehearth means **9**. Also, the optimum depth of glass in refiner means **6** will vary between 6" to 10", with the depth of the glass in the forehearth maintained at approximately 2¾" to 3¼". Preferably, therefore, the depth of the glass in the refiner means will be approximately twice that of the glass in the forehearth. The relative levels of glass in the refiner and forehearth may be maintained as desired.

Figure 3.11: Fiber-Forming Apparatus Utilizing Electric Furnace

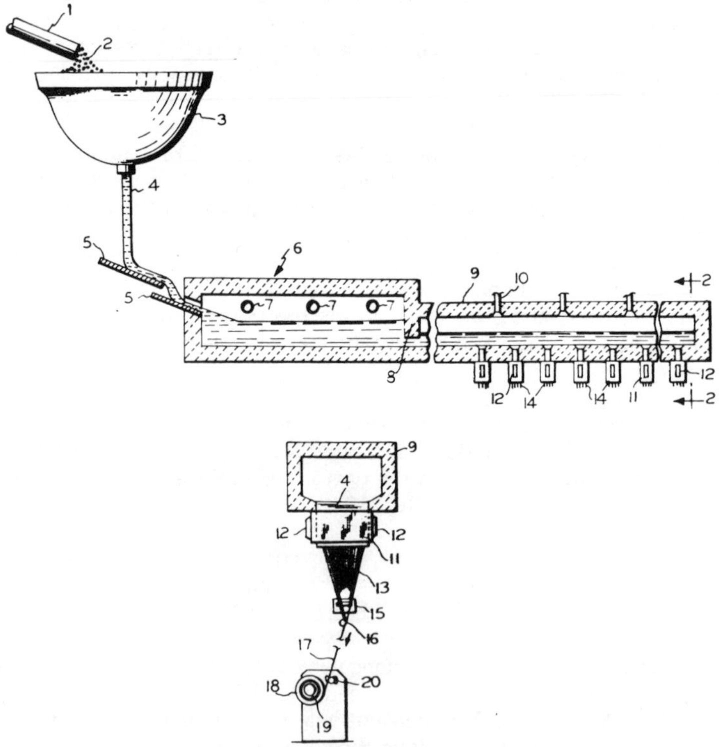

(a)   Schematic diagram of process
(b)   Side view along line **2–2** of Figure 3.11a

Source: U.S. Patent 4,146,375

Draw Forming Apparatus with Increased Production Rates

M. Ishikawa, T. Watanabe, M. Takita, S. Shikama and K. Nishimaki; U.S. Patent 4,197,103; April 8, 1980; assigned to Nitto Boseki Co., Ltd., Japan describe an improved method and apparatus for producing glass fibers by suitably modifying a conventional draw forming apparatus to achieve greatly increased production rates. Specifically the flow hole diameters in the nozzle plate are increased while maintaining the same spacing between adjacent nozzle tip peripheries, and cooling air is blown across the nozzle tips from a supply manifold in a direction parallel to the cooling fins at least during the initial start-up period of the filament-forming operation. After stabilized filament-forming conditions have been reached, usually within 5 to 10 seconds, the air flow is reduced or terminated.

Continuous filament withdrawal and winding can be maintained thereafter at a molten glass flow rate of at least 0.75 g/min/nozzle, whereby the glass fiber productivity rate is markedly increased.

The apparatus for draw forming glass filaments (Figure 3.12) includes a melt furnace **1** having a bottom plate **12** with a plurality of flow holes therein, each terminating in a conical nozzle tip **2**.

Figure 3.12: Modified Draw Forming Apparatus

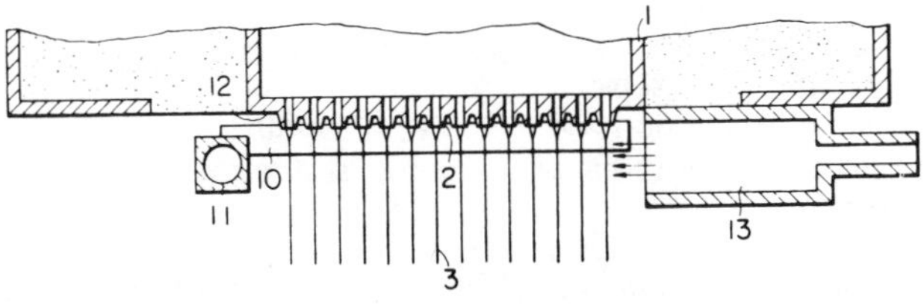

Source: U.S. Patent 4,197,103

Cooling fins **10** extend outwardly from a water pipe **11** and are disposed between the rows of tips, and additional cooling is provided at least during the initial start-up period by air blown across the nozzle tips from a manifold **13** disposed opposite the water pipe.

Example: The diameter of each nozzle tip was initially 1.80 mm, and the total molten glass flow rate was 300 g/min. The winding rate was set at 1,900 m/min to form glass fibers 10 μ in diameter. The forming was carried out using a conventional furnace having cooling fins but without any additional or auxiliary cooling means. Filament-forming could be carried out steadily and continuously. The diameter of each nozzle tip was then increased to 1.92 mm, and as a result the molten glass flow rate increased to 400 g/min. With such a flow rate the winding speed had to be increased to 2,500 m/min to form glass fibers 10 μ in diameter. The forming was again carried out using a conventional furnace with cooling fins but without any auxiliary cooling means, and it was found difficult to achieve stable forming conditions.

Next, an air flow was supplied across the nozzle tip surface in a direction parallel
to the rows of cooling fins at a rate of 0.7 m³/min and a velocity of 0.6 m/sec
until the forming and winding operation became steady, which took approximately 5 seconds. The air flow was thereafter suspended, and the forming and
winding conditions remained steady and continuous.

Environmentally Safe Fiber Collection Apparatus

In the conventional method of forming glass fibers utilizing hot, high velocity
gaseous blasts to attenuate the fibers during formation, the gaseous blasts with
entrained fibers and a large volume of inspirated process air are contained and
conducted by a forming tube and discharged into a collection chamber and onto
a moving perforated collection surface upon which the fibers are collected. A
suction means draws spent gas and air through the collection surface.

Emission control problems arise with such a known method, particularly with
the production of small diameter or microfibers (e.g., 0.05 to 2.60 micron diameter fibers and typically 0.1 to 0.7 micron diameter fibers) due to the difficulty of efficiently handling a large volume of moving gases. Furthermore, the
gas-entrained fibers tend to escape into the ambient surroundings, especially
in the regions adjacent the moving collection surface and the collection chamber.
The collection surface often becomes clogged with fibers which causes fibers to
be blown around the production area since a fiber clogged collection surface
prevents efficient exhausting of the gases.

This clogging problem necessitates replacing the collection surface, e.g., screen
material, frequently. This substantially diminishes the efficiency of the system
due to interrupted production and excessive downtime. In addition, it is often
necessary to install expensive emission control systems to avoid discharging fibers
into the atmosphere.

*R.E. Loeffler, S.R. Genson and J.L. Brunk; U.S. Patent 4,167,404; September 11,
1979; assigned to Johns-Manville Corporation* describe a method and apparatus
for collecting fibrous material which includes a collection chamber and a rotating
fluid-pervious collection drum preferably having a fine mesh screen positioned
around its peripheral surface. The collection drum is rotated along a path a
major portion, i.e., at least half, of which is within the collection chamber, and
a minor portion of which is located outside the collection chamber. Fibrous
material is drawn onto the peripheral surface of the collection drum moving along
the portion of the path within the collection chamber. The portion of the path
outside the collection chamber is sealed from the interior of the collection chamber.

The collected fibrous material is removed from the path while the collection
drum moves through the portion of the path outside the collection chamber.

Fiber-Handling Apparatus with Increased Tension

In a fiber collecting operation the fibers in an array are passed into contact with
a surface, which can be an applicator surface, and then collected, usually by being wound into a package on a winder. A comb-like splitter is often inserted
into the array of fibers just below the applicator surface to divide the fibers
into bundles and to produce a split-strand package. The splitter physically separates the array of fibers into bundles of fibers.

Heretofore the insertion of the splitter into the array has been a time-consuming task, requiring a manual separation of the array into bundles. In the array handling apparatus of the prior art, the fibers of the array generally do not assume a uniform alignment and are continually "dancing," or moving laterally relative to each other. This lack of uniformity of alignment of the fibers makes separation of the array into bundles even more difficult and it is especially difficult to obtain bundles with approximately equal numbers of fibers.

It has been discovered that increasing the tension of the fibers in the array reduces the dancing effect, makes insertion of the splitter easier, and allows the array to be divided into bundles of nearly equal size. The increase in tension can be effected by increasing the frictional drag force applied to the fibers as they contact the applicator surface. The fibers, under the influence of the increased drag force have a tendency to assume the alignment of the source of the array of fibers. For example, fibers pulled from a fiber-forming bushing will, under increased drag force, be more disposed to assume an alignment at the contact surface which corresponds to the alignment of fibers at the bushing.

J.E. Myers; U.S. Patent 4,170,459; October 9, 1979; assigned to Owens-Corning Fiberglas Corporation describes a process whereby an array of fibers is passed in contact with a surface to produce a first frictional drag force on the array, the array is divided into a plurality of bundles of fibers, the drag force of the surface on the bundles of fibers is modified to produce a second frictional drag force, and the bundles of fibers are collected.

In its preferred embodiment, the first frictional drag force is greater than the second. In its most preferred embodiment the surface is a rotating surface and the frictional drag force is modified from a first frictional drag force to a second frictional drag force by changing the speed of rotation of the surface.

Referring to Figure 3.13 there is shown bushing **10** comprising a chamber for holding glass mass **12**. The chamber is adapted with orifices **14** through which glass is emitted and attenuated into fibers **16** in the form of an array.

The fan of fibers is passed in contact with contact surface **18**, thereby creating a frictional drag force on the fibers. Below the contact surface the fibers are contacted by gathering member **20** which is suitable for gathering the fibers into a strand. The strand can then be collected on rotating collet **22** of winder **24**, which can be a conventional winder.

In order to divide the array of fibers into bundles of fibers, splitter **26** is inserted into the array. The splitter can be inserted manually. The comb-like splitter has projections to maintain separation between fiber bundles. It is usually desirable to divide the array evenly to obtain bundles containing substantially equal numbers of fibers. Subsequent to this splitting process, the bundles of fibers can be combined to form a single strand, as shown in Figure 3.13. The bundles can also be collected while the separation between bundles is maintained.

One method which can be used to modify the drag force exerted by the contact surface from a first drag force to a second drag force is to make the contact surface rotatable and to rotate the contact surface at different speeds. For example, in a typical fiber forming and collecting operation, a rotating size applicator surface is utilized to contact the fibers and apply a size.

Figure 3.13: Fiber Array Handling Apparatus

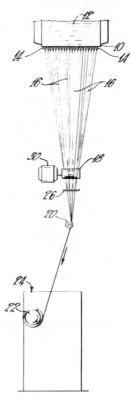

Source: U.S. Patent 4,170,459

By changing the speed of rotation of the applicator surface, the drag force on the fibers is modified. The rotation of the applicator surface can be accomplished by means of variable speed motor **30** shown.

In another embodiment the frictional drag force applied to the fibers by the contact surface is modified by changing the amount of lubrication applied to the contact surface. An increase in the amount of lubrication decreases the drag force on the fibers.

Glass Separating Device

H. Shono, K. Nakazawa and S. Ishikawa; U.S. Patent 4,171,212; October 12, 1979; assigned to Nitto Boseki Co., Ltd., Japan describe a method to mechanize the separation of glass, whereby workers may be freed from the manual and hard work and at the same time the separation failure may be reduced and productivity may be enhanced.

To this end the process provides a glass-fiber-forming apparatus of the type having an orifice plate with a plurality of closely spaced orifices and a flat undersurface, means for downwardly drawing glass cones formed under the orifices to form into glass fibers, and means for blowing cooling air against the undersurface of the orifice plate. Means is provided for separating a mass of molten glass adhered to the undersurface of the orifice plate in flooding condition into individual glass fibers. The separating means comprises an elongated stationary shaft standing on a floor spaced below the orifice plate at a position out of the stream of glass fibers and extending toward the orifice plate, a carrier movable along the shaft, means mounted on the carrier for releasably clamping the mass of molten glass, and a drive means for moving up and down the carrier to and from a position which may enable the clamping means to reach the mass of molten glass.

As shown in Figure 3.14, the separating device **15** has a clamp **17** consisting of a pair of gripping members made of a heat-resistant alloy and formed with saw-tooth blades at the leading ends thereof, and this clamp **17** is mounted on a slide box **18** movable up and down along the slide shaft **16**.

Figure 3.14: Glass Separating Device

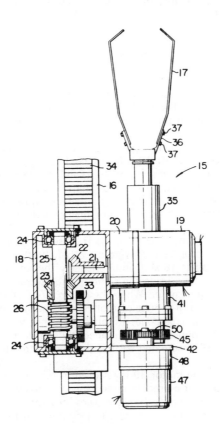

Source: U.S. Patent 4,171,212

Disposed at one side of the slide box **18** are a reversible variable speed motor **19** and a reduction gear **20** which is directly connected to the motor **19**. A bevel gear **22** is securely attached to the output shaft **21** of the reduction gear **20**, and a bevel gear **23** which is in mesh with the gear **22** is securely attached to a worm shaft **25** which in turn is supported by a pair of bearings **24** within the slide box **18**. A worm **26** provided on the worm shaft **25** is in mesh with a worm wheel securely attached to a counter shaft which is supported by a pair of bearings within the slide box **18** at right angle to the worm shaft **25**.

The slide box **18** has an integral cylindrical bracket having a bore surrounding the slide shaft **16** and is slidable along the slide shaft **16** through a metal bushing attached in the interior surface of the bracket and is prevented from the rotation by means of a key fitted into the slide shaft **16**. A spur gear **33** is securely attached to a counter shaft and the spur gear **33** is in mesh with a rack **34** formed on the slide shaft **16** through a slot formed through the bracket in opposed relationship therewith.

With the above arrangement, the rotation of the motor **19** is transmitted to the spur gear **33** through the reduction gear **20**, the bevel gears **22** and **23**, and the worm **26**.

As a result of the rotation of the spur gear **33** in mesh with the rack **34**, the slide box **18** moves up or down along the slide shaft **16**.

Each of the gripping members of the clamp **17** is connected with bolts **37** to a finger **36**, which is called a solenoid finger, of an electromagnetic automatic actuating switch **35** and when the switch **35** is energized, the clamp **17** is closed.

Next the mode of operation of the separating device with the above construction will be described. The slide shaft **16** is maintained at a predetermined angle of inclination by the operation of the worm shaft, and during normal spinning operation the slide box **18** is located on the slide shaft **16** at a relatively lower position thereof while a rotary arm is located at an angular position at which the clamp **17** will not interfere with the stream of filaments drawn from the orifices.

When the breakage of a filament or filaments is sensed by any suitable device, the slide box elevating motor **19** is energized so that the slide box **18** moves upwards along the slide shaft **16** and simultaneously the rotary arm turning motor **47** is energized so as to bring the clamp **17** to a position directed toward the center of the spinning furnace. When the clamp **17** reaches a position at which it may clamp the tip of a mass of molten glass hanging down from the orifice plate in a form of a large cone, a limit switch (not shown) is actuated so that the slide box **18** is stopped and simultaneously the electromagnetic automatic actuating switch **35** is actuated so that the mass of molten glass is clamped.

Next the slide box elevating motor **19** is reversed in rotation so that the slide box **18** moves downwards at a predetermined speed, thereby pulling down the clamped molten glass. As soon as the downward movement of the slide box has been started, the amount and pressure of air ejected out of the air nozzles and the temperature of the orifice plate are adjusted so that the separation of the molten glass to filaments is started.

When the clamp **17** is moved downwards to a certain position, the filaments are completely separated. When it is observed that filaments have been completely separated from each other, the electromagnetic automatic actuating switch **35** is de-energized and the lumped glass attached to the leading ends of the filaments is removed and the clamp **17** is returned to its initial waiting position. Thus the spinning of separated filaments is started again.

Recycling of Glass Fibers

M.A. Grodin, H.W. Barch, F.E. Harvey and P.C. Chardello; U.S. Patent 4,145,202; March 20, 1979; assigned to PPG Industries, Inc. describe a method which is suitable for reclaiming and recycling any of the numerous and varied types of glass filaments and/or strands produced in a continuous filament and/or strand manufacturing facility.

In Figure 3.15 the method is shown in block diagram form to illustrate the process variables that may be utilized. Thus, in the pretreatment of the scrap glass, recourse to all three of the pretreatment steps may be had or to none if desired. In draining the scrap as shown, it is desired to remove as much free water as possible to minimize the load on the drying step which is undertaken. Preferably the draining will be accomplished by placing the scrap glass in a suitable holding tank for a period of time to permit the free water present to be removed therefrom by gravity flow to the exterior of the zone in which the glass is held in any such vessel.

The scrap glass can also be placed on a foraminous chain and conveyed from a forming area permitting the free water or aqueous solutions of binder and/or size present to fall through the foraminous chain by gravity. Where desired or possible, recourse to suction boxes on the underside of foraminous chain conveyers can be used to assist in removing water. Similar physical aids can be used where the scrap glass is held in a container or vessel for draining. In instances where the scrap glass consists primarily of dried strands and/or filaments, this draining operation is unnecessary and can be eliminated. It can also be eliminated when the drying capacity is available to handle completely wetted material and for production purposes or other reasons when it is not desired to conduct a draining treatment.

The use of a metal detection and removal system is optional. The employment of these procedures will depend upon the quality of the scrap strand and/or filaments that the manufacturing operation is producing in a given area of the plant. Thus in a fabrication area of the plant producing a mat product where the contamination of the scrap glasses produced can be minimized or eliminated by good housekeeping practices, such a step may be unnecessary to provide good quality scrap to the thermal process.

It is preferred in feeding scrap glass and/or filaments to the thermal treatment zone to employ as feed glass strands and/or filaments which have been cut or otherwise reduced in size to lengths of 1 to 4 inches (2.54 to 10.16 centimeters) or more. This is particularly desirable when the thermal treatment zone is divided into physically separate compartments or where the scrap material is continuously moving through several distinct thermal treatment zones during the thermal treatment. It is, however, possible to employ bundles of scrap strands and/or filaments in the form they exist and subject them directly to the thermal treatment.

Figure 3.15: General Method of Scrap Reclamation

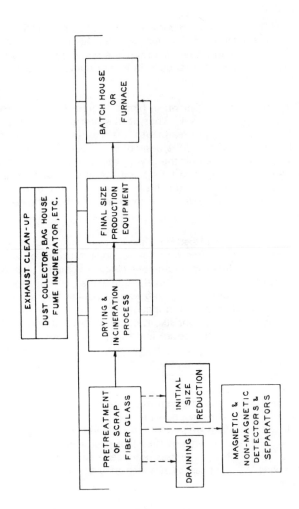

Source: U.S. Patent 4,145,202

In this latter form the strand and/or filaments are continuously agitated during the thermal treatment to insure that they receive thorough and adequate exposure to the thermal environment for the requisite time necessary to accomplish treatment being undertaken at a given temperature condition.

The scrap glass and/or filaments are treated in the drying and incineration zone under precise and controlled thermal environments to provide a glass product emerging therefrom that is chemically, and in some instances physically, capable of being employed as batch feed to a glass fiber forming furnace. This drying and incineration zone may be formed of a single thermal treatment unit or several. In instances where a single furnace or kiln is employed for the thermal treatment zone, the zone may be divided by different temperature regimens being maintained therein along its length or depth or the unit may be operated as a batch unit by employing different temperatures therein in stepwise timed sequence.

The important consideration in the operation of the thermal treatment zones is that the scrap glass strands and/or filaments be subjected to the requisite thermal treatments before they are removed therefrom.

Thus, scrap glass strands and/or filaments are subjected to a first thermal environment which involves holding them at temperatures typically from 200° to 350°F (93.3° to 176.7°C) for a period of time sufficient to drive off all free water contained on the strand. When observation indicates water is no longer being driven off from the strand, the strand is then heated to temperatures typically from 500° to 700°F (260° to 371.1°C) to remove the organic materials contained on the strand or filaments as coatings.

This thermal treatment for removal of organic materials present involves the use of sufficient oxygen as oxygen or air to support combustion of the carbon atoms present to CO and CO_2 and a sufficient period of time to insure removal of the volatile organic coating materials present. Upon completion of the evolution from the thermal treatment of organic constituents from the strands and/or filaments, which may be readily ascertained by observation and analysis of the gases emanating from that treatment, the glass strands and/or filaments are subjected to a final thermal treatment at elevated or incineration temperatures, typically at 900° to 1250°F (482.2° to 676.7°C) or higher but below the softening point of the glass to thereby remove any residues present on the glass surfaces.

This treatment is for a time sufficient to insure removal of such residues. For a given quantity of glass treated the time of treatment can be selected by observing the condition of the glass with sampling at specific time increments to insure residue-free glass and selecting a specific time for a specific quantity of glass at the temperatures employed based upon such sampling.

In removing the material from the thermal treatment zone or zones the glass is physically measured by suitable sieve techniques to determine the particle size range of the glass produced. If the glass removed is inherently in the range of 60 to 325 mesh (U.S. Sieve Series), the material may be fed directly into a glass-making furnace as cullet or fed to the batch mixing operation where it can be blended with the normal batch ingredients as feed to fiber glass furnaces. Should the particles of glass be too large in size, they may be subjected to comminution in suitable grinding equipment to reduce their size to the desired 60 to 325 mesh range.

Various methods may be utilized to remove volatile materials discharged from the thermal treatment zone and the furnace area prior to release of any gases to the atmosphere. Thus, dust collectors, such as bag houses, electrostatic precipitators and other units designed to purify gas streams may be employed to remove materials from the furnace flues and thermal treatment systems employed.

Glass Pellets from Fiber Glass Cullet

S.F. Brzozowski; U.S. Patent 4,188,228; February 12, 1980; assigned to PPG Industries, Inc. has found that, despite the low soda content of fiber glass batch ingredients, one can incorporate into pellets of fiber glass batch ingredients considerable quantities of fiber glass cullet and make acceptable pellets useful as feed to a fiber glass furnace. The cullet feed is a fiber glass composition which is in the form of ground or pulverized glass of short fibers, i.e., 0.0625 inch (0.158 centimeter) or less up to about 0.5 inch (1.27 centimeters). Fibers in a length above 0.5 inch (1.27 centimeters) are not in general conducive to good pellet formation.

Fibers for this process include individual filaments or strands which contain many filaments. The use of fiber glass cullet in pellets of fiber glass batch in accordance with this process enhances the pellets by providing a substantial quantity of glass in the pellet which assists in melting the raw batch ingredients such as SiO_2 present in the pellet.

The presence of significant quantities of ground or fibrous glass intimately adhered to other batch ingredients in the pellet also reduces the tendency of such recycled cullet to be carried away from the furnace proper in flue gases. Thus, the pellets prepared in accordance with this process make use of waste glass as furnace feed while providing a minimum of lost feed due to dust carryover in furnaces in which recycled glass is used.

In preparing pellets using fiber glass cullet and fiber glass batch compositions, the cullet may be employed in a range of from 5 to 35% by weight of the batch. Preferably the cullet is used in a range of 10 to 30% by weight.

In a typical operation the cullet is prepared by burning off the binder contained on fiber glass strand which has been sized by chopping into lengths of about ¼ inch (0.635 centimeter). The material cut in these lengths is incinerated at temperatures of 300° to 1200°F (148.9° to 648.9°C) and is preferably crushed on a roll mill to about 60 to 325 mesh (U.S. Sieve Series). This material is used as the cullet and has the following composition:

Cullet Composition

Ingredient	Weight Percent
Al_2O_3	13.56
SiO_2	54.31
K_2O	0.12
CaO	21.92
TiO_2	0.45
Fe_2O_3	1.08
MgO	0.35
Na_2O	0.78
B_2O_3	6.80
F_2	0.47

A standard boric acid fiber glass batch having the following composition is employed to supply the major portion of the pelletized batch:

Batch Composition

Ingredient	Weight Percent
Al_2O_3	13.64
SiO_2	54.41
K_2O	0.13
CaO	22.0
TiO_2	0.42
Fe_2O_3	0.22
MgO	0.38
Na_2O	0.75
B_2O_3	7.52
F_2	0.77

The batch composition described above is mixed with the cullet composition described above with the quantity of cullet being regulated to provide in the final batch 10% by weight cullet based on the total batch. This mixed batch is then fed to a disc pelletizer with water being fed thereto at a controlled weight to provide 15% by weight water to the batch ingredients in pellet form. The pellets are continuously formed until the total batch has been added. Upon completion of the pellet formation, the pellets are heated in an oven at temperatures between 220°F (104.4°C) and about 1000°F (537.8°C) until their free water content is below about 1% by weight. Pellets prepared in this manner are readily melted to provide a molten glass source from which glass fibers may be drawn.

Devitrification-Resistant, Amorphous Silica Fibers

G.B. Price and W.H. Kielmeyer; U.S. Patent 4,200,485; April 29, 1980; assigned to Johns-Manville Corporation describe a method of producing silica fibers uniformly and consistently having a sodium oxide and potassium oxide content of less than 0.01 weight percent by leaching the silica fibers in two steps or cycles, using an equal amount of leaching acid solution during each cycle for an equivalent amount of time. At the end of the first leaching cycle, the silica fibers are given an initial rinse with deionized water and then immediately subjected to a second leaching cycle.

In addition to producing the highly desirable result mentioned above, silica fibers made according to the process are uniform in properties, as evidenced by an acceptance level in excess of 90% of the lots or batches of fiber produced based upon a devitrification resistance for 4 hr at 2500°F of less than 5 weight percent crystallinity, preferably less than 2 weight percent, and most preferably less than 1 weight percent, as determined by x-ray diffraction in comparison with a 100% cristobalite control sample.

Fibers made according to the process typically have compositions as shown below, in percent by weight:

Chemical Analysis	Suitable	Preferred
SiO_2	99.6 min	99.7 min
Al_2O_3	0.20 max	0.025–0.16

(continued)

Chemical Analysis	Suitable	Preferred
$Na_2O + K_2O$	0.01 max	0.005 max
$CaO + MgO$	0.04 max	0.03 max
TiO_2	0.01 max	0.007 max

Most preferably the alumina content of the finished fibers is within a range of 0.03 to 0.11 weight percent.

As implied above, the method is applicable to those areas which require the use of an amorphous silica fiber having excellent resistance to devitrification at temperatures up to 2500°F and for exposure times at this temperature up to 4 hr.

FIBER COMPOSITIONS

Boron- and Fluorine-Free Glass Composition

The conventional glass compositions used in producing glass fibers into continuous glass fiber strands are E glass and 621 glass. Both of these glasses are calcium-aluminum-borosilicate glasses characterized by a low alkali oxide content usually calculated as sodium oxide (Na_2O). E glass is generally described in U.S. Patent 2,334,961. The glass composition of E glass is given below in Table 1 with the constituents being in weight percent:

Table 1

Ingredients	Weight Percent
SiO_2	52–56
Al_2O_3	12–16
CaO	16–19
MgO	3–6
B_2O_3	9–11

The 621 glass is a modification of a typical E glass formulation as is shown above in Table 1 and is typically devoid of magnesium oxide (MgO) and has a calcium oxide (CaO) content that is much higher than that usually found in an E glass. The 621 glasses are described in more detail in U.S. Patent 2,571,074. The glass composition of the 621 glasses is presented in Table 2 where the percentages shown are by weight percent:

Table 2

Ingredients	Weight Percent
SiO_2	52–56
Al_2O_3	12–16
CaO	19–25
B_2O_3	8–13

It is well-known in the art that both E and 621 glasses contain minor constituents which are typically F_2, Fe_2O_3, K_2O, Na_2O, SrO, and MgO, and on occasion BaO. In general the minor constituents are present each in amounts of less than 1% by weight in the glass.

In order to reduce the cost of producing glass fibers and to reduce environmental pollution during the production of glass fibers without increasing the cost of production, a glass composition is needed that does not contain F_2 and B_2O_3 but still retains the favorable properties of E glass such as softening point, liquidus temperature, and tensile strength.

H.E. Neely; U.S. Patent 4,199,364; April 22, 1980; assigned to PPG Industries, Inc. describes a process whereby the foregoing objects are achieved by a glass composition having critical amounts of silica, alumina, magnesia, lime, a low amount of alkali, and a small amount of barium oxide plus a minor amount of foreign materials such as ferrous oxide, and titania and the like. The glass composition is free of boron and fluorine, thereby being a glass composition that when melted does not release boron or fluorine to the environment.

In the preferred embodiment the glass composition comprises the ingredients listed below.

Table 3

Ingredients	Weight Percent
SiO_2	55.9-60.5
Al_2O_3	12.8-17.1
MgO	5.3-9.4
CaO	14.9-17.7
Na_2O	0.4-1.1
Li_2O	0.6-1.0
BaO	0.0-0.9

Also included are minor amounts of constituents such as Fe_2O_3, preferably 0.3 weight percent; TiO_2, preferably 0.6 weight percent; K_2O, in amounts of 0.1 to 0.5, typically 0.2, weight percent.

In an alternative embodiment the glass composition may be boron-free and may contain small quantities of F_2 up to 1 weight percent and preferably around 0.1 to 0.7 weight percent.

Glass Composition of Low B_2O_3 Content

In recent years the cost of boric acid and colemanite ore, another source of boric acid, has risen steadily causing overall batch costs in the manufacture of glass fibers to increase substantially. While all batch costs have in general increased, the boric acid constituent has had a particularly significant effect on the increased cost of glasses for the manufacture of continuous glass fibers. It is thus desirable to reduce the quantities of B_2O_3 in glasses used to make continuous glass fibers while still adhering to the properties of the E and 621 glasses.

H.E. Neely, Jr.; U.S. Patent 4,166,747; September 4, 1979; assigned to PPG Industries, Inc. describes a fiber glass composition which contains a substantially reduced B_2O_3 content when compared with conventional glasses used to manufacture glass fibers.

In general the glasses have the compositions shown in Table 1, the percent being percent by weight.

Table 1

Ingredients	Weight Percent
SiO_2	54–55
Al_2O_3	13–14
CaO	22–24
MgO	0.2–0.5
B_2O_3	5.5–6.8

In the preferred embodiments the glasses have the composition set forth in Table 2, the percents being percent by weight of the glass.

Table 2

Ingredients	Weight Percent
SiO_2	54.2–55
Al_2O_3	13–14
CaO	22.1–23.2
MgO	0.3–0.4
B_2O_3	6.5–6.8

In either of the above compositions listed in Tables 1 and 2, an F_2 content (wt %) between 0.3 to 1.4 may be used and preferably between 0.7 to 1.4. Other substituents may also be present in small amounts, typically below 1%. In general Fe_2O_3 is present in quantities of 0.1 to 0.5%, preferably 0.3; Na_2O is present at between 0.1 to 0.7%, preferably 0.5; and TiO_2 between 0.2 to 0.8%, preferably 0.6. K_2O in amounts of 0.1 to 0.5%, typically 0.2 may also be found.

Example: A glass batch composition was prepared by mixing the ingredients listed in Table 3 in the amounts shown:

Table 3

Ingredients	Grams
Silica	183.2
Clay	166.7
Limestone	181.2
Boric acid	57.0
Fluorspar	7.0
Sodium sulfate	4.8
Ammonium sulfate	1.3
Coal	0.5

The batch ingredients listed above in the amounts listed were mixed and then melted at temperatures between 2600° and 2700°F (1425° and 1480°C). The molten glass produced by this melting was fiberized in a single hole bushing at forming speed of 26.7 ft/sec (8.1 meters per second) at temperatures in the range of 2330° to 2350°F (1277° to 1288°C). Measurements of the softening point and liquidus of the glass were made and fiber tensile strength was measured. These measurements and the chemical composition of the glass produced, calculated on the basis of batch ingredients used, are listed below in Table 4.

Table 4

Glass Composition and Properties

Ingredients, wt %

SiO_2	55
Al_2O_3	13.6
Fe_2O_3	0.3
MgO	0.3
CaO	22.2
TiO_2	0.6
B_2O_3	6.8
Na_2O	0.5
F_2	0.7

Properties

Softening point, °F	1591–1596
Liquidus, °F	2170
Tensile strength of fiber, 10^3 psi	510

Basalt Glass-Ceramic Fibers

Basalt-type materials are widely available, and relatively inexpensive, natural materials. They are generally classified in mineralogy as basic volcanic rocks wherein the essential constituents are the minerals feldspar, pyroxene, and magnetite, with or without olivine, and a black basalt glass. It has been recognized that they are resistant to alkaline attack, a characteristic that makes them of particular interest in connection with alkaline environments such as are encountered in concrete work. Basalt materials are also easily melted and drawn as fibers, a further fact that makes them of interest as potential concrete reinforcement fibers.

H.L. Rittler; U.S. Patent 4,199,336; April 22, 1980; assigned to Corning Glass Works has found that vitreous basalt fibers, having a diameter not greater than about 250 microns, can be internally crystallized to form corresponding glass-ceramic fibers having a magnetite-clinopyroxene mixed crystal phase constituting at least 35% of the fiber and preferably over 50%.

Basically, production of the fibers contemplates three steps. First, crushed basalt rock, optionally containing various additives as modifiers, is melted in a crucible, pot, or continuous glass melting unit, depending on the quantity of product wanted and the manner of fiber formation to be utilized. Second, when a suitably homogeneous melt is attained, amorphous fibers are produced from the melt in any conventional manner such as drawing, spinning, or blowing. Since fiber orientation is important in reinforcement applications, fibers for this purpose will normally be drawn, wound on a drum, and subsequently cut into bundles.

Finally, the fibers are subjected to a heat treatment to convert them from the vitreous state to the glass-ceramic state by internal development of a polycrystalline phase. This heat treatment consists of exposing the fibers to a temperature in the range of 900° to 1250°C for a time not exceeding ten minutes, and preferably less than one minute.

Glass Composition Suitable for Rotary Process

A recent advancement in the rotary fiberization field provides for the manufacture of glass fibers having an average diameter below 7 microns without the necessity of using a relatively high temperature gaseous blast to attenuate the primary fibers. It was discovered that staple fibers having the desired diameter could be produced solely by passing molten material through the orifices of a rotor into a plurality of relatively cold and relatively low pressure gas flows, thus eliminating the hot gas blast used in the prior art and the fuel usage associated therewith.

Satisfactory production of fibers by the method described in the preceding paragraph requires that the glass compositions have certain defined characteristics. One of the primary requirements is that the glass have a relatively low melting and working temperature, so that it may be readily worked on a continuous and rapid basis with a minimum of energy required for melting it. The low working temperature is also needed to minimize corrosion and wear of the metal spinner. The glass must also have a liquidus temperature that is sufficiently lower than the relatively low working temperature, thus enabling low temperature rotary fiberization by the abovedescribed method without devitrification, i.e., the undesirable formation of crystals.

Additionally, the glass must have a viscosity at these relatively low working temperatures that is sufficiently low to permit high levels of fiber production. A final requirement is chemical durability. It is essential to use a glass composition that provides a fiber of extremely high weather resistance since the great amount of surface exposed per unit of weight renders the fibers sensitive to the corrosive influence of even such moisture as is present in the air.

Glass compositions heretofore used which meet the requirements of softening point, viscosity, liquidus and durability, have also been characterized by relatively high working temperatures.

L.V. Gagin; U.S. Patent 4,177,077; December 4, 1979; assigned to Johns-Manville Corporation describes glass compositions which are useful for the production of glass fibers by a rotary or centrifugal process which does not use the conventional external hot gas attenuating technique.

The glass compositions and glass fibers made from them fall within the following broad range of proportions:

Ingredient	Parts by Weight
SiO_2	54–57
Al_2O_3	3–4
CaO	10–13
Na_2O	16–19
B_2O_3	9–12
ZnO	1–3.5

Other ingredients which enter the composition either as impurities, because of economic reasons or as melting aids include the following:

Ingredient	Parts by Weight
Fe_2O_3	0–0.13
K_2O	0–2.7
MgO	0–0.6
SO_3	0–0.3

Glass compositions falling within this range of proportions have softening points of 1207° to 1221°F, liquidus points of 1700° to 1740°F, and temperatures at a viscosity of 1,000 poises of 1690° to 1730°F. The durability of the glass compositions in the form of fine diameter fibers to chemical attack by water is determined by submerging 5 micron monofilament samples in pure water for one hour at 100°C. Weight loss ranges between 2.1 and 3.0%.

The glass compositions and the glass fibers made from them have a preferred range of major ingredients as follows:

Ingredient	Parts by Weight
SiO_2	54.5–57.0
Al_2O_3	3.3–3.8
CaO	10.1–12.2
Na_2O	16.2–18.1
B_2O_3	9.2–11.1
ZnO	1.2–1.7

Specific glass compositions constituting the preferred embodiments of this process are described below.

	. . . . Example Number					
	1	2	3	4	5	6
SiO_2	55.0	55.2	55.2	54.5	55.3	57.0
Al_2O_3	3.6	3.6	3.6	3.8	3.3	3.6
CaO	11.5	11.8	11.7	11.0	10.8	10.1
Na_2O	17.8	16.5	17.0	16.3	17.1	17.2
B_2O_3	9.8	9.7	9.7	10.0	11.1	9.2
ZnO	1.5	1.4	1.4	1.5	1.3	1.5
Fe_2O_3	0.13	0.05	0.08	0.05	0.06	0.08
K_2O	0.4	1.2	0.7	2.7	0.7	0.7
MgO	0.2	0.3	0.4	0.2	0.3	0.3
SO_3	0.1	0.2	0.2	0.1	0.1	0.3
Softening point, °F	1215	1221	1215	1215	1207	1221
Liquidus temperature, °F	1740	1740	1720	1720	1710	1700
Weight loss, %	NM	2.1	2.3	NM	3.0	2.3
Temperature, °F at 1,000 poises*	1690	1730	1730	1710	1710	1700

Note: NM is not measured.

*Estimated.

Thermally Stable Quartz Glass Containing Cr_2O_3 and/or Mn_2O_3

M. Mansmann; U.S. Patent 4,180,409; December 25, 1979; assigned to Bayer AG, Germany describes homogeneous quartz glass of high thermal stability with a controlled tendency towards cristobalite formation up to a final, definite average crystallite size of about 100 to 500 Å, which is stable up to temperatures of about 1400°C with a modifying content of chromium and/or manganese of about 0.05 to 20% by weight, expressed as Cr_2O_3 and/or Mn_2O_3 and based on SiO_2, and at the very most mere traces of alkali metal and alkaline earth metal oxides, such as are inevitable in industrial processes. The process also relates to very finely crystalline cristobalite, obtained by heating the quartz glass to temperatures above about 1100°C, with crystallite sizes as determined by x-ray photography in the range of about 100 to 500 Å.

The quartz glass or the cristobalite according to the process is eminently suitable for use as quartz fibers or cristobalite fibers with improved mechanical properties resistant to high temperatures for the production of flameproof textile articles, for reinforcing and for insulating purposes.

Example 1: *Production of Pure SiO_2-Fibers* — 104 g of silicic acid tetraethyl ester were heated with stirring to 50°C with 24 g of water and 15 g of methanol, which contained 7.5 mg of hydrogen chloride to accelerate the hydrolysis reaction, the exothermic hydrolysis reaction taking place with an increase in temperature to 70°C. After 10 minutes, 35 g of a 3% methanolic solution of polyethylene oxide with a degree of polymerization of about 100,000 (Polyox WSR 301 Union Carbide Corp.) were added with stirring to the hydrolysis solution at a temperature of about 50°C, followed by stirring for 20 minutes at room temperature. The excellently spinnable solution contained 16.9% of SiO_2 and 0.59% of polyethylene oxide.

The solution was spun at room temperature through a 0.3 mm nozzle into a duct which had been heated to 120°C and through which air was passed in the same direction as the filaments. At the end of the duct, the fibers were wound onto a drum at a speed of 125 meters per minute.

The fibers were then heated in air to 300°C at a heating rate of 400°C per hour, from 300° to 500°C at a heating rate of 100°C per hour and from 500° to 1000°C at a heating rate of 500°C per hour, and subsequently removed from the furnace. The completely clear, colorless fibers had a diameter of 12 microns and a tensile strength of 100 kiloponds per square millimeter. X-ray examination showed that the fibers were amorphous.

Examples 2 through 8: *Production of Chromium-Containing SiO_2-Fibers* — A chromium(III) salt solution, which is highly compatible with the SiO_2-spinning solution prepared as described above, was obtained as follows: 150 g of $CrCl_3 \cdot 6H_2O$ were dissolved in 1.5 liters of water, followed by the addition with stirring at 80°C of concentrated ammonia solution up to pH 7.3. The precipitate was filtered off and washed until substantially free from chloride. The filter cake was stirred for 1 hour at 80° to 90°C in a solution of 90 g of oxalic acid in 300 ml of water, as a result of which it was almost completely dissolved.

After filtration, the clear violet filtrate was concentrated in vacuo at 75°C to 183 g using a Rotavapor. After dilution with 50 g of methanol, the resulting solution, which had an analytically determined chromium content of 15.0%, expressed as Cr_2O_3, was used for the production of chromium-containing SiO_2-fibers.

The spinning solutions were prepared in the same way as described in Example 1, except that before addition of the methanolic polyethylene oxide solution the chromium oxalate solution was added, corresponding to the required chromium content, and in addition a small quantity of citric acid was included.

Spinning and calcination up to 1000°C were carried out in the same way as in Example 1. The strength of the resulting fibers was as high as that of the chromium-free SiO_2-fibers. The fibers were x-ray amorphous. They were more or less green in color, corresponding to their chromium content.

Table 1 shows the compositions used for producing the chromium-containing SiO_2-fibers.

In order to assess their thermal stability, the fibers of Examples 1 through 8 were heated in air for 24 hours at 1300°C. Whereas the chromium-free fibers of Example 1 had undergone 45% linear shrinkage and were sintered solid, the chromium-containing SiO_2-fibers of Examples 2 through 8 showed minimal shrinkage (less than 5%) and had remained completely discrete, flexible and loose. The strength of the chromium-free fibers could no longer be measured, whereas the strength of the chromium-containing fibers had remained substantially unchanged.

Although an aluminum silicate fiber sample (Fiberfrax), which had also been tempered for 24 hours at 1300°C, for comparison, did not undergo significant shrinkage, it was sintered solid and, hence, also suffered loss of strength.

Examination of the fibers of Examples 1 through 8 by x-ray photography after tempering at 1300°C, showed that, in every case, devitrification had occurred with formation of cristobalite.

Table 1

	Additions to 143 g of Silicic Acid Ester Hydrolyzate			Composition of the Spinning Solution				Cr_2O_3 in SiO_2 Fiber
Ex. No.	Chromium Oxalate Solution	Citric Acid	Polyethylene Oxide Solution	SiO_2	Cr_2O_3	Citric Acid	Polyethylene Oxide	
	. . . . (g)			. . . . (%)				
2	0.2	0.06	35	16.9	0.02	0.03	0.59	0.1
3	0.5	0.15	35	16.8	0.04	0.08	0.59	0.25
4	1.0	0.30	35	16.8	0.08	0.16	0.59	0.5
5	2.0	0.60	35	16.6	0.17	0.33	0.58	1.0
6	6.0	1.80	35	16.1	0.48	0.97	0.56	3.0
7	11.5	3.0	35	15.8	0.79	1.58	0.56	5.0
8	20.0	6.0	40	14.3	1.43	2.90	0.57	10.0

The average crystallite size was determined x-ray photographically by measuring the integral half-value width with a counter tube goniometer according to Scherrer and Bragg. These values are set out in Table 2.

Table 2

Ex. No.	Cr_2O_3 (%)	Average Crystallite Size (Å)
1	0	620
2	0.1	295
3	0.25	255
4	0.5	225
5	1.0	230
6	3.0	235
7	5.0	260
8	10.0	220

Alkali-Resistant Glass Fiber Compositions

Y. Iizawa; U.S. Patent 4,142,906; March 6, 1979; assigned to Ikebukuro Horo

Kogyo Co., Ltd., Japan describes a glass composition for alkali-resistant glass fiber, which comprises:

   (1)   30 to 57% by weight of SiO_2;

   (2)   12 to 25% by weight of ZrO_2;

   (3)   14 to 26% by weight of R_2O (wherein R represents Na, K or Li);

   (4)   1 to 11% by weight of $R'O$ (wherein R' represents Ca, Ba, Mg, Zn or Co);

   (5)   0.1 to 6% by weight of CaF_2;

   (6)   0.1 to 10% by weight of $M_2(SiF_6)$ (wherein M represents Na, K or Li);

   (7)   0.1 to 12% by weight of B_2O_3;

   (8)   an effective amount of up to 3% by weight of TiO_2;

   (9)   an effective amount of up to 3% by weight of Al_2O_3; and

 (10)   an effective amount of up to 3% by weight of Fe_2O_3.

K. Ohtomo, T. Yoshimura and H. Fujii; U.S. Patent 4,140,533; February 20, 1979; assigned to Kanebo Ltd., and Nippon Electric Glass Company Ltd., both of Japan describe another alkali-resistant glass composition comprising the following oxides in a weight percentage composition of 57 to 64% SiO_2, 19 to 23.5% ZrO_2, 0.5 to 2.5% Li_2O, 11 to 18% Na_2O, 0 to 6% K_2O, 0.5% or less RO, 0.5% or less Al_2O_3 and 0.5% or less M_xO_y where R is an alkaline earth metal, M is a metal other than an alkaline earth metal, x and y are positive integers where (valence of M) multiplied by x = 2y, and in which the numerical values corresponding to the weight percentages of the above oxides satisfy the relationships defined by the following expressions (1) and (2):

    (1) $21 \geqslant Na_2O + Li_2O + K_2O \geqslant 15.5$

    (2) $52 - 2ZrO_2 \leqslant 4Li_2O - K_2O \leqslant ZrO_2 - 21$

Further, the oxides of the following formulation were especially suitable. 58 to 63% SiO_2, 20 to 22.5% ZrO_2, 1 to 2% Li_2O, 13 to 17% Na_2O, 0.5 to 2.5% K_2O, 0 to 0.5% RO, 0 to 0.5% Al_2O_3, and 0 to 0.5% M_xO_y, wherein R, M, x and y are as defined above, and the percentages are on a weight basis.

Radiant-Energy-Absorbing Glass-Ceramic Fiber

Heat-insulating materials consisting of formations of inorganic fibers have already been known. However, among such known ones, those of glass fibers and of ceramic fibers are particularly effective in preventing heat loss caused by heat conduction or convection, but they are not yet satisfactory as heat-insulating materials because they are pervious to the radiant heat energy emitted from a heat source at high temperatures.

S. Seki, T. Kobayashi, T. Kato, T. Suzuki and T. Matsubara; U.S. Patent 4,169,182; September 25, 1979; assigned to Honda Giken Kogyo Kabushiki Kaisha, and Ishizuka Carasu Kabushiki Kaisha, both of Japan describe a process which has improved remarkably the heat-insulating effect of the heat-insulating materials by either incorporating in the fibrous materials a substance that absorbs the radiant heat energy or coating the surface thereof with a film of such a substance, so that they can absorb and accumulate therein the radiant heat energy.

Effective substances for absorbing radiant heat energy are metals such as Cu, Co, Fe, Cr, Mn and Ni, and the foregoing effects can be obtained either by mixing in the glass raw materials for glass-ceramics any of those metals or their compounds such as oxides, etc., and thereafter forming the mixture into fibers, or by forming a coating film of such metals or metal oxides on the surface of the fibers obtained after the glass raw materials are formed into fibers or after the fibers are formed into a felt, blanket or the like, by any means such as depositing such metals or metal oxides thereon, or the like.

In general, manufacturing thereof in accordance with the following steps can provide homogeneous products and is economically advantageous:

(1) melting a batch of raw materials for glass which contains one or more of the above metals or metallic compounds and is capable of crystallizing;

(2) the metallic compound being used in an amount corresponding to 0.1 to 15 wt % thereof, calculated in terms of the oxide thereof;

(3) forming the molten glass into fibers in accordance with any conventional methods;

(4) forming the glass fibers into a product such as a felt, blanket, bulk, sheet, paper, board, blank, etc; and

(5) heating the above product.

By this treatment, the glass fibers are converted into glass-ceramic fibers in such a manner that the matrix glass in which the metallic ions are present is formed into predetermined crystallite substances.

It is essential to incorporate into the glass-ceramic fibers an appropriate amount of any of the metals Cu, Co, Fe, Cr, Mn and Mg. It has been confirmed that any of these metals is present as ions in the matrix glass and exhibits a geometrical effect, in the radiant heat energy absorption, with the radiant heat energy absorbing effect of the crystallites, and also exhibits an excellent effect in preventing the fibrous structure from changing due to high temperatures. Each of the metals as essential ingredient is used in the range of 0.1 to 15 wt %, calculated in terms of its oxide, and the optimum amount of each metal is shown below. Two or more of these metals may be used as an admixture.

Ingredient	Percent by Weight
CuO	1–10
Fe_2O_3	0.1–10
Cr_2O_3	0.5–5
MnO	0.5–12
NiO	0.1–8
CoO	0.1–10

FIBER COATINGS

Alkali-Resistant Coating

When glass fibers are used in alkaline environments, such as when they are used as fibrous reinforcement for Portland cement, they are rapidly attacked by the alkali and deteriorate rapidly. Since economic and environmental considerations have made the use of glass fiber for cement reinforcement much more attractive in recent years, there have been numerous efforts to render glass fiber less susceptible to alkaline attack. These efforts have been in two principal directions. First, much effort has been directed toward developing glass compositions which are themselves resistant to alkaline attack. For instance, one type of glass composition which is commercially used and considered moderately alkaline resistant contains large amounts of zirconia.

Fibers made of such compositions deteriorate much less rapidly in an alkaline environment than do fibers made of conventional glass compositions such as the well known E glass. The alkali resistant glass compositions, however, are quite expensive and thus cannot be readily used for such purposes as cement reinforcement without unduly raising the cost of the finished cement product.

The second approach to imparting alkali resistance to glass articles has been to develop coatings for the glass fibers which will prevent the alkaline components from contacting and attacking the glass surface. Use of such coatings is intended to allow the glass fiber to be composed of inexpensive and conventional materials such as those in the aforementioned E glass. Alternatively, use of such coatings on fibers of alkali resistant glass compositions would enhance that alkali resistance and significantly extend the useful life of the expensive alkali resistant fibers, thus improving the economics of their use for cement reinforcement and similar uses.

K.L. Jaunarajs; U.S. Patent 4,191,585; March 4, 1980; assigned to Johns-Manville Corporation describes a fibrous glass article which is resistant to an alkaline environment and which comprises glass fiber coated with a coating comprising zinc stearate.

The zinc stearate used as a glass fiber coating can be any zinc stearate of commerce. Zinc stearate is commonly considered to be the zinc salt of stearic acid with the formula: $Zn(C_{18}H_{35}O_2)_2$. When used herein it may be in a finely powdered form, it may be used molten at elevated temperatures (since its melting point is approximately 120°C) or it may be dispersed in water. It is particularly advantageous to dust zinc stearate powder onto the glass fibers to be coated and then heat the dusted glass fibers to melt the zinc stearate and allow it to flow and coat the glass fibers.

Alternatively, the zinc stearate can be suspended in water and sprayed onto the fiber, thus leaving the zinc stearate as a deposited coating when the water evaporates or is driven off by subsequent heating of the fibers.

To provide an adequate coating the zinc stearate must be present as from 1 to 50% by weight of coated fiber, preferably 3 to 10% by weight of coated fiber. Quantities less than this will not satisfactorily coat the fiber, while quantities in excess do not further enhance the alkali resistance and therefore are wasted. When used as a water suspension, the zinc stearate will normally be suspended in water in an amount of from 5 to 450 g/ℓ of suspension, preferably from 30 to 70 g/ℓ.

Coating Applicator System

G.M. Schmandt, R.N. Chappelear and R.D. Hand; U.S. Patent 4,192,663; March 11, 1980; assigned to Owens-Corning Fiberglas Corporation describe an apparatus adapted to apply a coating, such as a binder or a size, to advancing continuous filaments comprising a bracket; a base journaled at the bracket for rotation about a first axis; means for rotating the base; an applicator adapted to apply a coating to the advancing filaments; a carriage joined to the applicator; the carriage being adapted for sliding engagement with the base along a path substantially transverse to the first axis; and a first adjustment means adapted to engage the carriage to move the applicator along a second axis substantially transverse to the first axis.

As shown in Figure 3.16, feeder **10** is adapted to supply a plurality of streams of molten inorganic material, such as glass.

**Figure 3.16: Fiber-Forming Apparatus Incorporating
Coating Applicator System**

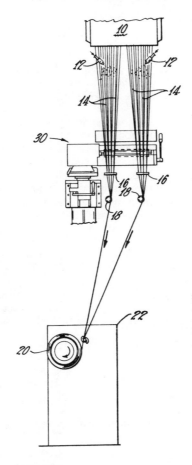

Source: U.S. Patent 4,192,663

The streams of glass are attenuated into continuous filaments **14** through the action of attenuating means or winder **22**. Just below feeder **10** nozzles **12** are adapted to apply a spray of water and/or lubricant to the advancing filaments which then pass through the region of applicator assembly **30**.

As is known in the art, the filaments then can directly advance to the gathering shoes **18**, or a splitter guide or guides **16** can be positioned intermediate the applicator assembly and the gathering shoes to better maintain the filaments in a spaced apart array at the applicator assembly **30**. The filaments are gathered into a pair of strands at gathering shoes **18** which are then wound into a package **20** by means of winder **22**.

The applicator assembly **30** is adapted to rotate the means for applying the coating to the filaments through a substantially horizontal plane or about a first axis to a position out of contact with the filaments **14** such that the coating is not applied to the advancing filaments. Among other things, this allows the operator greater access to the feeder area and saves size and/or binder that would otherwise be applied to the advancing filaments.

Fiber Size Composition

D.M. Walser; U.S. Patent 4,197,349; April 8, 1980; assigned to PPG Industries, Inc. has found that an improved fiber size composition is produced when a relatively narrow range of an increased amount of crosslinked starch that has been only partially cooked or gelatinized is used as the starch component of the fiber size composition.

The improved binder size composition comprises about 50 to 65 weight percent of the size composition based on nonaqueous components of a crosslinked starch which has been only partially cooked or gelatinized. The size composition also contains other ingredients such as a vegetable oil, as a lubricant; a fungicide of the metallo organic quaternary type, e.g., tributyltin oxide ammonium complex; coupling agents; softening agents; wetting agents and additional lubricants. Small amounts of additional film-forming ingredients other than the previously mentioned starch can also be included, such as for example, gelatin, polyvinyl alcohol, etc. to make a more durable strand.

Example 1: A fiber sizing composition having the following ingredients was prepared:

Ingredient	Amount per 1,000 Gallons
Amaizo 213 Hybrid Cornstarch, lb	352.4
Paraffin, lb	72.1
Soybean oil, lb	72.1
Tween 81 ester derivative (96 MS), lb	6.6
Cation-X reaction product, lb	48.0
Igepal CA-630 wetting agent, ml	500
Biomet 66 biocide, ml	68
Gamma methacryloxypropyltrimethoxy silane, lb	11
Acetic acid, ml	250
Versamid 140 resin, lb	44.1

(continued)

Ingredient	Amount per 1,000 Gallons
Acetic acid, ml	5,000
Water, gal	240

The final percent solids of the composition was 6.60±0.20% and the final pH was 6.0±0.2 and the final viscosity was 30 to 50 cp (150°F). The weight percent of crosslinked cornstarch in the composition was 57.5%.

The above composition was prepared by heating the starch in an aqueous solution to a temperature of 225±2°F in a jet cooker and then cooling to below 190°F. This constituted a partial cooking of the starch. In a separate tank, the silane is mixed with acetic acid and water. Also, an emulsion of wax, soybean oil and Tween 81 is prepared. The emulsion and silane are added to the partially cooked starch. Then, an aqueous dispersion of Cation-X and a mixture of Versamid 140 in acetic acid and mixture of Biomet 66 in water and a mixture of Igepal CA-630 in aqueous solution are added to the starch mixture. The starch mixture is then diluted to give the final volume of 1,000 gallons.

The size prepared as described was applied to individual glass fibers as they were drawn from orifices of an electrically heated, platinum alloy bushing containing molten glass to form fibers of 0.0003/6 inch in diameters. The size was applied to the fibers prior to the time they were grouped together to form a strand containing 200 fibers by means of a roller applicator which is partially submerged in the sizing solution contained in a reservoir. The fibers were grouped into strands by a gathering shoe and wound on a forming package rotating approximately 4,420 rpm to produce a strand travel speed of approximately 14,000 fpm.

The glass fiber strands wound on the forming package are then dried. This may be done by any number of known methods sufficient to reduce the moisture level to that appropriate for such processing (i.e., 5 to 10%). After the forming packages have dried, the strand is unwound onto a bobbin, the forming package and the bobbin being mounted on a twist frame. During the unwinding and rewinding step, a twist is imparted into the strand which provides integrity for subsequent processing.

When the twisted strand or yarn is to be used for textile applications, beaming is a standard process employed to prepare the yarn. This involves a plurality of bobbins being mounted on racks and the end of each bobbin being threaded through a tensioning disc and through a plurality of guide eyes over a separating comb and onto a beam which is a large cylinder. Then, the plurality of yarn ends are wound on this beam in parallel fashion.

Example 2: Glass fiber strands prepared with the fiber size composition of Example 1 were made into a 600-end section beam. The beam was then placed in a slasher and evaluated for high humidity setups. Approximately 2,000 yards of strand were run at the 85 to 95% relative humidity level with no gumming on an orientation bar such as a meir bar. Also, additional yardage holding the percent relative humidity at the 75 to 85% range was run. Again, the meir bar had a powdery buildup with no gummy deposits.

Recycling of Size Material

F.R. Paul; U.S. Patent 4,192,252; March 11, 1980; assigned to Owens-Corning Fiberglas Corporation describes an apparatus for applying liquid to advancing filaments which is comprised of a pair of nestable containers having specific liquid control portions for improved excess liquid capture and recycling.

As shown in Figures 3.17a and 3.17b, size applicator assembly **10** is adapted to supply liquid size and/or binder **6** to the array of advancing filaments **8** as is known in the art.

Applicator **10** is comprised of housing **12** having an application surface or roll **14** rotatably journaled therein at shafts **15** extending from roll **14**. It is to be understood that the application surface can be a belt-type applicator as well as a roll-type applicator.

Roll **14** is driven by motor **17** by means of pulleys **18** and belt **19**. Generally, the roll is driven such that the surface of the roll **14** in contact with the filaments **8** is moving in the same direction as the filaments. However, in some instances, the roll can be driven counter to the advancement of the filaments.

First container **22** is mounted with respect to housing **12** and roll **14** such that during operation the surface of roll **14** is partially immersed in the body of liquid **6** in container **22**.

First container **22** is comprised of a front wall **24** having an upper edge **25** extending along the length of roll **14**, a bottom wall **27**, sidewalls **29**, and a rear wall **31** suitably joined together to form a reservoir for holding liquid **6**. First container **22** also includes a vertically oriented liquid flow control channel means **33** comprising a plate **34** attached to rear wall **31** and a pair of bars projecting laterally beyond the exterior surface of plate **34** to form the channel. Channel means **33** is oriented such that the upper lip **36** thereof is positioned in a horizontal plane below the upper edge **25** of front wall **24**.

As shown in Figure 3.17b, a pair of channel means **33** are located at the rear wall **31**. In operation, as an excess of liquid **6** is supplied to first container **22** at liquid inlet port **38** located in one of the sidewalls **29**, the excess liquid flows over the upper lip **36** of channel means **33** and downwardly into settling zone **72** of second container **50**.

Second container **50** is adapted to slideably receive first container **22** and to capture the excess liquid from first container **22** and any spray thrown from the filaments or the like deposited on the front wall **24** of first container **22**. Second container **50** is mounted within housing **12** by positioning means **80**.

Second container **50** is comprised of a front portion **52** extending outwardly or forwardly from bottom portion **55** joined thereto. Second container **50** also includes side portions **57**, one of which has recess **58** located therein to accommodate inlet tube **85** joined at liquid inlet port **38** of sidewall **29** of first container **22**. Each side portion **57** has a projection **59** extending inwardly toward each other adapted to accommodate each of the sidewalls **29** such that securement means **68** positions the sidewalls **29** against projections **59** to adjustably locate first container **22** within second container **50**.

Figure 3.17: Recycling of Size Material

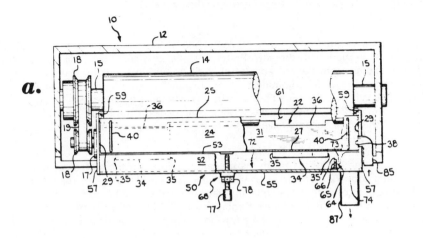

a.

b.

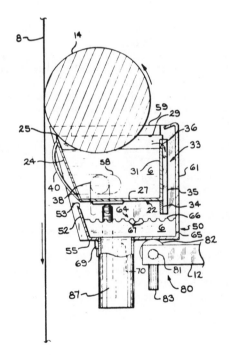

(a) Size applicator apparatus
(b) Side view of apparatus shown in Figure 3.17a

Source: U.S. Patent 4,192,252

Baffle means **64** which is comprised of plate **65** securely joined to front portion **52**, bottom portion **55**, and rear portion **61** of container **50** divides the lower region of the second container **50** into a settling zone **72** and an outlet region **73**.

Plate **65** has a serrated upper edge **66** comprised of a plurality of slots or recesses **67** along the length thereof. Preferably, the slots are approximately $\frac{1}{16}$" wide and extend from the top edge of plate **65** to within about $\frac{1}{8}$" from the bottom edge of plate **65** adjoining bottom portion **55** and are about ¼" apart along the length of plate **65**. This permits contaminants in the excess liquid delivered to zone **72** to collect in the settling zone beneath the lower edge of the slots **67** if the contaminants have a density greater than the density of the liquid **6** (i.e., sinkable).

During operation, floating foreign matter can also form on the surface of liquid **6** in settling zone **72**. The serrated upper edge **66** provides a means for retaining the floating matter within the zone and yet permit the liquid **6** to flow into the outlet region **73** to outlet tube **87** through outlet **74** located in bottom portion **55**. Thus, the sinkable and floatable foreign matter or contaminants are retained in the settling zone **72** for periodic removal by the operator. Thus, the foreign contaminants are removed from the excess liquid before being recycled.

First container **22** is positioned within second container **50** such that the upper rim **53** of front portion **52** extends outwardly beyond the junction of the front wall **24** and bottom wall **27** to collect any excess liquid flowing down front wall **24**. Also, first container **22** incorporates a handle means **40** comprising a pair of curved bars adapted to direct any liquid on the handle means along the front wall **24** and/or into the second container **50**. The handles **40** are adapted to provide the operator with a convenient means for removing, inserting and adjusting the first container **22** in second container **50**.

Securement means **68**, which urges the sidewalls **29** into projections **59** to frictionally retain the first container **22** in second container **50**, is comprised of a boss **69** on bottom portion **55** having a threaded hole **70** to receive screw **77** which is adapted to contact bottom wall **27** and urge first container **22** upwardly into projections **59**. It is preferred that the bottom wall **27** have a boss or landing section to receive screw **77**. Screw **77** is locked into place by any suitable means such as locking nut **78**.

As shown in Figure 3.17b, positioning means **80** for locating second container **50** within housing **12** is comprised of a shaft extending along the length of second container **50**. Shaft **81**, which is rotatably mounted in housing **12**, has an eccentric cam **82** thereon such that as arm **83** of shaft **81** is rotated, the surface of cam **82** urges the second container **50** into fixed engagement with housing **12**.

OTHER PROCESSES

Porous, Corrosion-Resistant Platelike Structure

As is known, the requirements of providing porous, dimensionally stable, heat-resistant and corrosion-resistant structures are encountered in a wide range of engineering. For example, in industrial plants, exhaust stacks for hot corrosive gases or vapors are frequently provided with exhaust hoods which must be corrosion-resistant to the corrosive substances contained in the exhaust gases, e.g., sulfuric acid vapors, as well as heat-resistant as the exhaust vapors frequently have high temperatures of 1000°C or more. In addition, various types of filters, which are used in industrial plants, require similar characteristics.

Packings for mass exchange and heat exchange processes also require structures with similar characteristics. Many packings are made of a fabric or netting of metal wires while other packings are made of asbestos or plastic. Generally, the packings made of a metal fabric or a metal netting use wires with a small diameter. These wires, which usually consist of steel, however, are not corrosion-resistant to a large number of substances to be treated. While asbestos packings are resistant to a relatively large number of substances, such packings require relatively large wall thicknesses because of the brittle nature of asbestos. This imparts a great deal of weight to the packing and, for a given diameter, effects a reduction in the free gap volume. This reduction, in turn, results in a larger pressure drop in the packing.

In the case of plastic packings, these have a disadvantage in that the packings can only be used up to a limited temperature. This reduces the application of the packings to a great extent.

Still other packings have been known which are made exclusively of ceramic. Although these packings are dimensionally stable and corrosion/heat resistant, they have certain disadvantages. That is, in order to be dimensionally stable, the packings require relatively thick walls, e.g., in the order of several millimeters. As a result, in addition to requiring a relatively great amount of material, the pressure drop of such packings is very large in material and heat exchange columns due to a relatively small gap volume.

V. Kubicek; U.S. Patent 4,157,929; June 12, 1979; assigned to Sulzer Brothers Limited, Switzerland describes a method of making a porous, dimensionally stable, heat-resistant and corrosion-resistant platelike structure. The method comprises the steps of obtaining a flat structure of glass filaments wherein the glass filaments are disposed in intersecting relation. Thereafter, a coating of a ceramic substance, which is capable of being fired, is formed on the flat structure of glass filaments and, subsequently, the coating is fired at a temperature sufficient to form a solid skin of the ceramic substance. This skin envelops the glass filaments and defines a skeletonlike supporting structure for the glass filaments. In addition, during firing, the temperature is at least in the range of the melting temperature of the glass filaments so as to cause the glass filaments to fuse together at intersecting points of contact.

In accordance with the method, the flat structure is shaped after the coating step and prior to the firing step. In this regard, the structure can be shaped into a corrugated platelike structure.

In order to effect the coating of the ceramic substance, a binder is applied to the glass filament structure prior to coating. The flat structure may then be shaped after application of the binder since the binder stiffens the structure to some degree.

Friction Material

Friction material, such as that used in brake lining, clutch pads, and the like, has severe performance requirements. The principal function of a friction element like a brake lining is to convert kinetic energy to heat and to absorb the heat or otherwise dissipate it while simultaneously through the agency of friction to reduce relative movement between the friction material and a part engaged by it. To achieve these objectives, it is necessary that the coefficient of friction between the friction material and the part so engaged be as reasonably high as possible, independent of variations in operating conditions, and accomplish the necessary energy conversion with a minumum wear of contacting parts. In particular, a friction material should not only have a relatively high coefficient of friction, but as well possess durability, heat stability, generate little or no noise while in rubbing contact with an engaging part, such as a rotor, and cause a minimum of wear on the engaged part.

In general, a friction material contains a matrix binder, such as a thermosetting resin or vulcanized rubber, a fibrous reinforcement, and a friction modifier which aids in imparting a desired coefficient of friction to the material. Often the friction material may also contain fillers or extenders which modify its physical characteristics and reduce its cost.

G.J. Roberts and J.H. Heasley; U.S. Patent 4,182,437; January 8, 1980; assigned to Ferro Corporation describe an amorphous glass which, in finely divided form, is adapted for use as a combined friction modifier and reinforcing agent for friction material and which, in spite of its amorphous character, does not substantially fuse or smear under heat generated by the friction material and, therefore, does not adversely affect performance of such material.

There are many glass compositions which can be smelted to form the unstable, amorphous glass of the process. One attribute of the glass is that it has a viscosity-temperature relationship enabling it to be fiberized when molten as the substantially amorphous glass, because rapid chilling of the glass is one technique of creating the unstable, amorphous character. This does not mean that the glass must be used in fibrous form although that is the preferred form. The glass can be employed in other forms, such as a particulate form. Glasses useful in the process include silicate, titanate, phosphate, aluminate, and borate glasses and mixtures thereof.

The finely divided glass is conventionally incorporated into a matrix for forming a friction material, such as a rigid thermosetting heat-resistant organic resin.

The glasses are conventionally smelted to a molten melt from the ingredients or from materials forming oxides and fluorides during the smelt in a manner known in the art. The melt is then fiberized by any known means, such as blowing or attenuating molten glass issuing from a bushing by steam or air, wheel spinning, flame impinging, mechanically drawing fibers from a bushing, and the like. The fibers may be either continuous or discontinuous. The diameters and lengths of the fibers are not at all critical and may vary widely. For example, a diameter may average from about 0.5 to 30 μ and usually is about 1.75 μ. Lengths, when continuous fibers are not used, may average from about 1 to 50 cm.

Fiberizing is an especially useful technique to obtain the desired unstable, amorphous glass condition, because the chilling rate during fiberization is so extremely high. It is estimated that fiberizing achieves a chilling rate upward of 1,000,000°C per second, such that fibers obtained in this way can be exceptionally unstable and eager for crystallization when their compositions favor this conversion. If a glass composition cannot be fiberized, it is unlikely that the composition in finely divided form will be amorphous and unstable in accordance with the process.

Any thermosetting, heat-resistant, organic resin capable of withstanding the heat generated by friction material may be used. As a rule, phenolic resins are best for this purpose, such as phenol-formaldehyde and phenol-furfural, but other thermosetting resins may also be used, such as melamine-formaldehyde, urea-formaldehyde, epoxy resins, diallyl phthalate resins, dioctyl phthalate resins, crosslinked alkyd resins, and the like. Phenol-formaldehydes of the novolak type are preferred.

Other additives, modifiers, fillers, extenders known in the art may be added to the thermosetting organic resin. Such other added ingredients include, for instance, barytes, graphite, talc, litharge, kaolin, rosin, waterproofing agents such as waxes like mineral, hydrocarbon and vegetable waxes including, for example, beeswax, Montan wax, paraffin wax, ceresin wax, and the like (such waxes also lubricating the mixing together of the components of the friction material), minor amounts of oxides such as lime, zinc oxide, lead dioxide, silica sand, and manganese dioxide, sulfur (when vulcanization of an elastomer is contemplated), and the like.

Proportions are not critical. In general, the friction material contains in parts by weight from about 20 to 80 parts of the resin, from about 5 to 40 parts of the present finely divided glass. When an elastomer is added, it may be used up to about 12 parts by weight. When filler material is used, it may be added up to about 30 parts by weight.

To prepare the friction material, one technique that may be used is the hot press method. Friction materials were prepared in this way: The resin, glass fibers of the process, and any filler material desired are dry blended in a mixer for about 15 to 45 minutes until a uniform mixture is obtained. The resulting mix is placed in a mold and compacted as by a ram, while being warmed to form a brake lining. The amount of mix and pressure used are adjusted to obtain a brake lining of a desired thickness and density. The brake lining is stripped from the mold and postcured, for example, at about 150°C for 30 minutes. The temperatures of the mold and postcure are far below those which trigger devitrification of the glass fibers of the mix.

During use, the friction material generates heat as, for example, in contacting a moving member to brake it, such as a brake drum. The heat generated is not uniformly distributed throughout the friction material, but tends to become concentrated at the interface between it and the moving member which may be considered to be the wear area. This phenomenon brings into play the amorphous-devitrification function of the glass.

Assuming for convenience of description that the glass is in fibrous form, as minute portions of the fibers reach the wear area (due to accumulated wearing away of the friction material) and are subjected to the generated heat, the glass of such minute portions releases its pent-up preference for crystallization, and the heated unstable portions quickly self-convert to a devitrified state which is brittle and dusts or flakes or otherwise separates from companion nondevitrified portions of the glass. The latter remain in the friction material as a combined friction modifier and reinforcing agent. In this manner, the fibers devitrify to the extent needed as the friction material wears away, that is, the fibers wear away in unison with the friction material.

GLASS-CERAMICS

GLASS-CERAMIC COMPOSITIONS

Glass-ceramics are produced in accordance with three general steps. First, a glass-forming composition, which customarily also contains a nucleating agent, is melted. Second, the melt is simultaneously cooled to a glass sufficiently quickly to prevent the occurrence of any substantial crystallization and an article of a desired geometry is shaped therefrom. Third, the glass is subjected to a heat treatment to cause the growth of crystals in situ. Commonly, the crystallization step will be carried out in two parts. Such involves initially heating the glass shape to a temperature within or somewhat above the transformation range thereof to cause the development of nuclei in the glass. After the nucleating step, the glass shape is heated to a higher temperature, commonly above the softening point thereof, to effect the growth of crystals on the nuclei.

Because the growth of crystals takes place on nuclei dispersed throughout the glass shape, a glass-ceramic will typically have a microstructure of fine-grained crystals randomly oriented, but uniformly dispersed, throughout a residual glassy matrix. In general, a glass-ceramic is predominantly crystalline, i.e., greater than 50% by volume crystalline, such that the physical and chemical properties exhibited thereby will be more comparable to the crystal phase present therein than to the properties of the original glass. Moreover, the composition of the small amount of residual glass will be quite different from that of the original glass inasmuch as the components making up the crystals will have been removed therefrom. Finally, because glass-ceramics are prepared via the crystallization in situ of glass articles, they will have smooth surfaces and be nonporous.

Anorthite Glass-Ceramic Composition

Anorthite, i.e., triclinic $CaO \cdot Al_2O_3 \cdot 2SiO_2$, has been known for its insulating properties. However, those properties are not sufficiently different from those of steatite and forsterite to warrant the added expense of manufacture. Nevertheless, it has been realized that a glass-ceramic body wherein the predominant and, preferably, the sole crystal phase is anorthite, could have significant practical application. Such bodies would have high dielectric constants and dc volume resistivities coupled with low loss tangents and, consequently, would be competitive with commercial electrically-insulating materials. Because of their mode of manufacture, such bodies could be produced at a rapid rate and would have the advantages of smooth surfaces and no porosity.

R.F. Reade; U.S. Patent 4,187,115; February 5, 1980; assigned to Corning Glass Works describes a limited range of compositions in the $CaO \cdot Al_2O_3 \cdot SiO_2$ system nucleated with TiO_2 which can be heat treated in a defined manner to form glass-ceramic articles wherein anorthite or an anorthite solid solution is the predominant crystal phase present. Solid solution can be had with divalent metal oxides such as MgO, SrO, BaO, CdO, and PbO, which articles will exhibit physical properties modified from those of the simple anorthite assemblage. As used herein, the term anorthite will include both the classic ternary composition and a solid solution thereof.

The articles are essentially free from surface distortion or wrinkling and exhibit high refractoriness, moderate coefficients of thermal expansion ($\sim$35-65 x 10^{-7}/°C over the temperature range of 20° to 300°C), high dielectric constants, high dc volume resistivity, and low ac dielectric losses. Such articles are highly crystalline and can be formed from glasses having base compositions, expressed in weight percent on the oxide basis, of 10 to 18% CaO, 29 to 35% Al_2O_3, 30 to 39% SiO_2, and 13 to 20% TiO_2.

The process comprises three general steps: First, a glass-forming batch coming within the abovecited composition ranges is melted; second, the molten batch is simultaneously cooled to a temperature at least within the transformation range of the glass (optionally to room temperature) and a glass shape of a desired geometry shaped therefrom; and third, the glass shape is heated to a temperature between about 1000° to 1400°C and maintained within that range for a sufficient length of time to cause the growth of anorthite crystals.

To insure the greatest uniformity of crystal size, the crystallization step will customarily be divided into two parts. Thus, the glass article will typically be heated to a temperature within and somewhat above the transformation range, e.g., between about 750° to 900°C, and held within that range of temperature for about 3 to 8 hours to foster good nucleation, after which the temperature is raised to between 1000° to 1400°C and maintained within that range of temperature for about 4 to 10 hours to cause the growth of crystals on the nuclei.

It will be appreciated that the rate of in situ crystal growth is dependent upon the temperature at which the crystallization is carried out, the rate being more rapid at higher temperatures. Consequently, whereas a time of only about one hour may be sufficient at the higher extreme of the crystallization range, 24 hours and longer may be required at temperatures around 1000°C to achieve high crystallinity. The use of dwell periods at specific temperatures is a matter of convenience only. All that is required is the maintenance of the glass within the range of crystallization temperatures for a period of time sufficient to induce crystal growth.

Beta-Spodumene Glass-Ceramic Materials

K. Chyung and J.E. Megles, Jr.; U.S. Patent 4,192,665; March 11, 1980; assigned to Corning Glass Works describe a method for producing a white, opaque, chemically durable, low-expansion glass-ceramic material comprising β-spodumene solid solution as the principal crystal phase which comprises the steps of:

(a) selecting for crystallization a glass having a composition falling within one of Ranges I and II below, wherein Range I includes compositions consisting essentially, in weight percent, of about 63 to 67%, SiO_2, 18 to 21% Al_2O_3, 3.0 to 5.0% Li_2O, 1.0 to 2.5% MgO, 0.4 to 2.5% ZnO, 4.0 to 5.0% TiO_2, 1.5 to 4.0% PbO, 0 to 1.0% B_2O_3, 0 to 1.0% Na_2O, 0 to 2.0% P_2O_5, and 0 to 2% total of B_2O_3 + Na_2O + P_2O_5, and Range II includes compositions consisting essentially, in weight percent, of about 68 to 71% SiO_2, 16 to 18% Al_2O_3, 3.0 to 5.0% Li_2O, 1.0 to 2.5% MgO, 0.4 to 2.5% ZnO, 4.0 to 5.0% TiO_2, 0 to 1.0% B_2O_3, 0 to 1.0% Na_2O, 0 to 2.0% P_2O_5, and 0.2 to 2.2% total of B_2O_3 + Na_2O + P_2O_5; and

(b) converting the selected glass to a glass-ceramic material utilizing a crystallization heat treatment comprising heating the glass to a nucleation temperature in the range of about 750° to 800°C for a time in the range of about 1 to 4 hours to develop crystal nuclei throughout the volume of the glass, followed by heating the glass to a crystallization temperature in the range of about 900° to 1000°C for a time not exceeding about 2 hours.

Applications for the products may be envisioned wherever usage at high temperatures or under adverse conditions of thermal shock is required. The glass-ceramics are strengthenable by lamination techniques, particularly differential densification strengthening, and could be used alone or in combination with other materials to provide strong, durable sheet, tubing, vessels for cooking or the like, and many similar articles.

Oxynitride Compositions

K. Chyung and R.R. Wusirika; U.S. Patent 4,141,739; February 27, 1979; assigned to Corning Glass Works describe a glass-ceramic article having a crystal content in excess of about 50% by volume, the composition of the article consisting essentially, by weight, of about 55 to 70% SiO_2, 20 to 30% Al_2O_3, and 3.5 to 15% N, wherein the predominant crystal phase consists of nitrogen-mullite solid solution.

Table 1 records a group of approximate compositions, expressed in weight percent on the oxide basis, illustrating the parameters of the process. Since it is not known with which cation(s) the fluorine and nitrogen are combined, they are merely reported as fluoride (F) and nitrogen (N) and the oxygen $\approx$ fluorine and oxygen $\approx$ nitrogen correction factor recited in accordance with conventional glass analysis practice. The actual batch ingredients can comprise any materials, either the oxide or other compound, which, when melted together, will be converted to the desired oxide in the proper proportions.

The fluorine will commonly be supplied utilizing such compounds as AlF_3, MgF_2, LiF, ZnF_2, and CaF_2, depending upon the compositions involved. In the following compositions, nitrogen was added as Si_3N_4. However, other nitrogen-containing compounds, such as AlN, Si_2ON_2, Li_3N, or Mg_3N_2, can be employed. Volatilization of nitrogen and fluorine from the melt is dependent upon the melting temperature and atmosphere utilized. Losses of up to 50% by weight can occur although, in general, such will run between about 10 to 40%.

Table 1

	1	2	3	4	5	6	7	8	9	10
SiO_2	70.4	70.4	70.4	70.4	70.4	70.4	70.4	70.4	65.4	65.4
Al_2O_3	22.2	22.2	22.2	22.2	22.2	22.2	22.2	22.2	22.2	22.2
MgO	7.0	5.0	5.0	5.0	5.0	5.0	5.0	5.0	15.0	–
F	1.4	1.4	1.4	1.4	1.4	1.4	1.4	1.4	1.4	1.4
N	4.8	4.8	4.8	4.8	4.8	4.8	4.8	4.8	4.8	4.8
Li_2O	3.0	–	–	–	–	–	–	–	–	–
SrO	–	5.0	–	–	–	–	–	–	–	–
BaO	–	–	5.0	–	–	–	–	–	–	–
ZnO	–	–	–	5.0	–	–	–	–	–	15.0
Na_2O	–	–	–	–	5.0	–	–	–	–	–
K_2O	–	–	–	–	–	5.0	–	–	–	–
Y_2O_3	–	–	–	–	–	–	5.0	–	–	–
Cs_2O	–	–	–	–	–	–	–	5.0	–	–
Total	108.8	108.8	108.8	108.8	108.8	108.8	108.8	108.8	108.8	108.8
$O{\approx}F$	-0.6	-0.6	-0.6	-0.6	-0.6	-0.6	-0.6	-0.6	-0.6	-0.6
$O{\approx}N$	-8.2	-8.2	-8.2	-8.2	-8.2	-8.2	-8.2	-8.2	-8.2	-8.2
	100.0	100.0	100.0	100.0	100.0	100.0	100.0	100.0	100.0	100.0

The batch ingredients will be compounded, ball-milled together to assist in obtaining a homogeneous melt, and the mixture then run into silica, molybdenum, or graphite crucibles. Since an inherently strongly reducing atmosphere is created by nitride, a platinum crucible cannot be utilized. And, since nitrogen solubility increases significantly in a reducing atmosphere (as much as five orders of magnitude compared to an oxidizing environment), it is most beneficial to melt the compositions in a nitrogen or other oxygen-deficient atmosphere.

The crucibles containing the batches are covered, placed in induction-heated furnaces operating at about 1500° to 1850°C, and the batches melted for about 1 to 16 hours. The melts are poured into steel molds to yield glass slabs and these slabs immediately transferred to an annealer commonly operating at about 600° to 700°C.

Table 2 records, in parts by weight, the actual batch materials utilized in the examples of Table 1. The melting temperature (°C), the type of crucible employed (SiO_2) and the furnace atmosphere employed are also listed. In the examples given, the atmosphere used was air.

Table 2

	1	2	3	4	5	6	7	8	9	10
SiO_2	55	55	55	55	55	55	55	55	50	50
Al_2O_3	21	21	21	21	21	21	21	21	21	21
Si_3N_4	12	12	12	12	12	12	12	12	12	12
AlF_3	2	2	2	2	2	2	2	2	2	2
MgO	7	5	5	5	5	5	5	5	15	–
Li_2O	3	–	–	–	–	–	–	–	–	–
SrO	–	5	–	–	–	–	–	–	–	–
BaO	–	–	5	–	–	–	–	–	–	–

(continued)

Table 2: (continued)

	1	2	3	4	5	6	7	8	9	10
ZnO	–	–	–	5	–	–	–	–	–	15
Na_2O	–	–	–	–	5	–	–	–	–	–
K_2O	–	–	–	–	–	5	–	–	–	–
Y_2O_3	–	–	–	–	–	–	5	–	–	–
Cs_2O	–	–	–	–	–	–	–	5	–	–
Temperature, °C	1650	1650	1650	1650	1650	1650	1650	1650	1650	1650

Crucible . . . . SiO_2 . . . .

pO_2 . . . . air . . . .

The molten glass batch is cooled to a temperature at least within the transformation range and then reheated to cause crystallization. (The transformation range is defined as the temperature at which a molten mass becomes an amorphous solid, that temperature customarily being deemed to lie in the vicinity of the annealing point of the glass.)

The rate of crystallization is a direct function of the temperature employed. Therefore, a brief exposure time, e.g., 0.5 hour or less, may be adequate to achieve substantial crystallization at the upper extreme of the crystallization range; whereas, in the cooler end of the crystallization range, much longer periods of time, i.e., up to 24 hours or more, may be demanded.

Although temperatures as low as 800°C can be operable, such are normally not utilized because the rate of crystal growth is generally so slow as to be unattractive from a practical point of view. Conversely, temperatures in excess of about 1300°C are avoided since grain growth of the crystals and deformation of the article are hazarded. Therefore, the more useful range of crystallization temperatures has been defined as about 900 to 1300°C.

If desired, a two-step heat treatment procedure may be followed. Thus, the glass article will first be heated to a temperature slightly above the transformation range, e.g., 750° to 850°C for a sufficient length of time, perhaps 1 to 8 hours, to cause extensive nucleation within the glass and initiate crystal growth. Thereafter, the nucleated article will be heated into the 900° to 1300°C range for about 2 to 10 hours to develop the desired fine-grained growth of crystals.

The products are conventionally highly crystalline, being greater than 50% by volume crystalline and, frequently, in excess of 75%, and the crystals are homogeneously dispersed within the minor residual glassy matrix.

Dental Restoration Material

J.M. Barrett, D.E. Clark and L.L. Hench; U.S. Patent 4,189,325; February 19, 1980; assigned to The Board of Regents, State of Florida, University of Florida describe a glass-ceramic suitable for use as a dental restorative material which combines the properties of high mechanical strength, good fracture toughness, high chemical durability in the intended physiological setting, good castability in conventional dental laboratory investment molds, biological compatability, and aesthetic properties resembling those of natural teeth.

A glass, suitable as a starting material in the production of the glass-ceramic, consists essentially of about 25 to about 33 mol % Li_2O, about 73.5 to about 52 mol % SiO_2, about 0.5 to about 5 mol % Al_2O_3 and about 1 to about 10 mol % CaO.

The process for preparing a glass-ceramic article substantially free of cracks arising from the local volume change of crystallization and consisting essentially of a fine-grained crystal phase uniformly dispersed within a vitreous matrix, comprises the steps of:

(A) preparing a uniform melt having the composition described above;

(B) cooling the melt to at least below the transformation range thereof and simultaneously forming it into a glass article of the desired shape;

(C) heat treating the glass article at a temperature of about 490°C to about 575°C to effect nucleation in situ of the crystal phase;

(D) heat treating the article resulting from step (C) at a temperature of about 600°C to about 700°C to effect growth in situ of the crystal phase; and

(E) cooling the resulting glass-ceramic article to room temperature.

The dental restorations can be provided with a broad range of colors to match the colors of the natural teeth of different patients. The color of the dental restoration may be very carefully and precisely controlled. Thus, for example, a glass frit containing 61.0 mol % SiO_2, 30.5 mol % Li_2O, 2.5 mol % Al_2O_3 and 6.0 mol % CaO is mixed with 0.2 to 0.7 weight percent niobium oxide, 0.3 to 1.1 weight percent $AgNO_3$ and a compound capable of generating 0.0033 weight percent Pt (the weight percents based on the weight of glass frit), and a glass prepared from the resulting mixture. The glass is thermally crystallized by heat treatment at 520°C for about 4 hours (nucleation) followed by heat treatment at 620°C for from about 0.5 hour to about 2 hours (crystal growth). This heat treatment is capable of providing a glass-ceramic material which is greater than about 70 volume percent devitrified, but still requires only about 4.5 to 6 hours for the two stages of nucleation and crystal growth.

The resulting glass-ceramic possesses a wide range of colors, depending on the precise $AgNO_3$ and Nb_2O_5 levels employed, matching those of natural teeth and has translucencies closely approximating those of natural teeth. Ceric oxide (about 1.5 to about 2 weight percent together with about 0.3 to about 0.5 wt % niobium oxide, both based on the sum of $Li_2O + Al_2O_3 + CaO + SiO_2$) has also been found to be a particularly useful coloring agent for the dental restorations. When ceric oxide is employed, a crystal growth heat treatment of from about 4 hours to about 15 hours at 620°C produces the best reproductions of the colors and translucencies of natural teeth.

Because niobium oxide causes the glass-ceramic to exhibit an unnatural whiteness and opacity and suppresses the effects of the inorganic coloring agents, dental restorations for teeth other than molars preferably contain from about 0.2 to about 0.7 wt %, based on the weight of $Li_2O + Al_2O_3 + CaO + SiO_2$, of Nb_2O_5.

DECORATIVE COATINGS

Brown Stain Decoration

V. Lupoi; U.S. Patent 4,192,666; March 11, 1980; assigned to PPG Industries, Inc. found that crystallizable glass articles having inhomogeneities which would normally lead to nonuniform coloration when stain-decorated with palladium can be successfully provided with uniformly colored stain patterns by carrying out the crystallizing and staining steps in separate, sequential heat treatment steps. Moreover, postcrystallization staining of glass-ceramics has been found surprisingly to produce not the usual gray or blue-gray colored stains, but rather a highly desirable family of brown colors.

In general, the crystallizable glass compositions of the process may be characterized as having essential inclusions of SiO_2, Al_2O_3, and Li_2O as crystal-forming constituents, ZnO as a melting aid, and TiO_2 or a mixture of TiO_2 and ZrO_2 as nucleating agents. The alkali metal content of the glass-ceramics is minimized, although a small amount of K_2O is typically included. Small amounts of melting and fining aids, such as fluorine, chlorine, antimony, or arsenic may also be included. The presence of arsenic and/or antimony has also been found to have a beneficial effect on the staining process. An example of a crystallizable glass composition is as follows:

Ingredients	Range	Preferred Embodiment
	....	.(% by wt)...........
SiO_2	67–71	70.22
Al_2O_3	18–21	19.21
TiO_2	1.4–5.0	2.15
ZrO_2	0–2.0	1.58
Sb_2O_3	0–1.0	0.38*
As_2O_5	0–1.0	0.01
Li_2O	2.5–4.0	3.99
Na_2O	0–1.0	0.30
K_2O	0–1.0	0.27
Cl_2	0–0.2	—
ZnO	0.5–2.0	1.59
F_2	0–0.5	0.24**
MgO	0–3.0	0
CaO	0–4.0	0
P_2O_5	0–1.5	0

*Sb_2O_5
**F^-

A glass of the above preferred composition may be melted from the following batch ingredients:

Ingredient	Parts by Weight
Silica	700
Hydrated alumina	296
Lithium carbonate	83
Zinc zirconium silicate	31
Zinc oxide	10.5
Titanium dioxide	15.0
Lithium fluoride	13.5

(continued)

Ingredient	Parts by Weight
Soda ash	4.0
Lithium sulfate	6.0
Potassium carbonate	2.5
Antimony oxide	4.0
Total	1,165.5

These materials may be melted on a continuous basis in a refractory melting chamber, from one end of which, in one embodiment, a ribbon may be withdrawn and formed into a flat sheet of glass by rolling in accordance with techniques similar to the plate glass method. Following forming, the glassy sheet is cooled and cut to the desired size. Optionally, the glassy sheet also may be ground and polished.

Crystallization of the formed glassy articles (e.g., plates) takes place in a heat treatment chamber into which the articles are placed at room temperature. The temperature is steadily raised over a period of several hours to about 1300°F (700°C) or higher, which temperature is maintained for several hours to initiate nucleation of crystallization sites within the bodies of the glassy articles. The temperature is then increased to about 1850°F (1010°C) to convert a major portion (at least 50%) of the glass to small, dispersed crystals.

The crystal which first forms is β-eucryptite, which is subsequently transformed by the heat treatment to β-spodumene crystals. Preferably, crystallization is carried out until the article is about 98% by weight crystal phase, with 2% remaining as a glassy phase. In its final state, the crystalline phase is a solid solution of β-spodumene and silica.

A specific example of a preferred heat treating schedule is as follows:

Raise temperature from room temperature to 1100°F (593°C) over 3 hours;

Hold at 1100°F (593°C) for 2 hours;

Raise temperature to 1285°F (696°C) over 2 hours;

Raise temperature to 1325°F (718°C) over 1 hour;

Increase temperature to 1385°F (752°C) over 6 hours;

Hold at 1385°F (752°C) for 2 hours;

Increase temperature to 1400°F (760°C) over 2 hours;

Increase temperature to 1650°F (899°C) over 1.5 hours; and

Hold at 1650°F (899°C) for 2 hours.

From this point crystallization is ordinarily completed by further increasing the temperature to 1850°F (1010°C) over 1.5 hours, holding at 1850°F (1010°C) for 3 hours and then cooling to room temperature over a period of about 3 hours. However, it is preferred to stop the crystallization heat treatment after holding at 1650°F (899°C), even though conversion to the β-spodumene crystal form is incomplete, since the subsequent heat treatment required for stain-decorating the glass-ceramic articles may be employed to carry crystallization to the desired point of completion. The result is a savings in thermal energy.

A specific palladium glass colorant which may be used is identified as Dark Brown A-1454 (Englehard Industries, Inc.) and includes a palladium resinate as the active staining agent, along with a small amount of bismuth in an organic carrier. The colorant contains about 2.26 weight percent palladium and 0.42 weight percent bismuth.

About 50 g of the colorant are mixed with about 450 g of titanium dioxide extender, which is preferably in the anatase crystal form as taught in U.S. Patent 3,816,161. The colorant and titanium dioxide may be mixed in a ball mill with a suitable liquid vehicle to establish the desired viscosity for use in the particular decorating technique to be employed. For use in the preferred silk-screen decorating method, pine oil is the preferred vehicle, for example Drakelene Oil (Hercules, Inc.). A viscosity of about 50,000 cp is typically considered suitable for silk-screening.

The stain mixture is applied to surface portions of the crystallized glass-ceramic articles in decorative patterns or to an entire surface. While silk-screening is the preferred method of mass producing identical patterns, any method of applying the stain mixture could be used, such as brushing, stenciling, or spraying.

The crystallized glass-ceramic articles with the stain material applied thereto are then returned to the heating chamber where they are heated to a temperature sufficient to drive the stain-producing palladium ions into surface portions of the glass. In the case of the specific heat treatment schedule set forth above, wherein crystallization was stopped short of completion, the second heat treatment consists of heating the heating chamber to a temperature of 1950°F (1066°C) over a period of about 10 hours (or approximately the maximum rate of the heating chamber) and holding at 1950°F (1066°C) for 1 hour, after which the heat is turned off and the heating chamber is permitted to cool over a period of several hours.

At the conclusion of the second heat treatment, the residue of the staining material is wiped from the surfaces of the glass-ceramic articles. The stain patterns which are formed are found to be a uniform brown color with a lustrous surface appearance and good abrasion resistance. When the crystallizable glass is selected from a production run known to have a high incidence of inhomogeneities which cause nonuniformity in the conventional gray stains, and is stained with palladium after crystallization in accordance with the process, the resulting brown stain is still found to be essentially free from nonuniformity.

Gray Stain Decoration

In a variation of the previous process, *M.J. Hummel, V. Lupoi and R.L. Cerutti; U.S. Patent 4,197,105; April 8, 1980; assigned to PPG Industries, Inc.* have found that it is possible to produce gray colored stain patterns on glass-ceramics which have already been crystallized. This is accomplished by heating a conventional stain-decorating composition as described in the previous process, to a temperature above the decomposition temperature of the organic constituents of the composition, reconstituting the residue of the heat treated stain composition by mixing with a suitable vehicle, applying the mixture in the conventional manner to crystallized glass-ceramic articles, and subjecting the articles to a second heat treatment to drive the stain into the glass surface.

The pretreatment of the stain composition is as follows: A substantial quantity of the stain composition is placed in a vessel and heated in a well ventilated facility to boil off the organic vehicle at a slow rate to prevent splattering. The residue is ground to -200 mesh, placed on a tray and heated in a furnace to a temperature at least sufficient to decompose any organic compounds present including the resinate portion of the palladium compounds. Although temperatures considerably lower may be adequate, it is preferred to heat the stain composition to a temperature similar to the maximum temperature obtained in the crystallizing heat process, i.e., about 1850° to 2000°F (1010° to 1093°C) for about 2 hours. A silica tray with the powdered material loaded to a depth of about one-half inch (one centimeter) has been found to be satisfactory for heat treating the stain material. The heat treated stain material is then reconstituted with a suitable vehicle, such as pine oil, for application to the glass-ceramic articles by the particular decorating technique chosen, preferably silk-screening.

The crystallized glass-ceramic articles with the heat treated stain material applied thereto are then returned to the heating chamber where they are heated to a temperature sufficient to drive the stain-producing palladium ions into surface portions of the glass.

The stain patterns which are formed are found to be a uniform neutral gray color closely matching that produced in glass-ceramics which are stain-decorated and crystallized simultaneously. The remainder of each crystallized article has an almost opaque, milk-white to grayish-white appearance. By employing other heat treatment schedules, glass-ceramics which are transparent or translucent may be produced.

MISCELLANEOUS PROCESSES

Conversion of Thin Glass Bodies to Glass-Ceramic Bodies

H.L. Rittler; U.S. Patent 4,201,559; May 6, 1980; assigned to Corning Glass Works describes a simple, convenient, and highly effective procedure for cramming certain glass bodies, i.e., converting such glass bodies to corresponding glass-ceramic bodies by thermally induced internal nucleation and crystallization. It is not limited to specific glasses, except as the glass must be amenable to the nucleated crystallization that is characteristic of the ceramming process. It is, however, limited to cerammable glass bodies having at least one dimension less than about 250 μ. This includes, but is not limited to, such shapes as fibers, filaments (both single and laminated), ribbons, microsheets, and coatings on such articles as glass, metal or ceramic cores or substrates.

The characteristic feature of the process is a crystallizing heat treatment which consists of a plurality of heating and cooling cycles. Instead of one continuous heat treatment of specified time to produce a crystallized glass body, the process contemplates a series of interrupted heat treatments of short duration. During each interruption the glass is cooled, and then reheated. Thus, the heat treatment will provide a total amount of time at the crystallizing temperature comparable to that of a prior single-cycle treatment. However, this time will be subdivided by cooling interruptions to provide several short applications of heat (pulses), rather than one long continuous application.

For example, the schedule for ceramming a particular crystallizable glass fiber is 21 seconds at 1250°C. In accordance with the process, the fiber might be exposed to seven separate heat treatments of three seconds each at the same temperature with cooling intervals intermediate each pair of successive exposures.

The process is based on the discovery that such an interrupted (pulsed) type of crystallizing heat treatment produces a crystal phase of finer grain size, i.e., smaller size crystals, than the comparable continuous heat treatment of equal total time and temperature. The finer grain size is manifested by a greater resiliency in the crystallized glass body and/or a degree of transparency. It was also observed, by way of powder x-ray traces, that the degree of crystallinity in the pulsed type process may be about the same or somewhat greater.

Neither the number nor uniformity of the cycles in a particular heat treatment, nor the length of the individual cycle, is particularly critical. As a matter of convenience, a heat treatment will normally consist of several uniform cycles of about 2 to 5 seconds each. For most purposes, a total heat treating time up to a minute or so is contemplated, but an even longer time of treatment, up to about 10 minutes, may be employed to advantage in some instances. The degree of crystallinity will normally increase, at least to some extent, with the total crystallization time. However, the distinct advantage of compatibility with a continuous operation becomes more difficult to achieve with longer time. Thus, where the crystallization heat treatment is incorporated into another continuous operation, such as the drawing of fibers, a high temperature, rapid ceramming schedule of less than a minute will normally be most convenient.

The pulse heat treatment may, in a simple form, be achieved manually. Thus, a number of fibers may be deposited on a suitable support with means to repeatedly move the support into and out of a heat treating furnace. For example, the fibers may be laid in grooves in a refractory support sheet resting in front of the opening for the heat treatment furnace. The furnace will be maintained at a suitable crystallization temperature, e.g., 1250°C. The sheet is then retractably pushed into the oven for a few-second interval, removed, reinserted for a second interval, and this cycle repeated, for example, several times.

In order to achieve maximum benefits from the process, however, it is usually desirable to automate the cycling heat treatment and to incorporate it into a continuous drawing or coating operation. For this purpose, a split laser beam is a particularly convenient source of heat.

As schematically shown in Figure 4.1, the heat treatment is applied to fibers **10** after they leave the bushing, and before they are gathered on the drum for storage. Thus, a split laser beam is generated to produce, in the present instance, four separate and distinct beams through which the fibers pass as they are drawn. As the fibers enter each laser beam, they are reheated to the desired temperature for internal crystallization (ceramming). Thereafter, as they leave the beam, they inherently cool before passing into the next beam.

Thus, in an arrangement such as illustrated in Figure 4.1, a laser beam is split into four separate beams which provide four separate reheating and cooling cycles for the fibers as they are drawn.

Figure 4.1: Apparatus for Production of Glass-Ceramic Fibers

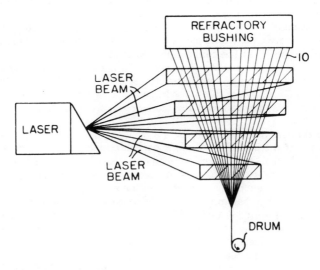

Source: U.S. Patent 4,201,559

Glass Envelope for Isostatic Pressing of Ceramic Articles

R. Grunke; U.S. Patent 4,199,339; April 22, 1980; assigned to Motoren-und Turbinen-Union München GmbH, Germany describes a method for the manufacture of a molded member from a ceramic material, for instance, of silicon nitride or silicon carbide, wherein the molded member, which is arranged in an evacuated capsule of a high temperature-resistant glass is compressed into a nonporous condition under a pressure acting on all sides thereof. The starting material of the glass capsule is applied onto the molded members in a nonvitrified state and exposed under a vacuum to its reaction temperature until there is formed a molded member-encompassing, cohesive glass layer which constitutes the capsule.

This method is generally and summarily designated as high-temperature isostatic pressing. The object of the method is to compress a ceramic member having a porosity of up to about 30% under a pressure uniformly acting thereon on all of its sides up to nonporosity in order to increase its strength.

The glass layer prevents the entry of air into the pores of the molded ceramic member beneath the glass layer and, thereby, oxidation at the pore edges in the molded ceramic member. In essence, this approach takes the pores and seals them with respect to the exterior by means of a durable glass layer or coating.

Described hereinbelow are two preferred approaches, having been found to be valuable approaches in experimentation. When reaction-sintered silicon nitride is specified, the assumption is that the numerously mentioned experimentations are carried out with silicon nitride. However, it may be assumed that other types of ceramic materials, when not a pure ceramic, will correspondingly perform, so that silicon nitride (Si_3N_4) or the similarly investigated silicon carbide (SiC) should be merely considered as safe forms of equivalent ceramics.

Approach 1: *Two-Step Heat Treatment Process* — Initially produced was a porous molded member of reaction-sintered silicon nitride. Vapor deposited thereon was metallic aluminum. In a first heat treatment step (annealing in air for about 12 hours at about 580°C) the aluminum was converted into aluminum oxide. Concurrently, silicon oxide is formed on the surface of the molded member. In a second heat treatment step (at about 1200°C for about 0.5 hour under a vacuum) silicon oxide and aluminum oxide formed a high temperature-resistant glass layer on the surface of the porous silicon nitride member. After the heat treatment, the vacuum remained intact interiorly of the pressing as a result of the glass layer.

Approach 2: *Single-Step Heat Treatment Process* — Initially, again produced was a porous molded member of reaction-sintered silicon nitride. Deposited on the surface thereof were the constituents of a high temperature-resistant glass such as, for example, oxides of the rare earths and silicon oxide, in the finest possible distribution. In a heat treatment operation at about 1200°C under a vacuum, the oxides were converted into a high temperature-resistant glass which covered the entire surface of the molded member and thus maintained intact the vacuum in the interior of the pressing.

Resulting from both approaches is a pressing or briquette with an envelope of a high temperature-resistant glass which can be further processed through high-temperature isostatic pressing into a dense molded member. During the compression of the molded member, the glass envelope will soften to such an extent that it no longer hinders the volumetric reduction, however, without allowing for the entry of gas into the evacuated envelope.

OPTICAL GLASS

LENSES

B_2O_3-ZnO-La_2O_3-Y_2O_3 Optical Glass

High refractive index, low dispersion optical glass heretofore has been formed of a basically B_2O_3-alkaline earth metal oxides-La_2O_3 system. In an optical glass of this type, a large quantity of alkaline earth metal oxides is required. Of the alkaline earth metal oxides, the use of appreciable amounts of MgO, CaO and SrO makes the resultant glass very unstable against devitrification. Also, the use of a substantial amount of the alkaline earth metal oxide, BaO, greatly diminishes the chemical durability of the resultant glass. Thus, high refractive index, low dispersion optical glass made of a system which is basically B_2O_3-alkaline earth metal oxides-La_2O_3 is unsuitable for mass production because of its tendency to devitrify and its poor chemical durability, and, therefore, its insufficiency as elements for optical instruments.

An improvement over such known glass is a high refractive index, low dispersion optical glass which has ZnO as one of the chief components, instead of alkaline earth metal oxides, thereby constituting an essentially three component system of B_2O_3-ZnO-La_2O_3. An example of this type of glass has a composition as follows, the ranges being present by weight percent:

B_2O_3	29-43
La_2O_3	16-27.5
ZnO	17-36
SiO_2	1-7
MgO + CaO + SrO	1-11
CdO	0-6.5

However, such optical glass is low in La_2O_3 content which serves to impart the high refractive index and low dispersion properties to the glass. Consequently, in order to maintain for this glass the high refractive index and low dispersion properties, it is necessary to increase the ZnO content which, among the chief or essential components, is next to La_2O_3 in the degree of contribution for imparting high refractive index and low dispersion properties to the glass. However, this glass which contains such a substantial amount of ZnO, which is a glass network modifying oxide, has a low viscosity; also, the glass may not be said to be sufficiently stable against devitrification. This is a serious disadvantage insofar as mass-producing the glass.

S. Matsumaru; U.S. Patent 4,144,076; March 13, 1979; assigned to Nippon Kogaku KK, Japan describes a high refractive index, low dispersion optical glass which has a refractive index ranging from 1.69 to 1.80 and Abbe number ranging from 43 to 55, which is stable against devitrification and excellent in chemical durability and which is suitable for mass production.

According to this process, ZnO is used instead of the alkaline earth metal oxides belonging to the B_2O_3-alkaline earth metals-La_2O_3 system. Then, Y_2O_3 is partially substituted for La_2O_3 in the three component system of B_2O_3-ZnO-La_2O_3. As a result, a four component system of B_2O_3-ZnO-La_2O_3-Y_2O_3 is provided to thereby enable mass production of high refractive index, low dispersion optical glass which is stable against devitrification and excellent in chemical durability.

The contents of the foregoing four requisite components may be in the following ranges, all by weight percentage:

B_2O_3	25-40
ZnO	11-30
La_2O_3	28-44
Y_2O_3	1-25

Also, the amounts to be added of alkaline earth metal oxides, SiO_2, Al_2O_3, PbO, ZrO_2 and Ta_2O_5 are in the following ranges:

SiO_2	0-5
Al_2O_3	0-5
PbO	0-15
ZrO_2	0-10
Ta_2O_5	0-10
Total alkaline earth metal oxides	0-20

Of the optical glasses according to this process, those having the following ranges of composition in percent by weight possess refractive indices of 1.69 to 1.75 and Abbe numbers of 50 to 55, and fall under a lower dispersion range among the high refractive index, low dispersion glasses.

B_2O_3	30-40
ZnO	11-30
La_2O_3	28-44
Y_2O_3	1-25
Alkaline earth metal oxides	0-20

The alkaline earth metal oxides are in the following ranges:

SiO_2	0-5
Al_2O_3	0-5
PbO	0-5
ZrO_2	0-5
Ta_2O_5	0-5

The optical glasses having the following ranges of composition, in percent by weight, are particularly stable against devitrification and suited for mass production.

B_2O_3	34-38
ZnO	12-16
La_2O_3	35-41
Y_2O_3	3-8
SiO_2	2-5
BaO	0-3

The optical glass of this process may be produced by using corresponding oxides, carbonates, nitrates, etc. as the materials for the respective components, taking them in desired amounts by weighing, thoroughly mixing them, placing the mixed compound in a platinum crucible which is put into an electric furnace and heated to 1200° to 1400°C, stirring and homogenizing after melting and fining, and thereafter casting the glass into an iron mold and gradually cooling.

Zirconium-Containing Borosilicate Glass

H. Bröemer, W. Huber and N. Meinert; U.S. Patent 4,200,467; assigned to Ernst Leitz Wetzlar GmbH, Germany describe a zirconium-containing borosilicate glass composition having a refractive index n_e in the range of between about $1.55 < n_e < 1.72$, an Abbe number ν_e in the range of between about $50 > \nu_e > 34$ and negative anomalous partial dispersion value $\Delta\nu_e$ between about –3.0 and –7.7 comprising the following components:

(a) From about 8.4 to 45.5% by weight SiO_2, and from about 0.9 to 33.0% by weight B_2O_3, wherein the sum of ($SiO_2 + B_2O_3$) amounts to between about 31.6 and 46.4% by weight;

(b) From about 13.5 to 18.9% by weight ZrO_2, wherein the sum of ($SiO_2 + B_2O_3 + ZrO_2$) amounts to between about 46.1 and 64.4% by weight;

(c) 0 to about 5.0% by weight Li_2O, 0 to about 22.3% by weight Na_2O, 0 to about 10.0% by weight K_2O, and 0 to about 24.0% by weight NaF, wherein the sum of the alkali metal oxides ($Li_2O + Na_2O + K_2O$) amounts to between 0 and about 21.8% by weight and the sum of (alkali metal oxides + NaF) amounts to between about 8.0 and 24.0% by weight;

(d) 0 to about 2.0% by weight BaO, 0 to about 2.3% by weight ZnO, 0 to about 0.5% by weight CdO, and 0 to about 5.0% by weight PbO, wherein the sum of the bivalent oxides (BaO + ZnO + CdO + PbO) amounts to between 0 and about 5.5% by weight;

(e) 0 to about 9.9% by weight Al_2O_3, 0 to about 24.8% by weight La_2O_3, 0 to about 23.1% by weight Sb_2O_3, and 0 to about 4.0% by weight Y_2O_3, wherein the sum of the trivalent oxides ($Al_2O_3 + La_2O_3 + Sb_2O_3 + Y_2O_3$) amounts to between 0 and about 26.9% by weight;

(f) 0 to about 5.0% by weight GeO_2, and 0 to about 1.0% by weight TiO_2, wherein the sum of the tetravalent oxides ($ZrO_2 + GeO_2 + TiO_2$) amounts to between about 13.5 and 20.6% by weight;

(g)   0 to about 5.0% by weight Nb_2O_5, and 0 to about 33.1% by weight Ta_2O_5, wherein the sum of the pentavalent oxides ($Nb_2O_5 + Ta_2O_5$) amounts to between 0 and about 36.1% by weight; and

(h)   0 to about 2.1% by weight WO_3.

The glass compositions may further comprise up to about 3% by weight of a modifying additive selected from SnO_2, P_2O_5 and mixtures thereof.

Also, described is a process for the production of a glass composition as defined above, comprising the steps of:

(1)   Heating the mixture of the aforementioned components in a platinum crucible to a temperature of between about 1345° and 1400°C to form a melt;

(2)   Refining the melt at a temperature of between about 1385° and 1450°C for a period of time of from about 7 to 15 minutes;

(3)   Homogenizing the melt with agitation at a temperature of between about 1360° and 1400°C for a period of time of from about 80 to 120 minutes;

(4)   Lowering the temperature of the melt down under agitation to a pouring temperature of between about 1135° and 1185°C within a period of time of from about 3 to 8 minutes; and

(5)   Pouring the melt into at least 1 preheated mold.

P_2O_5-PbO-Nb_2O_5 Optical Glass

K. Ishibashi and T. Ichimura; U.S. Patent 4,193,807; March 18, 1980; assigned to Nippon Kogaku KK, Japan describe an optical glass of high refractive index and high dispersion capability and which is excellent both in transmissivity and chemical durability.

The optical glass contains phosphoric acid P_2O_5, lead oxide PbO and niobium oxide Nb_2O_5 in the range of 11 to 32, 34 to 60 and 22 to 50, respectively, all being percentages by weight. As regards its optical constant, the refractive index and Abbe number range from 1.75 to 2.13 and 17 to 28, respectively.

The optical glass can be produced by obtaining component P_2O_5 from aqueous solution of orthophosphoric acid H_3PO_4, or a phosphate, for example, potassium metaphosphate, and using, as the other component, corresponding oxides, carbonates, nitrates, fluorides, etc. and, if required, adding a desired amount of defoaming agent such as arsenious acid As_2O_3 or the like, mixing them at a desired ratio to prepare a composition, placing the composition into a platinum crucible in an electric furnace heated to 950° to 1200°C, stirring the material after melting and homogenizing the same, and then casting it into an iron mold and gradually cooling the same. Fluorine is introduced as a fluoride having positive ions as a component.

The composition (% by weight), the refractive indices and Abbe numbers of embodiments of the optical glass according to this process are shown in the table below.

	.... Compositions (% by wt)								
	1	2	3	4	5	6	7	8	9	10
P_2O_5	20.54	16.79	16.45	12.92	20.12	16.14	20.39	25.04	20.63	20.19
PbO	53.82	58.06	46.57	50.79	42.19	35.52	49.31	34.25	49.01	37.66
Nb_2O_5	25.64	25.15	36.98	36.29	37.69	48.34	25.38	23.91	24.51	25.14
Li_2O	–	–	–	–	–	–	0.63	–	–	–
Na_2O	–	–	–	–	–	–	0.66	–	5.85	–
K_2O	–	–	–	–	–	–	3.63	16.80	–	7.01
TiO_2	–	–	–	–	–	–	–	–	–	10.00
n_d*	1.9753	2.0311	2.0468	2.1220	2.0181	2.0858	1.9030	1.7568	1.9041	1.9501
ν_d**	21.9	20.7	19.5	18.3	19.7	17.9	22.5	27.3	23.2	19.7

*Refractive indices.
**Abbe numbers.

Optical Glass with Low Nb_2O_5 Content

J.E. Boudot and H.J. Meyer; U.S. Patent 4,149,895; April 17, 1979; assigned to Corning Glass Works have found that glasses can be produced having indices of refraction varying between 1.675 to 1.720, densities within the interval of 3.1 to 3.4 g/cm³, Abbe numbers between 33 to 36.5, and which demonstrate very good chemical durability, as represented by a loss of weight of less than 0.01 mg/cm² when measured in the American Optical test, from compositions consisting essentially, as expressed in weight percent on the oxide basis, of:

Component	Weight Percent
SiO_2	40-46
Al_2O_3	0-3
$SiO_2 + Al_2O_3$	40-46
TiO_2	13-17
ZrO_2	4-11
MgO	0-2
CaO	3-7
BaO	8-16
SrO	0-4
CaO + MgO	4-8
CaO + BaO + SrO	15-25
Nb_2O_5	2-10
Li_2O	0-3
Na_2O	2-7
K_2O	3-8
$Li_2O + Na_2O + K_2O$	5-12

The American Optical test is the standard acid bath test for ophthalmic applications and consists in determining the decrease in weight of a glass disc, the physical dimensions of which have been carefully measured, after complete immersion thereof for 10 minutes in an aqueous 10% by weight HCl solution at a temperature of 25°C. (The test has been described in *Applied Optics Review*, 7, No. 5, p. 847, May 1968.) A weight loss of <0.5 g/cm² is deemed to successfully pass the test.

The abovecited limited range of operable compositions is particularly interesting from a practical point of view in that the content of Nb_2O_5 in the glass is low, thereby keeping the cost of manufacturing such glasses low.

Lead-Free Optical Glass

A number of glasses having optical constants expressed by a refractive index of 1.66 to 1.77 and an Abbe number of 30 to 50 have been known for a long time. Many of them, however, contain a relatively large amount of PbO, or both PbO and TiO_2. PbO has the defect of poor chemical resistance. The copresence of PbO and TiO_2 has the defect of causing a strong coloration. Moreover, such glasses tend to undergo greater phase separation as the TiO_2 content increases. In order to avoid phase separation, relatively large amounts of B_2O_3 and Al_2O_3 must be used in the glass. Consequently, this causes a reduction in the amount of SiO_2, and sufficient chemical resistance cannot be obtained.

H. Sagara; U.S. Patent 4,179,300; December 18, 1979; assigned to Hoya Corporation, Japan describes an optical glass which consists essentially of an SiO_2-Li_2O-alkaline earth metal oxide-TiO_2-ZrO_2 glass, and which has reduced coloration, superior chemical resistance, high hardness facilitating polishing, and a meltability and resistance to devitrification that are suitable for mass production. This optical glass comprises:

Component	Percent by Weight
SiO_2	30-45
B_2O_3	0-6
Li_2O	1-5
$Na_2O + K_2O$	0-4
BaO	20-40
MgO + CaO + SrO + ZnO	5-20
MgO	0-15
CaO	0-20
SrO	0-20
ZnO	0-15
ZrO_2	2-7
TiO_2	2-20
$La_2O_3 + Ta_2O_5 + Nb_2O_5 + WO_3$	0-10
La_2O_3	0-10
$Ta_2O_5 + Nb_2O_5 + WO_3$	0-4

The optical glass can be obtained by melting a mixture of materials, for example, silica powder, boric acid, lithium carbonate, sodium carbonate, potassium nitrate, barium carbonate, magnesium carbonate, calcium carbonate, strontium nitrate, zinc oxide, zirconium oxide, titanium oxide, lanthanum oxide, tantalum oxide, niobium oxide and tungsten oxide in a platinum crucible at about 1300° to 1400°C, stirring the mixture to homogenize the mixture and remove bubbles, casting the molten mixture into a mold preheated at a suitable temperature, and annealing the product.

Optical Element with Refractive Index Gradients

W. Jahn; U.S. Patent 4,177,319; December 4, 1979; assigned to Jenaer Glaswerk Schott & Gen., Germany describes a vitreous material in which an especially steep refractive index gradient can be produced by means of an ion exchange process, e.g., a salt bath.

As is known, glasses of this type may be utilized for optical systems in which the refractive deflection of a light beam is not (or not exclusively) produced by any curved surfaces but instead is produced by the medium itself. The material may be utilized in the form of rods and discs, as well as in the usual lens form.

This optical quality glass composition, suitable for the production of optical elements with a refractive index gradient Δn of at least 100×10^{-4}, is produced by utilizing an ion exchange, consisting essentially of the following synthesis composition, in percent by weight:

Component	Percent by Weight
SiO_2	56-78
MgO	6-15
Li_2O	5-14
Na_2O	3-14
K_2O	0-9

wherein the total of MgO plus Li_2O is 12 to 23%.

Particularly high refractive index gradients may be obtained with glasses of the following compositions (in percent by weight):

Component	Percent by Weight
SiO_2	65-75
MgO	9-13
Li_2O	8-12
Na_2O	4-10
K_2O	0-8

wherein the total of MgO plus Li_2O is 16 to 22%.

Example: 295.6 g of SiO_2, 41.6 g of MgO (as carbonate), 44.4 g of Li_2O (as carbonate), 18.4 g of Na_2O (as carbonate) and 1.2 g of As_2O_3, as purifier, are placed into a platinum crucible in an electric furnace at 1500°C, melted down and then maintained at this temperature for 30 minutes. The mixture is then purified for 40 minutes at 1520°C and subsequently stirred for 15 minutes while the melting temperature is reduced to 1380°C.

After the crucible has been removed from the furnace, the molten mass is poured into a preheated iron form and cooled to ambient temperature in the course of 20 hours. Glass rods with a 2 mm diameter are drilled from the glass block by means of a hollow drill. After finish grinding, these rods are cleaned in an ultrasonic bath and then subjected to ion exchange in a sodium salt bath composed of 90% by weight of $NaNO_3$ and 10% by weight of NaCl for 24 hours at a temperature of 480°C.

The refractive index gradient between the rod axis and periphery was measured as $\Delta n = 158 \times 10^{-4}$.

Infrared-Transmitting Glass

Various infrared transmitting glasses are known and are used in the lens systems of infrared, or thermal imaging systems. In these infrared imaging systems, the infrared, or thermal radiation from a scene is focussed onto a detector whose output is proportional to the change of radiation received. By scanning the scene in a raster manner, a picture of the scene can be displayed on a cathode ray tube.

The material used in the lens system is frequently germanium formed into a single lens and this is adequate for many purposes. However, for high resolution the small degree of chromatic aberration in germanium limits the optical performance of the lens. Therefore, the chromatic aberration in germanium must be corrected by means of an additional lens component made from another material. Such a material could be a glass from the GeAsSe ternary system as described in *Journal of Non-Crystalline Solids* 20 (1976) 271-283, P.J. Webber and J.A. Savage.

A.H. Lettington and J.A. Savage; U.S. Patent 4,154,503; May 15, 1979; assigned to The Secretary of State for Defence of Great Britain describe an infrared transmitting glass which has up to 30 atomic percent of telluride substituted for selenide in a glass having a composition within the following range: germanium, 10 to 35 atomic percent; arsenic, 9 to 45 atomic percent; and selenide, 45 to 70 atomic percent.

Preferably, the glass range is within the area defined and bounded by the lines joining the following compositions in the GeAsSe ternary diagram (Figure 5.1) expressed in atomic percentages $Ge_{22}As_9Se_{69}$, $Ge_{35}As_9Se_{56}$, $Ge_{35}As_{20}Se_{45}$, $Ge_{15}As_{40}Se_{45}$.

The addition of Te increases the refractive index and decreases the dispersion between 8 and 12 μm. Also, the optical absorption coefficient between 8 and 12.0 μm is reduced sufficiently that infrared transmitting fiber optic components may be made.

Figure 5.1: Preferred Area of GeAsSe System upon Which Substitution of Te Is Based

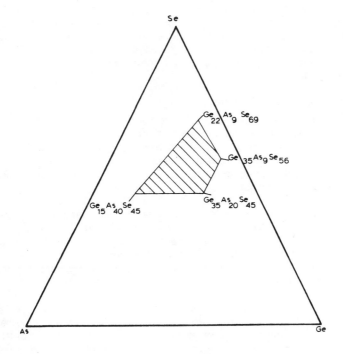

Source: U.S. Patent 4,154,503

Ion-Exchanged Antireflection Coating

L. Chang and J.D. Masso; U.S. Patent 4,168,113; September 18, 1979; assigned to American Optical Corporation have found that metal oxide films including TiO_2, CeO_2, ZrO_2, La_2O_3, Nd_2O_3, Al_2O_3, and SiO_2, when evaporated in a vacuum chamber by electron beam techniques, form hard optical coatings which are chemically stable at high temperatures. When thin films of these oxides are properly deposited to a quarter or half-wave thickness in the visible range (the geometrical thicknesses are on the order of 0.1 to 0.2 μ, or 40 to 80 microinches), they allow the potassium and sodium ions to penetrate the thickness of the film without changing the chemical composition thereof or the physical integrity of the bond between the coating and the glass substrate being ion exchange toughened.

One useful coating combination is a one-quarter wave bilayer antireflection coating. This coating design has been chosen because it allows the reflectance to be decreased to near 0% at the center wavelength and the total optical thickness of the coating combination is less than two quarter waves. For a conventional crown glass (n = 1.523) substrate, for example (the preferred embodiment), the antireflecting coating formulation is G 0.26H 1.33L air, wherein G denotes glass substrate, 0.26H denotes 0.26 quarter wave optical thickness of a high index material, and 1.33L denotes 1.33 quarter wave optical thickness of a low index material.

According to the preferred embodiment, the coating materials are CeO_2 (n = 2.34) as the high index material and SiO_2 (n = 1.46) as the low index material. Antireflective coatings made with this composition have demonstrated their optical function and ability to withstand the ion exchange process without being damaged. Other coating combinations using the abovementioned oxide filming materials can include various bilayer, trilayer and other multilayer antireflection coating designs.

Hardened Circular Lens Element

It has been found that it is not possible to harden round lens blanks and thereafter edge the hardened lenses, i.e., adapt them to the shape of the eyeglass (spectacle) frame, by removal of material from the edge. Upon such removal of material, the stress conditions which have been built up upon the hardening are drastically changed, leading, in by far the majority of cases, to the breaking of the lens during the edging.

For this reason, it has been necessary to effect the hardening of eyeglasses after the edging has been effected by the optician. This means either a high expense for investments for the optician since he must have in his shop apparatus for the hardening of eyeglasses which is generally utilized insufficiently, or a large amount of time is required if the optician gives the eyeglasses to the eyeglass factory for hardening after the edging.

B. Gänswein, E.J. Daniels, H. Schürle and K. Grosskopf; U.S. Patent 4,178,082; December 11, 1979; assigned to Carl Zeiss-Stiftung, Germany describe hardened, nonedged (i.e., circular) eyeglasses (lens elements) which enable the optician to effect edging by customary methods without the glass breaking and without reducing the strength achieved as a result of the hardening. This purpose is achieved in the manner that during the hardening the action of the hardening medium is limited to a central region of the uncut circular glass, the diameter of the region being smaller than the diameter of the ultimately desired edge contour of the eyeglass. The eyeglass is therefore hardened only in a central region which is surrounded by a nonhardened annular zone.

Upon the subsequent edging, only material in the nonhardened annular zone is removed, i.e., the actually hardened central region remains unaffected. In this way, a bursting of the eyeglass and a decrease in strength are avoided.

It is particularly advantageous to effect the hardening chemically. For this purpose, an annular layer of a masking substance which impedes ion exchange is formed on both sides of the eyeglass to be hardened so as to surround the central region; the glass is thereupon introduced as a whole into a bath of hardening salts and maintained therein for a predetermined period of time and then, after the removal of the glass from the bath, the hardening salt and the masking substance are washed off.

Molding Surface of SiC or Si₃N₄

G.E. Blair, J.H. Shafer, J.J. Meyers, and F.T.J. Smith; U.S. Patent 4,139,677; February 13, 1979; assigned to Eastman Kodak Company have found that glass elements, having high surface quality and high surface accuracy and, therefore, requiring no further preparatory operations such as grinding or polishing, can be prepared by molding glass against a molding surface formed from either silicon carbide or silicon nitride. The molding surface can be formed from a solid body of the particular material or from a layer of the material on a substrate.

In either case, the material must be of sufficient thickness that the molding characteristics of the molding surface are exclusively attributable to the silicon carbide or silicon nitride; preferably such a layer should be at least 10 μ thick. The use of a silicon carbide or silicon nitride molding surface is particularly necessary to the process because it has been found that certain other refractory materials, such as, for example, aluminum oxide or boron nitride, could not be satisfactorily employed as molding surfaces.

In one embodiment of the process, the molding process comprises the steps of placing a portion of heat-softened glass in a mold having silicon carbide or silicon nitride molding surfaces such as described above, pressing the glass against the molding surfaces until the glass conforms to the shape of the mold, cooling the glass and mold, and removing the glass element from the mold. In another embodiment, the molding process comprises the steps of placing a portion of glass in a mold having the abovementioned silicon carbide or silicon nitride molding surfaces, heating the mold to soften the glass, pressing the glass against the molding surfaces until the glass conforms to the shape of the mold, cooling the glass and mold, and removing the glass element from the mold.

In yet a third embodiment, the molding process comprises the steps of placing a portion of glass in a transfer chamber having silicon carbide or silicon nitride walls, heating the chamber to soften the glass, applying pressure so that the heat-softened glass is transferred through sprues into mold cavities defined by molding surfaces formed of silicon carbide or silicon nitride until the glass conforms to the shape of the mold cavities, cooling the glass and molding surfaces, and removing the glass element from the mold.

FIBER OPTICS

Method of Improving Cross-Sectional Circularity

Two fabrication processes are found to yield particularly pure fibers of loss as low as 2 dB/km. One process is commonly referred to as the soot deposition technique. The other is known as the modified chemical vapor deposition technique.

In both of these techniques, a hollow glass cylinder may be formed. Ordinarily the cylinder has at least two compositional regions. The interior region will ultimately form the core of the optical fiber through which the optical radiation will pass. The exterior region forms the cladding for the optical fiber. The remaining critical step involves pulling this relatively large diameter (5 to 25 mm) cylindrical preform into a relatively small diameter (5 to 100 μ) fiber. Prior to pulling the preform into a fiber, the preform is usually collapsed to a smaller diameter, or preferably into a solid cylindrical mass.

During both preform fabrication and preform collapse, noncircularities are introduced into the otherwise circular preform cross section. If these asymmetries are not removed before pulling the preform into a fiber, they will be reflected in the cross section of the resultant fiber yielding a noncircular optical wave guide cross section. Such asymmetrical fibers are difficult to splice to other optical fibers with different cross-sectional properties, and may yield a fiber with degraded pulse dispersion properties.

W.G. French and G.W. Tasker; U.S. Patent 4,154,591; May 15, 1979; assigned to Bell Telephone Laboratories Incorporated describe an improved technique for producing optical fibers with circular cross sections. In this process, a hollow cylindrical preform is collapsed while under slight positive pressure. This may be accomplished by at least partially sealing the preform at one end and statically pressurizing it through the other end. It was found that under appropriate conditions of positive pressure, the preform collapses upon heating into a circular structure rather than expanding. The technique not only prevents the introduction of additional noncircularities but, in addition, removes noncircularities that may have been present prior to the collapse.

Figure 5.2 is a schematic representation of an otherwise standard glass lathe modified so as to be applicable to the process. In Figure 5.2, **11** is the standard supporting structure associated with glass lathes; **12** is a hollow glass cylinder placed in the lathe prior to collapse. The cylinder is a preform which will subsequently be drawn into an optical fiber.

At **16**, one end of the glass cylinder is shown to be at least partially closed off so that an appropriate pressure may be developed within the cylinder during collapse. The tube may be closed off by applying heat to one end and allowing it to melt and fuse in the ordinary fashion known to those skilled in the glass blowing arts. However, other alternative techniques may be used to seal off the downstream end of the tube, and this process is not restricted to any particular method of sealing the tube.

Once the tube is sealed off, a positive pressure is introduced into the cylinder, for example, by means of an inlet at **17**. This positive pressure is measured by the gauge at **14** and may be lowered by allowing an appropriate escape of gas through the valve at **15**. The gas used to pressurize the cylinder is restricted only by obvious considerations. For example, it is clear that one requires a non-explosive gas, and, in the case of the collapse of optical fiber preforms, it is clearly advantageous to operate with a gas, such as oxygen, nitrogen, or the noble gases, that does not degrade the transmission characteristics of the preform material.

Figure 5.2: Glass Lathe Employing Positive Pressure

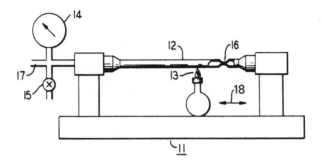

Source: U.S. Patent 4,154,591

The heat source **13** is applied to the cylinder during collapse. While a broad heat source which extends over the entire cylinder may be used, the heat source in the figure is of limited extent and is not maintained stationary but rather traverses the extent of the cylinder during the collapse. The means for traversal may be the standard screw mounted table associated with such lathes and indicated by **18**. However, any other means for providing relative translation between the heat source and the preform may be utilized. The heat source is shown schematically to be a flame burner; however, this too is not critical and any appropriate heat sources may be utilized, e.g., multiple burners, an electric resistive heat source, or an oven. Depending upon the composition of the cylinder, the heat source may traverse the cylinder one or more times in order to effect adequate collapse and maintenance of, or substantial improvement in, the circularity of the cylindrical cross section.

A central idea associated with this process, and which may serve to guide those who practice it, is that upon heating the glass cylinder will collapse despite the fact that it is under positive pressure. Of significant importance is that under the conditions described the cylinder collapses into a structure of approximately circular cross section even if the starting structure had serious cross-sectional distortions.

Drawing Filaments from Soot Preforms in Helium Atmosphere

A.C. Bailey; U.S. Patent 4,157,906; June 12, 1979; assigned to Corning Glass Works describes a method of forming glass filaments by the flame hydrolysis process. Briefly, this method comprises applying a first coating of glass soot to a surface of a substrate to form a soot preform. The preform is heated to a temperature that is sufficiently high to cause an end thereof to be consolidated while it is simultaneously drawn into a filament.

The method can be employed to form optical wave guides having relatively high numerical apertures. A second coating of glass soot is applied over the outside peripheral surface of the first soot coating, the refractive index of the soot of the second coating being less than that of the first coating. As this composite preform is consolidated and drawn, an optical wave guide filament having a solid cross section is formed.

Optical wave guide soot preforms are conventionally prepared in accordance with the following methods. A coating of glass soot is applied to a cylindrical mandrel by means of a flame hydrolysis burner. Fuel gas and oxygen or air are supplied to the burner to produce a flame. A gas-vapor mixture is oxidized within the flame to form a glass soot that leaves the flame in a stream, which is directed toward the mandrel. The mandrel is supported by means of a handle and is rotated for uniform deposition of soot. This method can be employed to produce either step index or gradient index wave guides.

A second coating of soot is applied over the outside peripheral surface of the first coating. In accordance with well-known practice, the refractive index of coating No. 2 is made lower than that of coating No. 1 by changing the composition of the soot being produced. This can be accomplished by changing the concentration or type of dopant material being introduced into the flame, or by omitting the dopant material. The mandrel is again rotated and translated to provide a uniform deposition of coating, the composite structure including the first coating and second coating constituting an optical wave guide soot preform.

In accordance with this method, the separate consolidation process is eliminated, and the overall process of forming a filament from a soot preform is greatly simplified as illustrated in Figure 5.3. The mandrel is preferably removed from the soot preform **30** and one end thereof is attached to a handle. The soot preform is then inserted into the top of a draw furnace, and a filament **38** is drawn therefrom. For the sake of simplicity, only the draw furnace muffle **40** is shown.

Handle **42** may consist of a low expansion glass tube **44** to which a short length of smaller diameter tubing **46** is attached. Quartz is a particularly suitable material for tubing **46** since it can withstand the 1600° to 1850°C draw temperature without excessive distortion, and since it does not add impurities to the blank. In one specific embodiment tube **44** was a ½" outside diameter tube of high silica content glass and tube **46** consisted of a 2" long section of ¼" diameter quartz tubing having two 0.5 mm bumps **48** of quartz flame-worked onto opposite sides of the end thereof that is inserted into the soot preform.

After the tube is inserted, it is rotated about 90° to lock it into the preform. Loose soot is then blown from the outside and inside surfaces of the soot preform with dry, filtered nitrogen. The soot preforms employed in this embodiment had an outside diameter of 32 mm in order to provide sufficient clearance between the preform and the muffle which had an inside diameter of about 44 mm.

Handle **42** is placed in a chuck above the draw furnace, and the preform is fed down into the furnace muffle **40** to a position just above the hottest zone. The top of the muffle is sealed. Muffle gas introduced at the top of the muffle flows down over the preform **30** as indicated by arrows **52**, as well as through the interstices of the preform to flush gases therefrom during consolidation of the soot. The muffle gas is exhausted at the bottom of the muffle. The muffle gas is one that will allow the soot to consolidate in a bubble-free manner and which will not interact with the glass soot constituents in a way that will harm the optical properties of the resultant filament.

Figure 5.3: Portion of Draw Furnace for Drawing Filament from Soot Preform

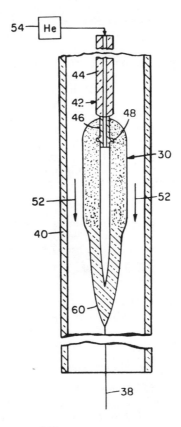

Source: U.S. Patent 4,157,906

Depending upon the particular dopants employed, an oxidizing or reducing condition may be desirable. The preferred muffle gas is one that is rich in helium which can readily pass through the interstices of the porous soot preform to purge residual gas therefrom. The phrase helium-rich atmosphere, as used herein, means one that contains at least 95% helium. Other gases such as oxygen, nitrogen, argon, neon and mixtures thereof may be employed to flush gases from the preform during consolidation. A halogen, preferably chlorine, is sometimes employed to dry the preform to an extent greater than that which can be achieved by the use of helium alone, combinations of helium and chlorine compounds being disclosed in U.S. Patent 3,933,454 for soot consolidation purposes. During this initial period of flushing gas from the preform and muffle, the feed mechanism is stopped so that the preform does not advance into the hottest zone.

During this time, flushing gas may also flow from source 54 through the handle and into the center of the preform to provide additional flushing thereof. Adequate flushing may be provided by flowing pure helium into the top of the muffle at a rate of 27 cfh for 15 minutes while flowing helium into the center of the preform at a rate of 3 cfh. Optimal flow rates of these flushing gases depend upon such factors as blank feed rate, consolidation temperature, specific flushing gas composition and the like.

The preform is then advanced into the hot zone of the furnace, whereby the soot begins to consolidate as indicated at region 60. The temperature in the hot zone is preferably such that the viscosity of the core glass is between 10^8 and 10^3 poises. The desired viscosity depends upon the filament draw rate. Draw rates between 0.75 and 100 fpm have been employed, but this method is not limited to this range. If helium source 54 is employed, it is turned off as soon as the tip of the preform begins to consolidate. The tip of the consolidated glass preform is contacted by the end of a low expansion glass rod to initiate drawing of the filament. The preform is continuously fed into the hot zone of the furnace where it continuously consolidates in a helium rich atmosphere and is drawn into a filament.

Since materials evolved from the aperture-forming surface must be permitted to escape, the end of tube 44 remote from the preform should be opened after source 54 is turned off so evolved materials can vent to the atmosphere. Flushing gas represented by arrows 52 continues to flow until the entire preform is consolidated.

In a similar process, *A.C. Bailey; U.S. Patent 4,154,592; May 15, 1979; assigned to Corning Glass Works* describes a method of drawing filaments which comprises flowing a helium-free gas through the muffle until about 10 cm of the draw blank remains, and thereafter flowing helium-containing gas through the muffle.

Drying of Glass Soot Preform

D.R. Powers; U.S. Patent 4,165,223; August 21, 1979; assigned to Corning Glass Works describes an effective and economical method of removing residual water from a flame hydrolysis-deposited glass soot preform during the consolidation process.

This process conventionally comprises the steps of depositing on a starting member a coating of flame hydrolysis-produced glass soot to form a soot preform, removing the starting member to form an aperture, consolidating the soot preform to form a dense glass blank, and drawing the blank to form a filament. The consolidation step conventionally comprises subjecting the soot preform to a temperature in the consolidation temperature range for a time sufficient to permit the soot particles to fuse and consolidate, thereby forming a dense glass which is free from particle boundaries. While the preform is heated to its consolidation temperature, a stream of an atmosphere including a drying agent flows into the aperture and through the porous preform, thereby removing water from the preform while the soot is consolidated. The process of fusing glass soot particles formed by flame hydrolysis is sometimes referred to as sintering even though no particle boundaries remain.

In accordance with this process, the drying gas mixture comprises chlorine and oxygen in amounts sufficient to substantially eliminate water from the preform while not removing excessive amounts of dopant oxide therefrom. During the drying process, the chlorine leaches some of the dopant oxide from the preform, especially from the central portion thereof adjacent to the aperture. By compensating for this leaching action by depositing an excessive amount of dopant oxide at the central portion of the preform during the soot deposition process, dry optical wave guide filaments are produced having a predetermined refractive index therethrough.

In accordance with a preferred embodiment, the refractive index of the soot preform decreases radially from a maximum value at the central portion thereof. The dry, dense glass blank, which is formed by the consolidation step, is heated to the drawing temperature and drawn to form a high bandwidth optical wave guide filament having a desired gradient refractive index.

Vitrification of Soot Layer in Inert Gas Atmosphere

T. Akamatsu and K. Okamura; U.S. Patent 4,149,867; April 17, 1979; assigned to Fujitsu Limited, Japan describe a method of forming a preform rod by the chemical vapor deposition process, characterized in that vitrification of the soot layer, resulting from oxidation of the starting materials in gas phase, is carried out in an atmosphere of an inert gas which is easily dissolved in a fused glass, such as a rare gas, preferably helium.

A preferable soot to be produced by the method comprises silicon dioxide, a first dopant for effecting an increase of the refraction index of the silicon dioxide and a second dopant for effecting reduction or lowering of the vitrification temperature of the silicon dioxide or soot.

A preferable first dopant is germanium dioxide and a preferable second dopant is phosphorus pentoxide.

A preferable method of this process comprises: bubbling at least one carrier gas through starting materials in liquid phase to generate the starting materials into a gas phase; introducing the gas phase starting materials with the carrier gas into a tube reactor of silica glass; introducing oxygen as a reactant gas into the reactor; heating the reactor at a local zone, the zone being shifted toward the output end of the reactor along the reactor length, thereby effecting reaction of the gas phase starting materials with the oxygen to produce a soot to be deposited in the form of a layer on the inner surface of the reactor and effecting vitrification of the resultant soot layer by the fusing of the soot.

The improvement of this process resides in the fact that an inert gas, which is easily dissolved in the fused glass, independent of the carrier gas, is introduced into the reactor. Preferably, the flow rate of the oxygen is set at a value as low as possible, but enough to effect the reaction with the starting materials in gas phase, and the gas mixture of the starting materials, the oxygen, the carrier gas and the inert gas is controlled in such a manner that it has a flow rate capable of providing a soot layer with a predetermined thickness.

Production of Glass Films by Thermal Decomposition

A cladded light-conducting fiber with a stepped index profile and a cladded light-conducting fiber with a gradient index profile are known. A stepped index profile means that the refractive index abruptly changes its value at the boundary surface of the core and cladding so that the core has a higher refractive index than the cladding. A gradient index profile means that the light-conducting fiber has a higher refractive index in the vicinity of the fiber axis and that this refractive index continuously decreases with an increasing distance from the fiber axis. Both types of fibers can have an additional synthetic material casing which is to protect the fiber particularly from mechanical damages. In each type of light-conducting fibers, light can be guided or conveyed over very long distances.

H. Aulich, H. Pink and J. Grabmaier; U.S. Patent 4,141,710; February 27, 1979; assigned to Siemens AG, Germany describe a method of producing light-conducting fibers, particularly cladded light-conducting fibers, with a low attenuation. This task is accomplished by providing a base member of an optical material, depositing at least one optical-material-forming layer from a liquid phase onto a surface of the base member, and subsequently transforming each of the layers into a film of optical material.

Advantageously, the method is inexpensive and has an additional advantage that it can be easily accomplished. The method may be used for providing light-conducting fibers having either a stepped index profile or gradient index profile.

In this process, an optical material is meant to include glass-like synthetic material as well as glass. When the optical material is glass, a glass base member may have glass-forming layers deposited thereon and these layers are then subsequently transformed into glass film by means of a thermal decomposition of the layers. When a glass-like synthetic material is used as the optical material, low polymers or monomers are dissolved in a solution medium and are brought to polymerization during the vaporization of the solution medium, for example, by means of increasing the temperature, ultraviolet radiation or gamma radiation.

The base member may be either a tube or rod. When the base member is a glass rod, the depositing proceeds on an external surface thereof and, when the base member is a glass tube, the deposition can also proceed on the interior surface of the tube. After applying the desired thickness of the film on the tube, it is collapsed into a rod. The composite rod and film are drawn into a light-conducting fiber such as a cladded light-conducting fiber.

In addition, the process may be used for applying a cladding directly on a core of a fiber. This is accomplished by drawing a glass fiber core and subsequent to drawing the core, depositing the layer of glass forming material and then transforming the material into a glass cladding.

Thus, light-conducting fibers formed of glass or of synthetic materials can be produced with either a stepped index or gradient index profile.

The preferred sample embodiment is described as follows. In one sample embodiment, a liquid phase consisting of water-free solution in which compounds of glass-forming oxides of elements selected from a group consisting of Se, Te, P, As, Si, Ge, Sn, Pb, Ti, B, Al, Mg, Ca, Sr, Ba, Li, Na, K and Rb are contained. The glass-forming layers are deposited from these solutions and the layers are then transformed into glass films by means of thermal decomposition.

In accordance with the glass composition desired, the solution to be utilized is to contain the desired amount of one of the abovementioned elements as:

(a) Pure compounds with low monocarboxylic or dicarboxylic acids, wherein the number of carbon atoms in these acids is not to be higher than four;

(b) Metal halides whose anions are partially substituted by the acid radical selected from a group consisting of monocarboxylic and dicarboxylic acids; or

(c) Metal halides whose anions are partially substituted by acid radicals selected from a group consisting of monocarboxylic and dicarboxylic acids and additionally partially substituted by radicals selected from a group consisting of hydroxyl, and alcoholate radicals of lower alcohols which have less than four carbon atoms per molecule.

A solution medium for the production of the solutions, which contain the compounds under (a) through (c) as mentioned above, is selected from a group consisting of lower ketones, lower alcohols, esters of lower alcohols with lower carboxylic acids and combinations thereof.

Apparatus Employing Rotating, Cylindrical Crucible

In a variation of the previous process, *H. Aulich, J. Grabmeier and H. Pink; U.S. Patent 4,173,459; November 6, 1979; assigned to Siemens AG, Germany* describe a method comprising providing a rotating, heatable crucible having a cylindrical interior wall, successively depositing optical-material-forming layers from a liquid phase with the first layer being deposited on the interior wall of the rotating crucible and the subsequent layers being deposited on the previously deposited layers, transforming each layer into a film of optical material to form a workpiece having a plurality of concentric, cylindrical films of optical material; and subsequently pulling a fiber from the workpiece.

Referring to Figure 5.4, a rotating crucible 20 is provided. The crucible 20 can consist of either platinum or iridium and other materials are also possible; however, it is necessary that the materials of the crucible do not enter into any undesirable reaction with the optical materials which are disposed therein. The crucible can be heated inductively as, for example, by an induction heating coil 5. The crucible 20 is also sheathed with a heat insulation layer 21 and will rotate about an axis such as the axis of pin 19.

To form a layer on an interior wall of the crucible 20, a solution, which is described in the previous process and is contained in a supply container 2, is guided through a valve 3 and a nozzle so that it is applied on the rotating interior wall of the crucible 20. A thin liquid film or layer will be formed by the rotation of the crucible on the total interior wall with the thickness of this layer being essentially determined by the rotational velocity of the crucible, the wettability of the material-forming layer, and the viscosity of the solution.

The rotation of the crucible will cause the layer to have an even thickness. To ensure a coating or layer with an even thickness, it is desirable to feed a surplus of the solution in container 2 into the crucible 20. The extra solution or surplus is collected in a container 4 after it flows through holes such as 22 in the crucible base. The collected surplus in the container 4, if desired, can be reused. As soon as a sufficiently thick glass-forming layer is deposited on the interior wall of the crucible 20, the valve 3 is closed and the transformation, which may be by thermal decomposition, of the layer on the interior wall of the crucible proceeds so that a glass-like film is formed on the interior wall.

**Figure 5.4: Glass-Forming Apparatus Having
Rotating, Heatable Crucible**

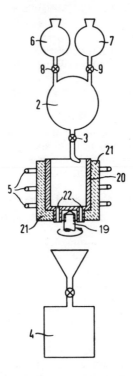

Source: U.S. Patent 4,173,459

Prior to starting the step of thermal decomposition, the solvent of the layer should be removed. If the crucible has been placed in a vacuum chamber, the chamber will be temporarily evacuated in order to vaporize the solvent from the deposited layer. It is also possible to blow oxygen into the crucible 20 after closing the dosage valves 3 so that the solvent remaining in the layer is oxidized and removed therefrom.

Whether the solvent was removed by a flow of oxygen or by evacuation, after its removal, the layer is heated by means of the induction coils 5 so that the thermal decomposition can occur. The temperature produced by the induction coil 5 is selected in such a manner that glassy or vitreous blister-free glass films are formed on the interior wall of the crucible during the thermal decomposition of the layer. It is also desirable to maintain the rotation of the crucible during the thermal decomposition to guarantee that the glass-forming layer retains its even thickness.

After producing the first layer, a second layer is then formed on the first layer by repeating the above described operations of depositing the glass-forming layer and then transforming the layer into the film.

The coating process described hereinabove can be repeated in multiple operating cycles so that the glass films of arbitrary thickness or multiple glass films composed of various individual glass films can be produced.

The pulling or drawing of a glass fiber from the workpiece can proceed in the following manner. The crucible 20 is placed in rotation and heated to a temperature at which the glass fiber can be stretched or pulled. Due to the rotation of the crucible, the individual concentric glass layers of the workpiece will not alter their position. A glass fiber can now be directly drawn or pulled out of the rotating crucible by means of a drawing mandrel.

Core Material of Graded Composition and Thickness

M.M. Ramsay and P.W. Black; U.S. Patent 4,140,505; February 20, 1979; assigned to International Standard Electric Corporation describe a method of making a glass optical fiber including the steps of making an optical fiber preform and of drawing the preform into fiber. The core material of the preform is formed by a thermally induced chemical vapor reaction under conditions providing a variation in both core material thickness and core material composition along the length of the preform chosen such that the fiber drawn therefrom has a central core region of substantially constant core thickness and composition and graded core end regions where the core is smoothly graded in thickness and composition to enlarged core thickness.

The composition grading is matched with the thickness grading in such a manner as to provide the fiber with a normalized frequency held within limits which provide the central core region and the two graded core end regions with the same single bound nonradiating mode or the same set of bound nonradiating modes. Preferably, the grading is such that the normalized frequency is held substantially constant over the three portions of the fiber.

A fiber produced by this method can be designed to combine the advantage of a large cored small refractive index difference fiber so far as coupling efficiently between butt jointed fibers is concerned, while benefiting, over the length of the central portion, which may be very long compared with the graded portion, from the relatively small radiative bending losses of a small cored large refractive index difference fiber.

A parabolic relationship between the core diameter and refractive index difference is required to maintain a constant normalized frequency. Thus, for instance, if the core thickness is to be increased by an order of magnitude from about 3 μ at the central core region to about 30 μ at the ends of the graded core end regions, the refractive index difference must decrease in a corresponding manner by two orders of magnitude, that is from a difference of about 1 to 0.01%. In principle, the grading can be over a distance of no more than a few centimeters, but, in practice, since the fiber is drawn from a preform, and this normally involves an extension of at least a thousand-fold, the grading is longer, and typically extends over several meters or even several tens of meters. Within the graded core end regions the fiber will have the sensitivity to bending losses appropriate to the core size in those regions, but since these graded core end regions will normally represent a very small proportion of the total length of the fiber, the overall sensitivity to bending losses of the fiber considered as a whole will approach the lower limit set by the size of the core in the central core region.

Since both the temperature and the rate of relative motion affect the deposition rate, the refractive index and thickness of the deposit may be controlled by suitable manipulation involving the variation of either or both of these two parameters. The particular way in which the variation in core thickness and composition is realized is not critical. What is important is that the core thickness be increased as the ends of the preform are approached and that the core composition be varied in relation to the thickness so as to maintain a substantially constant normalized frequency.

Optical Wave Guide with Diameter Variations

T.C. Kurth; U.S. Patent 4,163,370; August 7, 1979; assigned to Corning Glass Works describes a process whereby an optical wave guide having variations in diameter is produced by controlling the drawing rollers. Predetermined diameter variations along the length of an optical wave guide reduce dispersion in multimode wave guides. These diameter variations are produced by varying the rotational speed of the pulling rollers, by varying the diameter of the pulling rollers, by moving the pulling rollers axially or by moving a tensioning roller during the drawing process.

Glass Rod Formed by Gaseous Deposition onto Rotating Base Plate

K. Fujiwara, G. Tanaka and S. Kurosaki; U.S. Patent 4,135,901; January 23, 1979; assigned to Sumitomo Electric Industries, Ltd., Japan describe a method of manufacturing glass rods for optical wave guides. Its basic principle consists in producing a glass rod in a vitreous state or a powdery state while rotating it on a base plate. The distribution of the refractive index in the rod can be varied in the radial direction in accordance with a prescribed optimum distribution for the specific application purpose of the optical fiber.

In order to produce such a cylindrical rod, several gaseous mixtures of glass-forming compounds and oxidizing gas are fed into a high temperature flame.

Glass is formed through chemical reaction in the mixtures at the prescribed high temperature. The composition of the mixtures to be ejected by the nozzles of the burners varies in the radial direction of the glass rod, resulting in the radial variation of the glass composition. This glass composition is deposited on the top or end of the glass rod while it is rotating around its axis, and which is grown on the rotating base plate. The top of the rotating glass rod is heated by a clean flame such as plasma or hydro-oxygen or a clean photo-energy source such as a CO_2 laser, or ultraviolet rays, or infrared rays.

The oxidation takes place in the mixtures on the heated top of the glass rod, forming a new glass material to be deposited on the top of the glass rod. This glass is piled up in a vitreous state or in a powdery state on the top of the previously formed glass rod. Thus a glass rod is continuously formed on the base plate.

Single Polarization Optical Fibers

I.P. Kaminow and V. Ramaswamy; U.S. Patent 4,179,189; December 18, 1979; assigned to Bell Telephone Laboratories, Incorporated have found that orthogonally polarized waves are more efficiently decoupled in a wave guide that is fabricated in a manner so as to deliberately enhance stress-induced, or strain birefringence. This behavior is accomplished by introducing a geometrical and material asymmetry in the preform from which the optical fiber is drawn such that the resulting strain birefringence Δn is advantageously greater than 5×10^{-5}. The resulting beat period, L, for such a wave guide is less than 20 mm at 1 μm wavelength and less than 10 mm at 0.5 μm, where $L = 2\pi/\Delta\beta$, and $\Delta\beta$ is the difference in propagation constants for the two orthogonal directions of wave polarization of interest.

The strain birefringence Δn is given by

$$\Delta n = (n^3/2)(p_{11}-p_{12})(\alpha_1-\alpha_2)\Delta T$$

where n is the refractive index, p_{11} and p_{12} are the photoelastic constants of the wave guide material. $\Delta T = T_a - T_b$, where T_a is the operating ambient temperature, and T_b is approximately equal to the softening temperature of the material, and α_1 and α_2 are the thermal expansion coefficients of the jacket and wave guide regions, respectively. For simplicity, α_1 and α_2 are assumed to be independent of temperature when making estimates.

Example 1: For a 5 mol % B_2O_3-SiO_2 cladding, the calculated Δn is 1×10^{-4}, where $n \approx 1.5$, $(p_{11} - p_{12}) \approx 0.15$, $(\alpha_1 - \alpha_2) \approx -5 \times 10^{-7}/°C$ and $\Delta T \approx -850°C$.

Example 2: For a 25 mol % GeO_2-SiO_2 cladding, the calculated Δn is 4×10^{-4}, where $n \approx 1.5$, $(p_{11} - p_{12}) \approx 0.15$, $(\alpha_1 - \alpha_2) = -1.6 \times 10^{-6}/°C$ and $\Delta T \approx -1000°C$.

Example 3: For a 12 mol % P_2O_5-SiO_2 cladding, the calculated Δn is 4×10^{-4}, where $n \approx 1.5$, $(p_{11} - p_{12}) \approx 0.15$, $(\alpha_1 - \alpha_2) = -1.4 \times 10^{-6}/°C$ and $\Delta T \approx -1200°C$.

In each of the preceding examples, the core and cladding are assumed to have approximately the same thermal properties.

Elliptical Core Single Mode Fiber

In the utilization of single mode fibers for high band-width information transfer or for phase modulation applications, it is desirable to maintain the plane of polarization of polarized light or to force the polarization of unpolarized light. A single mode fiber with an elliptical core will maintain the plane of polarization of the injected light.

M.S. Maklad; U.S. Patent 4,184,859; January 22, 1980; assigned to International Telephone and Telegraph Corporation describes a method for fabricating elliptical core single mode fibers by first preparing a substrate hollow silica tube to remove contaminants which may be present both on the inside of the tube's bore and outside of the tube. The tube is then fire polished inside and out while the bore is under positive pressure to smooth out and remove the moisture from the bore of the tube. This is accomplished by mounting the tube or substrate on a rotating lathe while accomplishing fire polish inside and out while maintaining positive pressure on the inside of the tube. While the tube is rotating, a barrier layer of silica is deposited on the bore by passing a mixture of dry $SiCl_4 + O_2$ while heating the tube with a traversing oxy-hydrogen flame to produce a glassy film barrier layer on the bore of the tube, for example, about 2 μm thick barrier layer, at the rate of about 1 μm per pass.

After the barrier layer of pure SiO_2 has been deposited, a cladding layer of borosilicate glass doped with phosphorus pentoxide (P_2O_5) is laid down in approximately 50 passes to produce 50 layers, for a cladding layer thickness of about 1 mm. This is followed by depositing the core composition of germania-doped silica. During this period of time, starting with the mounting of the tube on the lathe through completion of the deposit of the core layer, the tube is rotated by the glass lathe and the flame or heat source is traversed from one end of the tube to the other.

The next step begins a collapsing process to produce a preform which can be used to produce a single mode optical fiber having an elliptical core. First, the inside of the tube is subjected to a partial vacuum and rotating of the lathe stopped. One side of the tube is subjected to heat, for example, an oxy-hydrogen flame, so that the heat will be concentrated at the interface between the heat source and the side of the coated tube closest to it. The heat source is such that the collapsing temperature is reached at that area of the coated tube closest to it, following which the flame or other heat source is longitudinally traversed at a speed allowing for a partial collapse of one side of the coated tube until the desired significant collapsing has been achieved all along one side of the coated tube. At this point, the coated tube is rotated 180° and the opposite side is subjected to the effect of the heat source on the side of the tube adjacent it to achieve the desired substantial degree of partial collapse.

Once the two sides of the coated tube have been partially collapsed, then the lathe is started up to rotate the partially collapsed coated tube and heat is supplied so that the entire periphery of the tube is heated to cause complete collapse of the bore and to produce a cylindrically shaped solid optical fiber preform having an elliptically shaped core. Preferably, the conditions of partial collapse are associated with the composition of the core and cladding layers so as to produce a single mode core having a substantially elliptical cross-sectional shape in which the major axis to minor axis ratio is about 2:1.

Next step is to draw the fiber in the usual manner thus producing an optical fiber of the single mode type which has a substantially elliptical shaped core.

High Temperature Internal Cladding Method

G. Gliemeroth and L. Meckel; U.S. Patent 4,199,335; April 22, 1980; assigned to Jenaer Glaswerk Schott & Gen., Germany describe a process which makes it possible to obtain preforms with exactly circular cross sections without having to separate the cladding step from the collapsing step. This process comprises precollapsing the tubular substrate during the cladding of the tube prior to the final collapsing step. The process utilizes, preferably and in contradistinction to the conventional internal cladding methods, tube temperatures of above 1850°C. Thus, it is termed the Schott High Temperature CVD Process.

The basic aspect of this process is based on the realization that it is possible to avoid deformations of the tube if certain dimensional conditions are maintained, and that this possibility exists in spite of the use of high temperatures of above 1850°C and the concomitant local softening of the silicate glass tube. Assuming an inner tube radius of R_i prior to cladding, this inner radius will change under the process conditions to an inner tube radius of r_i after the cladding step and prior to the actual collapsing procedure.

Preferably, tubes are used as the starting material which have a ratio of $R_o/R_i > 1.15$, e.g., 1.15 to 1.3, most preferably 1.16 to 1.24, and preferably have a precollapsing ratio of $R_i/r_i > 1.7$, e.g., 1.7 to 2.5, most preferably 1.75 to 2.0.

In the CVD method, the desired metal oxide glass layers are deposited on the silica substrate by decomposition of a gaseous starting material such as a halide of the metal cation. One apparatus which can be used is schematically illustrated in Figure 5.5.

Figure 5.5: Apparatus for High Temperature Internal Cladding Method

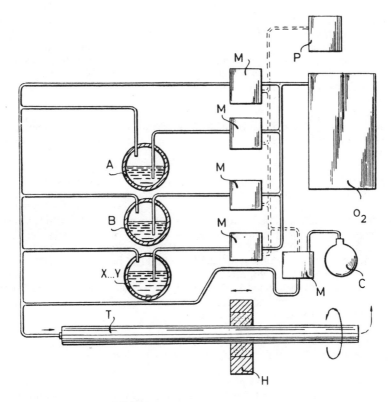

Source: U.S. Patent 4,199,335

Figure 5.5 shows a gas bottle containing a carrier gas, O_2, from which a stream of O_2 flows through a flow volume regulator **M**. The correctly measured dose of oxygen is then conducted, either directly or through charging or doping vessels, into a glass tube **T**. The charging or doping vessels **A, B, X...Y** are filled with liquid halide compounds so that the oxygen carrier gas stream can be doped or charged with molecules of such halide compounds before it enters into the tube **T**. The flow-volume regulators **M** may be controlled by a programmer **P** to provide the correct dosage in each case. Such volume or flow regulator apparatus may also be provided to control gaseous as well as, where appropriate, compressed components **C**.

The appropriate relative flow rates corresponding to any desired coating layer composition can be easily determined by fully conventional considerations. The gaseous streams are combined forwardly of the tube **T**, which is mounted on a synchronous lathe made for use with glass, flow through the tube and pass into the effective temperature zone of a suitable heat source **H**, i.e., a burner travelling along the longitudinal axis of the tube on its outer periphery. It is here that decomposition takes place, namely, pyrolysis to form metallic oxide glass layers.

The excess oxygen leaves the tube at the opposite end. The heat source is slidingly displaced to and fro over the length of tube parallel with the tube axis to ensure an even layer coating along the inner wall surfaces of the tube. By stepped variation of the concentration of the metallic halides which are carried along in the stream of carrier gas, and with the aid of the flow volume regulator **M** and the programmer **P**, it is possible to vary the composition of the resulting glass from layer to layer in the course of the coating process. When a sufficiently thick layered interior coating of varying concentration has been precipitated, the tube is collapsed by conventional methods to form a rod, or preform, which is subsequently drawn out into a fiber.

To avoid undesired condensation of high temperature vaporizing components, the entire tube-deposition system is normally enclosed in a conventional heating jacket.

In general, the ranges of appropriate values for the significant process parameters are as follows. The most important feature, of course, is the heretofore never used high temperature range of 1850° to 2150°C, preferably 1950° to 2050°C, and especially 2000° to 2050°C for the tube substrate. To successfully utilize this temperature range, the following burner parameters are suitable: velocity, 0.15 to 3 m/min, preferably 1.5 to 2.0 m/min or also 25 to 35 cm/min or 28 to 33 cm/min; burner temperature, 2000° to 2300°C, preferably 2100° to 2150°C; and burner width, 1 to 8 cm, preferably 3 to 5 and 4 to 5 cm. Suitable silica tube characteristics include: rotation speed, 50 to 200 rpm, preferably 70 to 100 rpm; width of heated zone, 2 to 5 cm, preferably 3 to 5 or 3 to 4 cm; and length, 1,000 to 1,500 cm.

Suitable outer diameters include 10 to 30 mm, preferably 14 to 20 mm; and wall thicknesses 0.8 to 2 mm, preferably 1.4 to 1.7 mm.

Both carrier gas (O_2) and decomposable metallic halides should be of the highest purity to ensure a high purity of the final product fiber and close control over the composition of the glass layers and correspondingly, the refractive index thereof, e.g., p.a. (pro analysis) grade reagents can be used. Generally, the flow rate of the carrier gas varies between 100 to 2,000 ml/min, preferably 500 to 700 ml/min; and of the decomposable gases between 10 to 2,000 ml/min; preferably 50 to 500 ml/min. Generally, the overall gas pressure in the tube is 800 to 1,000 mm Hg and the overall flow rate is 100 to 2,000 ml/min. Suitable carrier gas temperatures are 20° to 30°C and halide or other decomposable metallic compound temperatures 30° to 70°C.

In general, 20 to 200 layers, preferably 70 to 100 layers are deposited in the light-conducting core. These layers are of 3 to 10 μm, preferably 4 to 7 μm thickness and generally all have about the same thickness. Preferably, the composition of each layer differs from that of the preceding layer so that the index of refraction increases from the first layer to the next, i.e., inwardly towards the center of the tube. The profile of the refractive index as a function of radial distance is usually a parabola but any desired profile is attainable.

After deposition of the layers is complete, the tube is conventionally collapsed and drawn. Collapsing is effected in stages by raising the tube temperature to greater than 2150°C, e.g., greater than 2150° to 2350°C. This is most easily accomplished by decreasing the burner velocity to values of 10 to 12 cm/min typically. Generally, 1 to 6 stages are used to fully collapse the tube, in each stage, the burner velocity being decreased by approximately 3 to 10 cm/min, i.e., the tube temperature being increased by 100° to 150°C in each stage. In the first stage, a temperature of 2150° to 2200°C is used.

During each stage, a stream of pure oxygen is passed through the tube. Initially, the flow rate of the oxygen is 5 to 500 ml/min. This flow rate is decreased in each step. Just before the last stage, i.e., elimination of the last internal cavity (a capillary), the oxygen flow is eliminated. The O_2 flow rate decrease for each collapsing step is in more or less linear fashion from initial value to zero flow rate.

The collapsed fiber can be drawn by fully conventional techniques. Typical optical fiber diameters are 50 to 290 μm, preferably 60 to 150 μm.

Optical Fibers Having High Infrared Transmittancy

G. Gliemeroth and L. Meckel; U.S. Patent 4,188,089; February 12, 1980; assigned to Jenaer Glaswerk Schott & Gen., Germany describe conveniently manufactured light-conducting fibers, having particularly good transmittancy in the infrared spectral region, comprising glass cores whose anions are predominantly ions of S, Se and/or Te. The cations in these light-conducting cores, on the other hand, are preferably ions of Ge, Si, P, B, As, Sb and/or Ti.

Such light-conducting fibers can be produced by precipitation of halides of S, Se, Te on the one hand, and halides of Ge, Si, P, B, As, Sb, Ti on the other, from the vapor phase by glass-decomposition processes to form light-conducting glass cores. These processes are conventional except that the conventional propellant oxygen must not be used.

In its place, inert gases and preferably Cl_2 is used as the propellant. It is also possible to incorporate considerable quantities of halides (up to approximately 40 atomic %) in the resulting glass or vitreous material. These can also be precipitated from the vapor phase.

The apparatus used in the example is the same as in the previous process, Figure 5.5, with the exception that the propellant is Cl_2 and not O_2 as shown.

Example: By making full use of the facilities presented by the apparatus shown in Figure 5.5, a refractive-index gradient profile fiber was produced in the manner and under the conditions specified below.

The doping vessel **A** contained $GeCl_4$, and vessel **B** contained SCl_2. An additional doping vessel **X** contained liquid PCl_3. Chlorine gas was used as propellant. The flow volume regulators were set in such a way that chlorine gas flowed at the rate of 175 ml/min through the vessel containing germanium chloride, at the rate of 245 ml/min through the vessel containing SCl_2, and at the rate of 280 ml/min through the vessel containing PCl_3.

In the course of the interior layer-coating process, the concentration was stepwise modified up to a final concentration corresponding to a rate of flow of 350 ml/min of chlorine gas through the vessel containing $GeCl_4$, 245 ml/min through the vessel containing SCl_2, and 105 ml/min through the vessel containing PCl_3.

In the resulting vitreous internal coating in the tube, the first layer, i.e., the layer deposited directly on the inner wall surface of the initial glass tube, was composed of 25 atomic % Ge, 35 atomic % S, and 40 atomic % P, while the last vitreous layer which was nearest to the center of the tube had a composition of 50 atomic % Ge, 35 atomic % S and 15 atomic % P. Preliminary experiments showed that a constant temperature of $790° \pm 10°C$ provided by the heat source throughout the duration of the process was the most favorable processing temperature. The internal coating process involved, in all, the application of 52 layers of relatively different composition with a mean layer thickness of 7 μm. The subsequent operations of collapsing the tube and drawing the fiber were executed under conventional conditions at a temperature which was higher than the plastification point of the initial sodium-calcium-silicate glass tube. The refractive index profile of the resulting light-conducting fiber is shown in Figure 5.6.

Figure 5.6: Refractive Index Profile of Infrared-Transmitting Optical Fibers

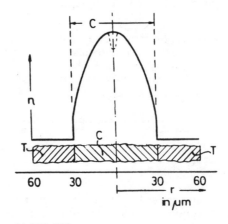

Source: U.S. Patent 4,188,089

For an overall fiber thickness of 120 μm, the light-conducting core had a diameter of 60 μm surrounded by an outer wall of 30 μm of the tube material **T**. The refractive index in the outer wall **T** is constant and corresponds to that of the sodium-calcium-silicate glass used. In the light-conducting core **C** in the fiber interior, the refractive index increases parabolically, after an initial jump at the interface with the outer wall. In the center of the fiber core, there is a small minimum due to evaporation loss which, however, does not affect the quality of the fiber with regard to transmittancy because by far the major part of energy is transported in the parabolic region of the fiber profile.

Three-Layer Optical Wave Guide

G. Gliemeroth; U.S. Patent 4,194,807; March 25, 1980 describes an optical fiber wave guide with a refractive index gradient profile, the optical fiber consisting of three units: an outermost mantle consisting of a silicate multiple component glass, an inner mantle and a core of the fiber which are both silicon dioxide-free.

The wave-guiding process occurs to the greatest extent inside the fiber in the core which has a refractive gradient profile through that intended unit. The inner silicon dioxide-free mantle contributes only marginally to the wave-guiding process; the outer silicate mantle has no part in the wave-guiding process.

The object of the process is a signal fiber whose linear thermal expansion coefficient in all of the fiber elements (mantle and core material) lies above $15 \times 10^{-7}/°C$ and in which the expansion coefficients in all fiber elements (mantle and core elements) are so arranged with each other that the lowest possible tension differences exist between these elements. This is true for the tension between the hollow tube and the inner coating as well as for the tension on the inside of a preform or a fiber.

An additional object is a wave guide fiber with a refractive index gradient profile in the area of the core which has a high as possible refractive index in the core material which is significantly and clearly higher than the refractive index of the silicon dioxide glass which is 1.458. The silicate glass hollow tube has the following composition:

Component	Weight Percent
SiO_2	51-92
$\Sigma \ Al_2O_3 + ZrO_2 + La_2O_3 + TiO_2 + B_2O_3 + P_2O_5$	1-40
P_2O_5	0-5
B_2O_3	0-26
Al_2O_3	0-28
ZrO_2	0-5
Alkali and alkaline earth oxides	2-40
BaO	0-7
CaO	0-10
MgO	0-9
PbO	0-6
ZnO	0-8
La_2O_3	0-6
Na_2O	0-12
K_2O	0-8
Li_2O	0-4

This silicate glass hollow tube has a linear thermal expansion coefficient between 14 and 120 x 10^{-7}/°C.

The optical fiber wave guide is in the wave guide conducting area silicon dioxide-free and consists of two parts, the inner mantle and the core. The inner silicon dioxide-free mantle has the following composition:

Component	Weight Percent
GeO_2	50-100
P_2O_5	0-45
B_2O_3	0-20
Al_2O_3	0-12

The linear thermal expansion coefficient of the inner silicon dioxide-free mantle corresponds to the linear expansion coefficient of the silicon dioxide glass hollow tube in the outer mantle with a tolerance of ±5 x 10^{-7}/°C. The core, also silicon dioxide-free, consists of a mixture of germanium oxide and at least one other component whereby the amount of the germanium oxide lies above 50% by weight. The composition of this core is as follows:

Component	Weight Percent
GeO_2	50-90
Sb_2O_3	0-30
Al_2O_3	0-15
B_2O_3	0-15
As_2O_3	0-30
BaO	0-10
PbO	0-15
Alkali oxide	0-15
Alkaline earth oxide	0-15
La_2O_3	0-15
SnO_2	0-20
TiO_2	0-20
WO_3	0-5
ZnO	0-10
ZrO_2	0-5
Ga_2O_3	0-14

In the glass composition for the silicon dioxide hollow tube, for the inner mantle and for the fiber core additional elements can be contained therein. For example, Ba, Rb, Cs, Sn, As, Sb, Bi, lanthanides as well as anions, acids and even halogens find use therein.

The silicon dioxide-free core is compositionally arranged so that, because of the concentration changes on the inside of the core, a refractive index change is steadily achieved. Thereby the concentration is changed in such a way that, after the collapsing step, a parabolic refractive index profile is obtained whose exponent for the parabolic equation lies between 1.7 and 2.1.

The optical fiber wave guide has in the area which results from the inner coating of the hollow tube a refractive index gradient greater than 1.55. This refractive index increases in the area of the core from the outside to the inside.

Also, the core is adjusted in the inner part with its linear thermal expansion coefficient as well as the linear expansion coefficient of the outside and that of the inner mantle so that, in spite of the refractive index gradient, no greater deviation than $\pm 12 \times 10^{-7}/°C$ is to be observed.

The outer silicate mantle can be pulled from a melt and consists of a silicate multicomponent glass whose linear thermal expansion coefficient of $15 \times 10^{-7}/°C$ lies clearly above the linear thermal expansion coefficient of silicon dioxide glass.

The inner silicon dioxide-free mantle and the silicon dioxide-free core can be manufactured by the coating of this glass hollow tube according to the deposition process from the gas phase whereby the inner coating after a collapsing to a preform and subsequently a pulling of the preform to a fiber is the wave-guiding fiber element.

The finished fiber wave guide is characterized by the lower transmission loss and higher transmission capacity as well as the large refractive index difference between the mantle and core material through a high aperture which lies above 0.25.

Fusion of a Particulate Tubular Structure

J.B. MacChesney, P.B. O'Connor and A.D. Pearson; U.S. Patent 4,191,545; March 4, 1980; assigned to Bell Telephone Laboratories, Incorporated describe a technique useful in the fabrication of optical fibers. According to the process, the glass tube used during the formation of the preform from which the fiber is drawn is fabricated from a tubular structure comprising amorphous, powdery particles which are fused to a unitary transparent glass tube while being constrained against shrinkage in both the radial and longitudinal directions. The fusion occurs in response to the application of heat to the structure while the structure is constrained by any appropriate technique. While the intermediate particulate tubular structure may be formed with the assistance of a mandrel, the mandrel and the structure are separated prior to fusion.

The process is more easily visualized with reference to Figure 5.7, which represents schematically a specific embodiment. In this figure, **14** is the particulate tubular structure, usually of wall thickness from 2 mm to 5 cm, previously formed from amorphous, powdery particles. The structure has been removed from any mandrel used during its formation and has been connected to appropriate glass tubes **11** by applying heat to the regions **13** of the structure, thereby simultaneously melting the ends of the structure and fusing it into a glass which becomes integral with the supporting tube **11**.

In other embodiments, the supporting tubes may be placed on the mandrel prior to particle deposition, and the deposition caused to overlap the tubes. A tapered section **17** may form during deposition and, depending on the requirements of the practitioner, may later be discarded. The rotation means **12** may conveniently comprise a glass lathe. In such an embodiment, the glass tubes **11**, to which the particulate structure has been fused, are securely installed in appropriate chucks which rotate synchronously. This arrangement simultaneously provides the means for constraining the structure against substantial longitudinal shrinkage during the fusion step to follow and means for rotating the structure during fusion to obtain more uniform heating.

A means **18** is also provided to prevent substantial radial shrinkage during the fusion step. In this embodiment, the radial constraint means comprises a means to internally pressurize the tube. In conjunction with this pressurization, the opposite end of the tube may be closed. The internal pressure will usually be monitored, and will most often be maintained between 0.1" and 1.0" of water for wall thicknesses between 2 mm and 5 cm when in the particulate state. The amount of pressure required to maintain the tube internal diameter will depend on the wall thickness, and for other wall thicknesses may fall outside this range. Ideally, an inert gas is used for internal pressurization. However, some other gas may be used whose reaction with the tubular structure during the fusion step to follow is not deleterious and is even perhaps beneficial.

Figure 5.7: Fusion of a Particulate Tubular Structure

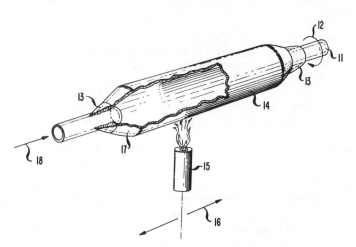

Source: U.S. Patent 4,191,545

The particulate structure is fused to a unitary transparent glass tube by traversing the structure, utilizing an appropriate traversal means shown as **16**, with a heat source, such as a flame, shown as **15**. In the figure, the region to the left of the flame has already been fused. The temperature to which the structure must be raised to obtain appropriate fusion depends upon the composition of the structure and the rate of traversal. Pure silica requires a temperature of from 1400° to 1900°C when a flame of from 2 to 4 cm in width traverses the tube at a rate of 3 to 7 cm/min. (All temperatures measured here are based on optical pyrometer measurements which are accurate to within 200°C.) Doped silicates, such as germania-silica, borosilicate, germanium borosilicate or phosphosilicates, may be fused at lower temperatures with otherwise similar fusion parameters.

Example: In this example, particulate material is deposited on an aluminum oxide mandrel, 18" long, which has a taper of 0.024"/ft of length and a maximum diameter of 1.6 cm. The mandrel is mounted in a glass lathe which is used to rotate the mandrel at approximately 60 rpm during deposition. The mandrel is slowly preheated prior to deposition by passing an oxy-hydrogen torch along the length of the mandrel. In this way, the mandrel is raised to a temperature greater than 500°C prior to deposition. In this embodiment, the glass precursor particles were formed in an oxidizing burner. The flame was supported by methane fed to the burner at between 6 and 7.5 ℓ/min and oxygen simultaneously supplied to the burner at between 6 and 7.5 ℓ/min. Glass precursor vapors were fed into the center of the flame in the form of silicon tetrachloride at 10 g/min, boron tetrachloride at 212 cc/min and oxygen at 400 cc/min. The glass precursor vapors may be supplied by bubbling the oxygen through appropriate liquids. The particles, which are formed by the reaction of the vapors in the flame, are directed to the mandrel and collect on it to form a cylindrical structure as the flame traverses back and forth over the length of the mandrel. The particulate deposition proceeds at an average rate of 1.3 g/min and at a temperature of approximately 1000° to 1050°C. The flame temperature is adjusted so as to obtain the desired density of collected particles.

At the temperatures used in this embodiment, a 5 mol % B_2O_3-SiO_2 powder is obtained with a porosity of 65%. After a sufficient volume of particles is collected, the reactant stream is shut off and the porous tube is allowed to cool to room temperature. In this embodiment, 100 passes of the torch resulted in a structure approximately 12" long, with a wall thickness of 1 cm, and an inside diameter of approximately 1.6 cm.

Subsequent to cooling, the tapered ends of the porous tube are discarded, and the tube is removed from the mandrel. The porous tube is then mounted in a glass-working lathe by inserting glass tubes of appropriate size into the ends of the particulate tube, and fusing the ends of the porous tube to the glass tubes by application of heat. The glass tubes, when so fused to the porous structure and mounted in the glass lathe, serve to constrain the porous tube from shrinkage in the longitudinal direction during fusion. One support tube was stoppered, while a slight positive pressure of from 0.3" to 0.6" of water of a suitable gas, such as oxygen, helium or mixtures thereof, was established within the tube via the second support tube.

A high temperature heat source of approximately 1700°C, such as an oxy-hydrogen flame or a graphite resistance furnace, is gradually traversed from one end of the tube to another, thus causing the porous material to fuse into a transparent glass tube. Additional material is then deposited on the inner portion of the tube as in the MCVD (modified chemical vapor deposition) process to yield a preform from which the optical fiber is drawn. The optical fiber was measured to have an optical loss of 3.8 dB/km at 0.82 μm.

Method of Reducing Absorption Losses by UV Radiation

K.G. Hernqvist; U.S. Patent 4,157,253; June 5, 1979; assigned to RCA Corporation describes a method for reducing the absorption losses in an optical fiber by the irradiation of a heated fused quartz or fused silica optical fiber in a vacuum for a time sufficient to dissociate and outgas the hydroxyl radicals contained therein as hydrogen.

The effective wavelength range of the UV (ultraviolet) radiation necessary for dissociation is from about 3000 to 4000 Å. The intensity of dissociation of the hydroxyl radical is not accurately known, but studies of this process indicate that a radiation intensity of about 100 W allow the processing of an optical fiber at reasonable speeds. The greater the wattage of the UV radiation, the more quickly and completely will be the dissociation of the hydroxyl radicals.

As an illustration, the method is carried out according to the following procedure: an optical fiber with a radius of about 5×10^{-3} cm coiled on a supply spool is passed through an oven and attached to a take-up spool in a vacuum jar. The vacuum jar is then evacuated to a pressure of about 1×10^{-7} torr and the oven is heated to about 1000°C. A 100 W high intensity UV lamp emitting UV radiation from about 3000 to 4000 Å wavelength is focused on the fiber optic. The take-up spool is turned so as to permit the UV radiation to irradiate the heated optical fiber for about 10 minutes to dissociate the hydroxyl radicals contained therein.

Optical Fibers with a Radial Refractive Index Gradient

A. de Panafieu, M. Villard, C. Baylac, and M. Favre; U.S. Patent 4,165,222; August 21, 1979; assigned to Thomson-CSF, France describe a process for the manufacture of optical fibers having a radial refractive index gradient, yielding a glass blank having a larger diameter than the fiber and destined for the production of optical fibers, comprising:

(1) A first stage consisting in mixing starting materials capable of entering into a glass composition showing the phenomenon of separation into two interconnected and continuous solid phases of different composition, the starting materials comprising at least oxides of boron, silicon, sodium and one oxide of a group consisting of germanium, titanium, phosphorus and aluminum oxides, the starting materials having an impurity level of under 10^{-5} calculated with molar proportions of oxides of so-called transition metals taken into account, the first stage subsequently comprising the preparation of a bath of molten glass;

(2) A second stage comprising a first step of drawing the blank at a predetermined rate from the bath of glass through a cooling system having a predetermined temperature gradient, and a second step comprising at least one thermal annealing treatment; and

(3) A third stage comprising at least one leaching treatment to eliminate the phase containing most of the impurities and a heat treatment for consolidating the blank.

In the embodiment described hereinafter, the following oxides are used as starting materials in the molar proportions indicated (in %): SiO_2, 35 to 70%; B_2O_3, 17 to 42%; and Na_2O, 4 to 15% and at least one of the following complementary or doping oxides: Al_2O_3, 0 to 5%; TiO_2, 0 to 10%; P_2O_5, 0 to 10%; and GeO_2, 0 to 15%.

The first stage of the process comprises melting the mixture of oxides as defined above. It will be recalled that these oxides must contain less than 10^{-5} of troublesome impurities, these troublesome impurities essentially consisting of so-called transition metals (iron, copper, nickel). The other impurities, such as water or the organic products, may be in a higher proportion.

Melting may be carried out by any conventional means which does not introduce any troublesome impurities. It is preferred to use a crucible of platinum alloyed with rhodium which is placed in a furnace capable of reaching a temperature of 1400°C. During melting, an oxidizing atmosphere is maintained above the starting materials, for example, by passing a stream of oxygen through the furnace.

The second stage comprises the step of shaping a blank and the thermal annealing step. The glass is shaped into elongate blanks: solid or hollow rods (radial dimensions on the order of 5 mm representing either the diameter of a solid blank or the thickness of a hollow blank). The thermal annealing step is carried out at a temperature of from 500° to 600°C over a period ranging from 1 to several hours.

The third stage comprises the following steps: (A) Extraction, rinsing and drying, wherein the annealed rod is cooled and then immersed in an acid solution (e.g., a 3 N aqueous hydrochloric acid solution with a temperature of 85°C) to eliminate the soft phase. The depth of penetration of the acid is typically on the order of 2 mm in 24 hours, although to obtain complete extraction it is preferable to continue the attack for a longer period, for example, for 48 hours for a solid cylindrical rod 5 mm in diameter.

During the attack of the soft phase by the acid, a gel is formed which can create sufficient stresses to break the rigid skeleton formed by the hard phase. Fractures such as these are avoided by starting from precise glass compositions (to be determined by trial and error) and by empirically determining the best time-temperature compromise for the thermal annealing step.

The rod, which has become porous as a result of the extraction step, has to be rinsed very carefully with deionized water. The blank is then dried by being kept for 3 to 4 hours in a gas stream at around 100°C. In order to eliminate the water present, the rod is heated in vacuo for around 20 hours at a temperature on the order of 550°C.

(B) Consolidation, the object of which step is to close the pores of the rod by collapse of the walls of the interstices left by leaching of the soft phase. The consolidation treatment is carried out by heating the glass to a temperature of from 100° to 900°C over a period ranging from 1 to several hours. The temperature selected should be as low as possible and compatible with the required effect in order to avoid deformation of the rod. In effect, the aim is to obtain a uniform reduction in the volume of the rod. This consolidation treatment is preceded and followed by the following optional steps.

(a) Heat treatment before consolidation treatment—The impurities existing in the hard phase are primarily in their reduced state after the stage of drying in vacuo. Depending upon the nature of these impurities, it may be desirable to modify their degree of oxidation. For example, the ferrous ion Fe^{++}, which has an absorption maximum for wavelengths on the order of 11,000 Å, is particularly troublesome to the optical fibers, while the ferric ion Fe^{+++} absorbs very little light for wavelengths ranging from 6000 to 11,000 Å.

By virtue of the large surface-to-weight ratio (on the order of 100 m^2/g) of the porous skeleton, it is easy to modify the degree of oxidation of the ions existing in the hard phase in known manner by carrying out heat treatments in controlled atmosphere.

(b) Treatment of tubular rods after consolidation treatment—In the case of a tubular rod, the tube may be closed by removing the inner wall by heating it in a flame. The advantage of the tubular rod during the second stage is due to the fact that it enables a greater mass of glass to be treated without any danger of breakage, the thickness of the glass between the outer wall and the inner wall being on the same order as the diameter of the solid rod. It is possible to increase the total length of optical fiber which can be drawn from a blank. For example, it is possible to reach lengths of several kilometers.

In order to obtain a glass fiber from the rod, one of the following two variants may be adopted:

(1)   The fiber is directly drawn from the rod and the product obtained retains the radial chemical composition gradient and therefore the refractive index gradient of the rod. The fiber can transmit light without cladding by virtue of its self-focusing properties; or

(2)   In order to provide the fiber with mechanical and chemical protection, the rod is placed in a tube of glass rich in silica (on the order of 90%) of which the internal diameter is adjusted to the diameter of the rod. The glass of the envelope has physical properties very similar to those of the rod, in particular, a drawing temperature of approximately 1400°C. Accordingly, a fiber cladded by the glass of the envelope is obtained by drawing. The glass of the envelope does not have to be purified because it is not used for the transmission of light, the light being guided in the core of the fiber by self-focusing.

The advantages of the process include the reduction in cost attributable to the use of less pure starting materials (impurity level approximately 10 times higher than that tolerated at the final stage) for the same result, i.e., a given attenuation of the light signal transmitted at the glass fiber stage. A further reduction in cost arises out of the relative simplicity of the process by virtue of the fact that the radial decrease in the refractive index is obtained without any special treatment during drawing of the rod through a cooler. Finally, the transparency of the glass thus obtained is promoted by the fact that no doping agent is added after the extraction stage.

Melting of Core Material Within Glass Tube

D.A. Krohn and S. Merrin; U.S. Patent 4,163,654; August 7, 1979; assigned to Exxon Research & Engineering Co. describe a method for making optical fibers in which the core has a graded refractive index such as to minimize dispersion.

According to the process, a tube made of cladding material is closed at its lower end. A commercially available, high purity, fused silica tube may be used as the tube. The components, which are to form the multicomponent core glass, are next introduced into the tube as batch material. For example, batch material may comprise SiO_2 (68% by weight), BaO (22%), and K_2O (10%). Batch material can (and should) be of high purity and substantially less refractory than the tube so that the tube can serve as a crucible in which the batch material can be melted. Heat is then applied by a furnace or other heating device to melt the batch material and form a glassy liquid within the tube.

The glassy liquid is then fined, i.e., heat treated sufficiently to make it homogeneous and bubble-free. The tube must be sufficiently refractory to serve as a crucible during the fining step which is an important part of the process if a low loss core glass is desired.

For production purposes, resistance heating, induction heating and/or laser heating devices (for example) can be used as the furnace.

In the final step, the temperature is further elevated until the fused silica tube softens and a fiber can be drawn. The resultant fiber will have low loss and a numerical aperture defined by the selected batch materials.

Removal of Substrate Layer

A.R. Asam; U.S. Patent 4,199,337; April 22, 1980; assigned to International Telephone and Telegraph Corporation describes a process for finishing high-strength multilayer optical preforms capable of being drawn into long optical fibers useful for light wave communications. The preforms are of the type prepared by sequential chemical-vapor-deposition of the various desired glass layers within a tubular glass substrate that is then collapsed into a solid cylindrical preform comprising a cylindrical light-transmitting core surrounded by a concentric cladding layer and one or more additional layers within the collapsed substrate.

In order for the outermost deposited additional layer to constitute a thin high-compression layer on the surface of the completed optical structure, it is necessary first to remove substantially all of the substrate layer from the preform in such a manner as to leave the adjoining high-compression layer intact. This is accomplished by selecting a glass for the substrate layer that is capable of being etched away faster than the high-compression layer, and then using controlled preferential etching to remove the substrate layer without penetrating or damaging the high-compression layer.

The silicate glass (Vycor) was found to be well suited for this purpose. This glass composition is a sodium-borosilicate glass composed of 96 to 97% silica with a 3 to 4% mixture of boron oxide and sodium oxide, e.g., 1 to 2% of boron oxide and the balance sodium oxide. It is well-known that this sodium-borosilicate glass is slightly softer than pure silica glass and that it has an etching rate in hydrofluoric acid that is approximately three times that for pure silica glass. Etching rates can, of course, be increased or decreased to some extent by altering the concentration and temperature of the etchant. Also, other glasses similar to Vycor can be formulated by introducing suitable modifiers to enhance the preferential etching rate with respect to silica glasses.

The final etching step may readily be controlled with precision by regulating the etchant concentration, temperature and immersion time. The etchant is also preferably stirred magnetically by known techniques to achieve a uniform etch. The etching time is adjusted in relation to the thickness of the substrate layer so that all of the substrate layer is removed, leaving only the high-compression layer **12** (Figure 5.8) as the outer layer of the finished preform.

This method makes it possible to provide a thinner and more effective outer high-compression layer on the surface of the completed preform, which is of uniform radial thickness regardless of any nonlinearity or nonconcentricity of the core, enclosing layers, or substrate layer. An optical fiber drawn from the completed preform will then retain a uniform high-compressive stress in its outer layer that contributes materially to its tensile strength, durability, and fiber life.

Figure 5.8: Partially Completed Optical Preform

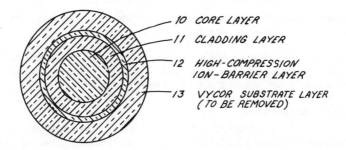

10 CORE LAYER

11 CLADDING LAYER

12 HIGH-COMPRESSION
   ION-BARRIER LAYER

13 VYCOR SUBSTRATE LAYER
   (TO BE REMOVED)

Source: U.S. Patent 4,199,337

High Tensile Strength Fiber Preform with Protective Layer

H. Schneider and E. Lebetzki; U.S. Patent 4,184,860; January 22, 1980; assigned to Siemens AG, Germany describe a method of producing a glass fiber light conductor of increased tensile strength. The following is an exemplary example of the improvement in the process in which a fiber preform receives a protective layer utilizing a flame hydrolysis.

Initially, a rod-like fiber preform is produced by a chemical vapor deposition process of depositing an internal coating in a glass tube and then subsequently collapsing the internally coated tube. Here a quartz glass is used for the glass tube and silicon dioxide-germanium dioxide glass is used for the inner coating. The dimensions are selected to be such that following the collapse of the tube to form the preform, the glass rod or preform has a silicon dioxide-germanium dioxide glass core with a diameter of approximately 6 mm and has a quartz glass casing or cladding which has an outer diameter of approximately 12 mm. Since the preform is produced by collapsing of an internally coated tube and thus was automatically flame polished, the glass rod has a surface which is subjected to low disturbances.

This rod or preform is then clamped horizontally in a turning lathe and rotated about its axis. The rod is coated with a protective layer by utilizing an oxyhydrogen flame or a burner, which is positioned at a distance of approximately 3 cm below the rotating rod. An excess of oxygen in the burner produces oxidizing conditions in the flame. The overall oxygen flow through the burner amounts to 1,300 N ml/min with 100 N ml/min being passed through a vaporizer vessel, which is filled with silicon tetrachloride at a temperature of 30°C and another 200 N ml/min being passed through a vessel containing titanium tetrachloride which is heated to 57°C. The chlorides transported by the gas flow are oxidized in the flame to form oxides which are deposited on the rod and form a white, porous oxide layer.

To enable the rod or preform to be uniformly coated, the burner is moved along the rod. The first burner is preferably accompanied by a laterally offset second burner which will melt the oxide layer deposited by the first burner at a higher temperature so that it will be a clear, bubble-free glass. At a burner speed of

20 cm/min, eighty layers are melted onto the preform and will result in a preform obtaining a glass layer of approximately 0.4 mm as the protective layer. This glass layer consists of silicon dioxide-titanium dioxide glass, which contains, approximately 6 wt %, titanium dioxide.

In order to avoid obstructions of the burner nozzle due to titanium tetrachloride being prematurely reacted to form a solid material in the nozzle, it is advantageous to use a burner wherein the flame is lifted or boosted from the burner by an additional gas flow. An example of additional gas flow is a flow of nitrogen, which emerges from a second nozzle.

The glass rod, which has been provided in this way with the protective layer is then drawn in a fiber drawing machine at a normal drawing temperature of approximately 2000°C to form the fiber. The furnace atmosphere must contain a high proportion of oxygen so that the silicon dioxide-titanium dioxide glass layer will not tend to decompose to form low grade color titanium oxides. Advantageously, the drawing is carried out in a pure oxygen atmosphere.

When the probability of breakage of the glass fiber produced in this way is determined, there is both an increase in the average tensile strength and a reduction of the distribution width of the probability of breakage in comparison to fibers without a protective layer.

As a result of the increase in the titanium dioxide concentration and the result of the optimization of the concentration profile and the thickness of the protective layer, the tensile strength is also further improved. In order to additionally increase the stability of the fiber following the drawing process and winding of the fiber, a synthetic layer is applied in a normal manner to form an outer layer.

The compressive stresses in the protective layer can be further increased in accordance with a further embodiment of the process. In the further embodiment, prior to the application of the protective layer, such as onto the fiber preform, an intermediate layer is applied to the cladding of the preform or to the fiber. The intermediate layer consists of a glass material having a higher coefficient of thermal expansion than the coefficient of thermal expansion for the material of the cladding layer. For example, the stress in the titanium dioxide-silicon dioxide protective layer can be increased by means of utilizing a silicon dioxide-germanium dioxide intermediate layer, which has a higher coefficient of thermal expansion than the quartz glass of the cladding layer.

The intermediate layer can also be produced by applying a silicon dioxide-titanium dioxide layer, which has a higher titanium dioxide content. This is then coated with the silicon dioxide-titanium dioxide layer having a lower titanium dioxide concentration. Here the TiO_2 concentrations are selected to be such that during the following tempering, which can be carried out as soon as the outer protective layer was applied, the intermediate layer is devitrified and partially crystallized, although the outer layer will remain clear.

As a result of the devitrification and recrystallization, the coefficient of thermal expansion of the intermediate layer increases and becomes greater than that of quartz glass. Typical concentration values for the titanium oxide TiO_2 are, for example, 15 wt % for the intermediate layer and 8 wt % for the protective layer. One of the advantages of this process consists in the fact that the intermediate layer is prevented from cracking or moving out of position, which may occur in a production method involving temperature fluctuations.

Relative Motion Between Tube and Plasma-Producing Apparatus

D. Küppers, H. Lydtin and L. Rehder; U.S. Patent 4,145,456; March 20, 1979 have found that in the reactive deposition of the core material from a gas which is passed through the tube onto the inner wall of the tube by means of a plasma zone, while a relative motion is effected in the axial direction between the tube and a plasma-producing device, the rate of precipitation is increased without impairing the quality of the core material coat, the reactive deposition being effected at a pressure of from 1 to 100 torrs and a temperature zone being superimposed on the plasma zone.

Referring to Figure 5.9a, a tube 1, for example, made of quartz, is moved through a heating device 2, for example, an electric heating coil, in the direction indicated by arrows. The heating device 2 is enveloped by a resonator 3 by means of which a plasma 4 can be produced in the gas mixture passed through the quartz tube 1.

In the reactive deposition, a coating 5 is directly formed on the inner wall of the tube 1.

Example 1: *The Deposition of Non-Doped SiO_2* — A gas mixture consisting of $SiCl_4$ and oxygen was passed through a quartz tube 1 (length, 150 cm; outer diameter = 8 mm; and inner diameter = 6 mm) at a throughput of 545 cm³/min. The mixture consisted of 7 vol % $SiCl_4$ and 93 vol % oxygen. The pressure in tube 1 was 12 torrs. The wall temperature was kept at 1000°C. The tube 1 was passed at a speed of 0.17 cm/min through the device, formed by heating device 2 having a length of 500 mm and resonator 3 having a length of 30 mm, while a plasma 4 was produced by a 2.45 GHz generator.

An SiO_2 coating having a thickness of 130 μm was formed directly on the tube wall. A gas phase reaction together with the formation of soot-like particles did not take place. The reaction efficiency in the plasma 4 is then almost 100%. The coating formed adheres well and is homogeneous. The gas mixture was measured in scm³ (standard cubic centimeters). 1 scm³ is one cm³ of the gas, where P = 760 mm and T = 0°C.

Example 2: *The Deposition of an SiO_2 Coat Doped with GeO_2* — A mixture of $SiCl_4$ and oxygen, consisting of 4 vol % $SiCl_4$ and 96 vol % oxygen was used to which increasing linearly with time, $GeCl_4$ was added until the content of $GeCl_4$ was 0.4% by volume. The pressure was 10 torrs. The wall temperature was kept at 960°C. The throughput was 40 scm³/min and the duration of the test was 2 hr. A well-adhering SiO_2 coat doped with GeO_2 was obtained. The coating consisted of 940 single layers of an increasing GeO_2 content. The resonator 3 was moved forward and backward along the tube in this test at 60 cm/min.

Example 3: A mixture of 0.4 vol % $AlCl_3$, 4 vol % $SiCl_4$ and 95.6 vol % oxygen was passed through the quartz tube at a throughput of 42 scm³/min (length and diameter as in Example 1). The pressure in the tube 1 was 15 torrs. The wall temperature of the tube **1** was kept at 950°C. A plasma **4** as in Example 1 was produced (power, 180 W; and frequency, 2.45 GHz). The reaction efficiency was approximately 100%. The tube was passed through the device **2–3** at a speed of 60 cm/min, while the resonator **3** was moved forward and backward along the tube **1**. A homogeneous, adhering coat **5** was obtained. The total thickness of the coating was 150 μm.

Figure 5.9b shows the total attenuation in dB/km as a function of the wavelength in micrometer of a fiber optic light conductor which was obtained by drawing at 1900°C of an internally-coated tube according to Example 2. The core diameter was 25 μm and the fiber diameter was 100 μm. The difference in the refractive indexes was approximately 0.5%.

Figure 5.9: Apparatus for Internal Coating of Glass Tube

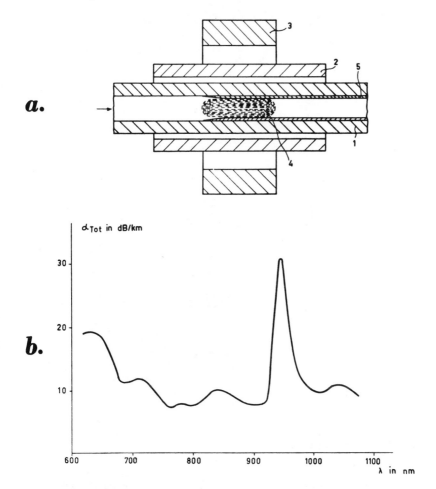

Source: U.S. Patent 4,145,456

Laser Drawing Apparatus

R.C. Oehrle; U.S. Patent 4,135,902; January 23, 1979; assigned to Western Electric Co., Inc. described a method of drawing optical fibers whereby a laser beam (Figure 5.10) is reflected from a first oscillating galvanometer controlled mirror **14** onto a second oscillating galvanometer controlled mirror **15**. The laser beam is reflected from the second mirror as a pattern **29** having an annular cross section by individually controlling the amplitude and phase relationships of the oscillations of the mirrors. The annular beam is directed at a frustoconical reflector **21** which reflects the annular beam radially inward to heat a portion of a glass preform **25** positioned along the axis of the reflector to form a melt zone therein from which an optical fiber **36** is drawn.

Figure 5.10: Laser Apparatus for Drawing Optical Fibers

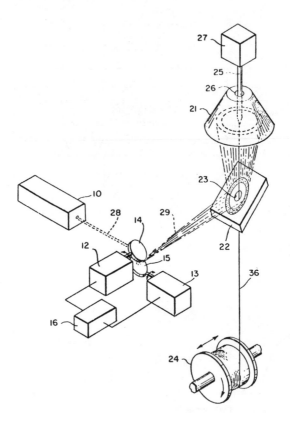

Source: U.S. Patent 4,135,902

Additionally, the amplitude of the mirror oscillations are modulated to vary the diameter of the annular beam to cause the beam to reciprocate along a portion of the length of the preform to expand the size of the melt zone.

Manufacture of Continuous Fibers Without Need for Preforms

J.E. Goell and M.S. Maklad; U.S. Patent 4,198,223; April 15, 1980; assigned to International Telephone and Telegraph Corporation describe a process for manufacturing continuous optical fibers whereby glassy oxides are deposited on a heated mandrel using vapor phase oxidation of the glass components. The degree of taper and the rotation rate of the mandrel provide for a gradient of material compositions to effect a step or graded index optical fiber. Plasma torches ensure that the glass components fuse upon deposit and induction heating and/or a high temperature laser beam maintains the mandrel tip at the optimum fiber drawing temperature.

Figure 5.11 is one apparatus for carrying out the method. A rotating tapered mandrel **10** is enclosed within a deposition chamber **11** containing a plurality of deposition nozzles **12-15** for depositing the various glass materials and an exhaust **16** for removing the gaseous by-products. The mandrel is heated by a plurality of RF coils **17** surrounding the enclosure **11**. The tip **18** of the mandrel is heated by separate RF coils **17'** to heat the deposited materials to their melting temperature.

Figure 5.11: Apparatus for Producing Continuous Fibers

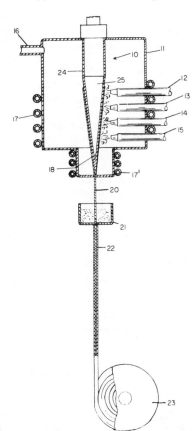

Source: U.S. Patent 4,198,223

The deposition nozzles 12-15 are located relative to the taper of the mandrel 10 so that core materials are deposited furthest from the mandrel point 18 whereas the cladding and outer layer materials are deposited respectfully at closer distances to the mandrel tip 18. This assures that the core material is innermost within the fiber and the cladding material is sandwiched with a glass outermost layer which protects the core and cladding in the drawn fiber 20.

The main RF heating coils 17 are adjusted to heat the mandrel at a temperature sufficient to cause the materials to melt upon deposit and the auxiliary RF coils 17' are adjusted to an optimum to create a suitable temperature for drawing a particular glass composition. Auxiliary heating can also be achieved by means of a high power laser.

The plastic coating applicator 21 is located immediately subjacent to the mandrel tip to insure that the fiber 20 immediately becomes coated with a plastic material 22 to keep airborne dust particles from gathering on the fiber surface 20.

The fiber drawing drum 23 is adjusted in speed to pull the glass materials into a continuous fiber 20 at a rate directly proportional to the rate of deposition of the glass-forming materials upon the mandrel.

The materials used for forming the optical fiber core comprise a mixture of germanium and silicon chlorides which coat the region furthest from the mandrel tip and is deposited by nozzle 12. The materials used for forming the cladding comprise a mixture of boron and silicon chlorides which mixture is deposited from nozzle 14 next closest to the mandrel tip and the material used for forming the outermost layer is generally silicon tetrachloride and is deposited therefore at nozzle 15 closest to the tip of the mandrel.

The mandrel is kept rotating during the deposition process to insure both uniform heating and uniform deposition within the enclosure.

The material of the mandrel comprises a platinum layer on a high temperature graphite rod 25. Alternatively, it can be platinum or a platinum alloy. The mandrel provides a susceptor for receiving RF heating from RF coils 17. The platinum coating 24 for the compound susceptor prevents the glass materials from reacting with the mandrel. The degree of taper of the mandrel determines the relative thickness of the corresponding core, cladding and outermost layers.

Multimember Crucible Apparatus

T. Yamazaki and K. Koizumi; U.S. Patent 4,145,200; March 20, 1979; assigned to Nippon Sheet Glass Co., Ltd., Japan describe a method for producing by high-speed spinning optical glass fibers of low loss while maintaining the temperature of the crucible at a relatively low temperature, and therefore inhibiting the diffusion of impurities from the material constituting the crucible into glass melts.

The method for producing optical glass fibers comprises cospinning glass melts of different kinds through coaxially disposed discharge nozzles of a multimember crucible composed of two or more crucible members having a discharge nozzle at their bottom; wherein the outermost nozzle has a length of at least 30 mm and is heated so that at least a part of it is maintained at a temperature equal to, or higher than, the temperature of the crucible thereby to increase the speed of spinning.

In the production of step-type optical glass fibers, a glass composition having a relatively low refractive index is used as a material for a cladding layer, and a glass composition having a higher refractive index is selected as a core-forming glass. Glass compositions usually employed in the art can be used as materials for the cladding and core. Examples of the cladding glass are silicate glass, borosilicate glass, and soda-lime-silicate glass which have a refractive index of 1.49 to 1.54. Suitable materials for the core glass are silicate glass, borosilicate glass, and soda-lime-silicate glass having a refractive index of 1.50 to 1.59.

For the production of focusing-type optical glass fibers, silicate glass containing an alkali metal ion having a low degree of contribution to refractive index, such as at least one of Li, Na, K, Rb and Cs, for example, is used as a material for a cladding layer. As a core-forming glass, glass having a metallic ion with a high degree of contribution to refractive index, such as a thallium ion, for example, is used.

During the spinning, the Tl ion in the core glass diffuses into the cladding glass, and the alkali metal ion in the cladding glass diffuses into the core glass, both at the boundary between the core glass and the cladding glass. In the inside of the core of the resulting optical fiber, the Tl ion concentration decreases progressively, and the alkali metal ion concentration progressively increases, from the center toward the radial direction, and because of this concentration distribution, the inside of the core has such a refractive index distribution that the refractive index decreases progressively from the center toward the radial direction.

Of the alkali metals, a Cs ion has a greater degree of contribution to refractive index than other alkali metal ions. For this reason, a focusing-type optical glass fiber can also be produced by using a Cs ion-containing glass as a core glass and a glass containing at least one of Li, Na, K and Rb as a cladding glass.

Generally, platinum of high purity is used as a material for the crucible and nozzles. Other highly heat-resistant materials, such as platinum-iridium alloy, quartz glass, alumina, tungsten and molybdenum can also be used.

Figure 5.12 shows an example of producing low-loss optical fibers by an improved three-member crucible method in accordance with the process. The three-member crucible consists of an inner crucible 401 into which a low-loss glass 402 for a core is fed, an intermediate crucible 403 into which a cladding low-loss glass 404 is fed and which surrounds the inner crucible, and an outer crucible 405 surrounding the intermediate crucible and into which a glass 406 for a protective layer is fed continuously. The three crucible members are fixed so that the inner nozzle 407, the intermediate nozzle 408 and the outer nozzle 409 provided at the bottoms of these members are located coaxially.

The three-member crucible and nozzles are maintained at certain high temperatures by an electric furnace 410, and by the melting of the material glasses 402, 404 and 406 fed from above; the glass melts within the crucible are maintained at certain water head values. These glass melts are associated at the nozzle portion, and drawn into an optical glass fiber having a three layer structure.

Figure 5.12: Spinning Furnace Employing Three-Member Crucible

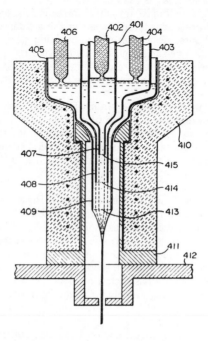

Source: U.S. Patent 4,145,200

Multilayer Optical Isolation Zone

G. Gleimeroth; U.S. Patent 4,148,621; April 10, 1979; assigned to Jenaer Glaswerk Schott & Gen., Germany describes a process for producing an optical fiber comprising an optical isolation zone and a light-guiding core having a radial refractive index profile. The process comprises depositing a plurality of layers of oxides inside a sheath to form the optical isolation zone. The oxides comprise silicon dioxide and boron oxide, wherein the boron oxide is present in an amount such that its mixture with silicon dioxide has a refractive index lower than that of silicon dioxide.

Then there is deposited a plurality of layers of oxides on the optical isolation zone to form the light-guiding core. These oxides comprise silicon dioxide and at least two oxides selected from the group consisting of and including germanium dioxide, phosphorus pentoxide, and boron oxide and an additional oxide. This additional oxide has a refractive index greater than 1.58. This additional oxide also has a cation diffusion coefficient measured at 860°C in silicon dioxide of at least 3×10^{-8} cm^2/sec.

The depositing of the layers to form the optical isolation zone can be conducted over widely varying temperatures at which deposition of the oxides takes place, and generally from 1700° to 2200°C and preferably 1800° to 1950°C.

According to a preferred embodiment, the average index of refraction in the optical isolation zone is at least 0.001 and preferably at least 0.003 less than the index of refraction in that portion of the core nearest the optical isolation zone. This can be effected in any convenient manner. According to a preferred embodiment, this is effected by depositing one to four layers of oxides consisting essentially of those of silicon, boron and phosphorus.

The depositing of oxides to form the core can occur at widely varying temperatures but is generally effected at 1750° to 2200°C and preferably at 2050° to 2100°C. These abnormally high temperatures are needed to give a glassy, flawless deposition, nonporous and without voids, and to allow diffusion of the specified component. The number of layers can vary widely but generally comprises from twenty to sixty. Silicon dioxide is an essential oxide in the core.

Furthermore, the core must contain at least two of the three oxides of germanium, phosphorus, and boron. When boron is present, it must be present in an amount such that its mixture with silicon dioxide has a refractive index greater than that of silicon dioxide and generally greater than 12%. The weight ratio of germanium dioxide to silicon dioxide is generally 15:100 to 200:100 and is preferably 25:100 to 100:100. The weight ratio of phosphorus pentoxide to silicon dioxide is generally 2:100 and 40:100 and is preferably 5:100 to 30:100.

The weight ratio of boron oxide is generally 12:100 to 100:100 and is preferably 12:100 to 50:100.

According to the process there is provided a fourth oxide in the core. The fourth oxide is produced by evaporating the corresponding halide at atmospheric pressure and reacting it with oxygen. Therefore, the preferred fourth oxide is one whose halide has a melting temperature at atmospheric pressure of less than 700°C, which is the highest temperature that can presently be conveniently maintained in the production apparatus. Likewise, the preferred halides are those which exhibit a vapor pressure of 760 mm Hg at somewhere between –100° and 700°C.

Example: A T08 silicon dioxide tube (Amersil) is used. The tube is 1.5 m long and has an internal diameter of 16 mm. The tube is rotated at 70 rpm on a lath. To prepare the tube for deposition, the hollow tube is first cleaned with hydrofluoric acid diluted 1:10, distilled water, and methanol and dried with clean nitrogen gas. The tube has oxygen passed through it at a temperature of 2100°C to prepare the tube for the glassy deposition. This is done with two passes of the burner. In all layering depositions, the rate of the burner is constant and is about 10 to 25 cm/minute. Subsequently, each traversal of the burner along the length of the rod with the concomitant feed of the components and oxygen will result in the deposition of one layer, which is generally approximately 7 μ in depth.

The optical isolation zone is composed of the layers deposited first on the inside wall of the prepared hollow tube. The oxygen is bubbled through vessels containing $SiCl_4$ and $POCl_3$. BCl_3 is added from another vessel. An additional oxygen stream is added to the tube which has helium entrained with it. 270 cc of oxygen per minute with an addition of 0.7 to 6.0 vol % of helium is used. Approximately twelve glassy layers are deposited, which form the optical isolation zone. The parts per weight of the P_2O_5 and SiO_2 are constant and the parts per weight of the deposited B_2O_3 are varied by monitoring means regulated by a preset programmer. Thus the parts of B_2O_3 are increased stepwise from 0 to 6 parts compared to 45 parts of SiO_2 to a plateau of seven constant layers and reduced to 0 at the twelfth layer. The temperature is 1825°C. This results in lowering the index of refraction relative to the refractive index of SiO_2. The 45 parts by weight of SiO_2 referred to above are equivalent to 72 ml/minute of oxygen bubbling through the vessel containing $SiCl_4$.

In order to form the glassy deposition of the core the parts of SiO_2 deposited remain constant and every additional modifier and glass-former component is increased stepwise, regulated by the monitoring system on command from the programmer. Thus in this example at layer 13, 45 parts of SiO_2 were introduced with 2 parts P_2O_5 and 3 parts Sb_2O_3. At layer 14 the GeO_2 is introduced at 14 parts. Vessels containing $SbCl_3$ and $GeCl_4$, which have oxygen bubbling through them, supply the antimony and germanium cations which react with the oxygen to form the respective oxides.

With each increasing layer, the deposited SiO_2 content remains at 45 parts, the deposited P_2O_5 content varies stepwise and linearly from 2 to 12 parts, the deposited Sb_2O_3 content varies stepwise and linearly from 1 to 20 parts, and the deposited GeO_2 varies stepwise and linearly from 14 to 45 parts. The vessel and tubing to the tube is heated and the oxygen is bubbled through the vessel. Auxiliary oxygen is also supplied and the tube is also heated. From layer 14 up to layer 50 the temperature is stepwise and linearly increased from 1850° to 2050°C.

When the 50 layers, each layer being of approximately 7 μ, have been deposited, the gas connection to the rod is closed and the temperature of the burner is increased to 2200°C. The rate of the burner traversing the tube is reduced to 4 to 6 cm/minute and the tube collapses in three passes of the burner over the tube and the rod is formed. After cooling the rod is drawn into a fiber. The numerical aperture was 0.35; the impulse dispersion was 2.7 ns/km. The optical loss or attenuation at 860 nm was 2.5 dB/km and at 1,060 nm was 1.7 dB/km.

Deposition of Variously Doped Layers Upon Tube Bore

C.P. Sandbank and J. Irven; U.S. Patent 4,155,733; May 22, 1979; assigned to International Standard Electric Corporation describe a method of optical fiber manufacture wherein glass tubing is fabricated around a vertical gas feed, which tubing is lowered through a hot zone where it is softened, its bore collapsed and where it is drawn into fiber, and wherein the gas feed is arranged to discharge vapor deposition reagents into the tubing in a zone above the region where the tubing bore is collapsed and to deposit upon the bore one or more layers, the composition of which is chosen to provide the drawn fiber with optical waveguiding properties.

The type of vapor deposition reaction that may be used can be a hydrolysis reaction, but where the material of the tubing is sufficiently refractory, as is the case when silica tubing is used, it is generally preferred to use an oxidation reaction. This is because hydrogen and hydrogen-containing compounds can be excluded from such a reaction, and thus preclude the formation of water vapor which is detrimental to optical transmissivity if incorporated into the deposit as −OH groups. The gas feed **19** of Figure 5.13 is preferably water-cooled, rotated, and at its outlet is provided with one or more jets directed towards the wall of the tubing **14**. In the most simple form of the apparatus the gas feed has only one mixture of vapors supplied to it in which case the mixture is such as to produce a glass deposit of higher refractive index than that of the tubing. This glass forms the core of a step index fiber while the tubing forms the cladding.

Figure 5.13: Apparatus for Depositing Layers of Reagents upon Collapsed Tube

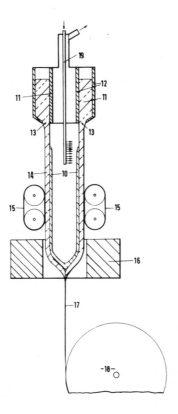

Source: U.S. Patent 4,155,733

In an apparatus where there are two or more jets supplied from separate ducts in the gas feed with different mixtures of vapors, these jets are located at different heights in order to produce a layered deposit in which the composition is graded from material derived from the top jet at the outermost portion of the deposit to material derived from the bottom jet at the innermost. Such an arrangement can be used to build up simple step index fiber in which both the core and the cladding are provided by vapor deposited material. This is a significant advantage in that it allows the tubing to be made of poor optical quality material. More complicated step index structures of optical fiber can also be made with the apparatus, such as W-guide. Furthermore, with a multifeed arrangement it is possible to make graded index structures.

Preferred reagents for making fiber in silica tubing are silicon tetrachloride, germanium tetrachloride, boron tribromide, and oxygen. To produce graded index fiber, the topmost jet or jets are supplied with a mixture to produce the lowest index deposit. This deposit is formed as the product of a halide oxidation reaction which proceeds at a high temperature promoted by a furnace (not shown) surrounding the tubing 14 in the region of the jets. If the endless belt drive is located above the deposition zone, this furnace may be an extension of the furnace 16 that is required to soften the coated tubing sufficiently for drawing.

The deposition reaction proceeds satisfactorily at a lower temperature than that required for drawing, and the deposition reaction temperature for depositing pure silica is somewhat lowered by the inclusion of other halide vapors, such as germanium tetrachloride. For this reason, in order to provide comparable deposition rates and conditions for both high and low refractive index material, it is generally preferred to include boron trichloride together with silicon tetrachloride and oxygen in the vapor mixture used for making the low index material.

The deposition product is silica doped with boric oxide. The boric oxide doping produces a glass with a refractive index slightly less than that of pure silica. Successively lower jets are supplied with vapor mixtures including silicon tetrachloride and oxygen, together with progressively increased proportions of germanium tetrachloride, and optionally progressively reduced proportions of boron trichloride. The germanium tetrachloride produces germania doping in the silica deposition product, and this has the effect of increasing its refractive index.

Other dopants, particularly phosphorus pentoxide, but also including alumina, and antimony trioxide, may be used in place of or together with, germania as refractive-index-increasing dopants. In the case of phosphorus pentoxide doping, the dopant is preferably provided by the inclusion of phosphorus oxychloride into the vapor mixture.

Joint Doped Porous Glass Fibers

The use of porous glasses as substrates for the molecular deposition of selected materials has shown great promise in the production of materials with selected physico-chemical properties and selected property variations. This process, called "molecular stuffing" or doping has been described in detail in U.S. Patents 3,938,974; 4,110,093; and 4,110,096.

P.B. Macedo, J.H. Simmons and M. Samanta; U.S. Patents 4,188,198; Feb. 12, 1980; and 4,183,620; January 15, 1980 have found that when the dopant in molecular stuffing is composed of certain combinations of lead and/or bismuth with cesium, rubidium, and/or potassium, then a remarkable and unexpected decrease in the scattering loss occurs. Further, it was discovered that certain of these combinations which give low scattering losses may be used to obtain high numerical apertures and/or high prestressing levels as well.

The desired combination of dopants leads to glass articles with the following final composition: at least 85 mol % SiO_2 with improvements which comprise at least 7 wt % of at least one member selected from Group I consisting of PbO and Bi_2O_3 and at least 1.5 mol % of at least one member selected from Group II consisting of K_2O, Rb_2O, and Cs_2O.

Even though the maximum dopant concentration is limited by the concentration of SiO_2, the broad limits are a maximum of 25 wt % for Group I and a maximum of 9 mol % for Group II. The preferred range covers at least 2 but not more than 9.5 mol % B_2O_3 and at least 7 but not more than 20 wt % of Group I and at least 1.5 but not more than 7 mol % of Group II.

Examples 1 through 3: Porous glasses are used as substrates for the deposition of the selected dopant combinations. Any porous glass preform is satisfactory. In these examples it is preferred to describe a specific method for forming the porous glass substrate by phase separation, although other processes are just as useful. (See for example U.S. Patent 3,859,073.)

An alkali borosilicate glass of composition 57% SiO_2, 36% B_2O_3, 4% Na_2O and 3% K_2O is melted in a Pt crucible in an electric furnace at temperatures between 1300° and 1450°C. The melt is homogenized by stirring with a Pt stirrer, and is then pulled in the form of rods $5/16$" diameter by 4' length. These rods are then cut into rods 4" in length which are heat-treated at 550°C for 1½ hours to induce phase separation and subsequently leached in a 3 N HCl acid solution. During the phase separation heat-treatment, the homogeneous glass decomposes into two phases, one with high silica content and one with high boron and alkali content and lower silica content. These phases are interconnected sufficiently that exposure to the leaching solution completely removes the alkali rich phase leaving behind a high silica porous glass substrate. The porous glass is washed with deionized water.

The porous glass substrate is immersed in a solution containing the desired concentrations of dopants (see Table 1) for 3 hours or longer to allow the solution to fill the pores completely. The dopant compounds are then precipitated from solution.

Table 1

Dopant Solution	Dopant Stuffing Temp. (°C)	Time (hr)	Temp. (°C)	Solvent Solution (%)	Refractive Index Core	Clad	Numerical Aperture
			..... Dopant Unstuffing*				
Example 1							
150 g $Pb(NO_3)_2$ +		(a) 2	0	50/50 methanol-water	1.527	1.459	0.45
150 g $CsNO_3$**	137	(b) 1	0	100 methanol			
Example 2							
100 g $Pb(NO_3)_2$ +		(a) 2	0	50/50 methanol-water	1.512	1.458	0.40
100 g $CsNO_3$**	107	(b) 1	0	100 methanol			
Example 3							
80 g $Pb(NO_3)_2$ +		(a) 2	0	50/50 methanol-water	1.499	1.4558	0.36
200 g $CsNO_3$**	100	(b) 1	0	100 methanol			

*Unstuffing is done in two steps: solution (a) followed by solution (b).
**Per 100 ml of aqueous solution.

In following this process it was found desirable to achieve precipitation of the dopants by thermal means; that is, lowering the temperature of preform and solvent to a point where the solution within the pores becomes supersaturated with the dopant causing the dopant to precipitate in the pores. The sample is then transferred to an unstuffing solution of solvent whereby some of the dopant is allowed to diffuse out of the pores yielding a sample with graded dopant concentration. This step is necessary in both fiber optics to achieve high numerical apertures and in strengthening to achieve high surface compressive stresses.

When graded properly, the dopant concentration is nil near the surface of the object thus yielding a low refractive index and/or a low thermal expansion coefficient in the cladding region. The unstuffing step is often conducted sequentially in two different solutions to insure maximum dopant removal from regions near the surface of the object [see Table 1 steps (a) and (b)]. The sample is kept at 0°C where it is exposed to vacuum for 24 hours and then heated at 15°C/hour up to 625°C under vacuum and sintered between 830° and 850°C.

Table 1 reports details (concentration of solution, temperature and times) of the preparation and measurements of refractive index in the core (central) and cladding (outer) regions of the objects made by using lead nitrate and cesium nitrate as dopants. Corresponding numerical apertures are also listed. Table 2 lists the compositions at the center of the final glass articles. A fiber was pulled from Sample 2 and scattering losses were found to be less than 20 dB/km in each case.

Table 2

Ex. No.	SiO_2 (mol %)	(wt %)	B_2O_3 (mol %)	(wt %)	Cs_2O (mol %)	(wt %)	PbO (mol %)	(wt %)
1	86	64	3	3	5	17	6	16
2	90	72	3	3	3	11	4	14
3	87	67	3	3	6	21	3	9

Incorporation of Index- and Stress-Modifying Dopants

P.B. Macedo, R.K. Mohr and P.K. Gupta; U.S. Patent 4,181,403; January 1, 1980 describe a glass dielectric fiber waveguide with a composition profile varying radially from the center to the surface, the composition profile causing a surface layer which is in compression, this composition profile being caused by (a) at least one dopant used substantially to create a stress profile wherein the surface is in compression, and (b) at least one dopant used substantially for the purpose of creating an index of refraction profile.

General guidance may be given for the selection of materials and relative radial dimensions for the surface (clad) and interior (core) of the preforms. In order to produce a preform with surface compression, it is required that the interior of the preform contract during manufacture after the surface has passed through its glass transition and that the interior contract relatively more than the surface, resulting in net compression at the surface.

To facilitate the achievement of large compression in the clad, the radius of the core should be more than a factor of 10 and preferably more than a factor of 50, larger than the thickness of the clad. In other words the surface layer should have a predetermined thickness which is less than 10%, preferably less than 2%, of the radius of the core. It would further facilitate the achievement of relatively more compression in the preform surface if the glass transition temperature T_g of the clad is at least 100°C higher than that of the core, preferably 250°C or higher.

To produce an optical fiber it is required that the surface layer have a relatively lower refractive index than the interior. The thickness of the optical clad depends on the loss desired and on the numerical aperture of the fiber. For fibers having attenuation below 30 dB/km and numerical apertures above 0.1, several microns clad thickness is usually recommended. It is not required that the optical index profile and mechanical stress profile be the same.

Although it is understood that a dopant may have some effect on both index and stress profiles by suitable manufacture the radial distribution of dopants which provide the desired optical profile and the distribution of dopants which substantially provide the mechanical stress profile may be different. This can be of considerable advantage. Because of the manner in which optical fibers are normally manufactured and used, deep scratches are avoided and therefore the mechanical clad can be very thin (i.e., less than 1 micron in a 100 micron fiber), while allowing the optical clad to be several microns thick which is needed for low loss light guidance; or allowing the optical profile to be parabolic while the mechanical profile has a step variation.

Example: *Melting* – A glass having the composition in mol % of 3.6 Na_2O, 3.4 K_2O, 32.8 B_2O_3, and 60.2 SiO_2 was melted using the following procedure. The raw materials Na_2CO_3, K_2CO_3, H_3BO_3 and SiO_2 are mixed and charged into a platinum crucible at 1400°C. After charging the glass was stirred between 1250° and 1450°C using a platinum/rhodium stirrer until homogenization and fining result.

Rod Forming – The well-stirred and fined glass is transferred to a furnace at 900°C. The glass is allowed to cool to 900°C during one hour.

Rods are drawn from the top surface of the melt having diameters of 0.7 to 0.8 cm. The rods are drawn through the center of a cooling cylinder.

Heat Treatment – The drawn rods were heat treated at 550°C for 1½ hours to cause phase separation.

Etching Before Leaching – The rod was etched for 10 seconds in 5% HF followed by a 30 second wash in water.

Leaching – The rods were leached at 95°C with 3 N HCl containing 20% NH_4Cl by weight. The leaching time for the rod was in excess of 30 hours. The time was chosen from previous trials to be sufficient for the rate of weight loss to be almost nil. During leaching, by providing a cold spot at 40°C, the boric acid concentration in the leaching agent was kept below 50 g/ℓ, thus speeding up leaching and avoiding possible redeposition of boron compounds in the pores of the matrix.

The leached material is washed with deionized water. The washing is conveniently carried out at room temperature using 10 volumes of water to 1 volume of glass. The water is changed 6 to 8 times during 3 days.

Stuffing – The stuffing solution was prepared by mixing $CsNO_3$, B_2O_3 and water in the following amounts: 124 g $CsNO_3$, 16 g B_2O_3 in 54 cc water for 100 cc of solution at 102°C. The porous rod remained for 3 hours in the stuffing solution. The index profile was produced by unstuffing the rod in a solution prepared by mixing 17 g B_2O_3 in 82 cc water for 100 cc of solution at 99°C. The rod was unstuffed for 10 minutes to produce a graded index profile. Precipitation of the dopants and establishment of the stress-producing profile was accomplished by replacing the first unstuffing solution by pure acetone at 0°C.

The rate of B_2O_3 removal is strongly dependent on B_2O_3 concentration in the solvent. The time for removing B_2O_3 is thus best determined by observing the clear unstuffed region formed as the B_2O_3 is removed. When the unstuffed layer reaches the desired thickness the acetone is exchanged for pure ethyl ether at 0°C. The rod is left in ethyl ether for 18 hours after which time the ethyl ether is removed and the rod is dried.

The rod was dried by exposure to vacuum at 0°C for 2 days followed by rate heating at 15°C/hour to 625°C under vacuum. The rod was then sintered under a ⅕ atmosphere of oxygen at 825°C.

Fiber Drawing — The preform was drawn down into a 170 μ diameter fiber using gas oxygen torches with the flames well-centered on the preform.

Fiber Properties — The unstuffed cladding thickness was 3% of the radius.

The optical attenuation of the fiber was measured by standard transmission methods to be less than 30 dB/km at 0.85 and at 1.05 μ.

The numerical aperture of the fiber was greater than 0.2.

Preform Manufacture Using Volatile Dopant

J. Irven and A.P. Harrison; U.S. Patent 4,165,224; August 21, 1979; assigned to International Standard Electric Corporation describe a method of silica optical fiber preform manufacture.

The material that is to form the core of the preform contains a volatile oxide and is provided as a layer of doped silica lining the bore of a silica tube, wherein the bore of the tube complete with its lining is first shrunk and then finally collapsed by repeatedly traversing a hot zone along the tube while it is rotated about its axis, and wherein during the shrinking of the bore a slight overpressure is maintained in the bore by a gas mixture containing oxygen and a halide or oxyhalide of the element having the volatile oxide, passing the halides mixture of the volatile oxide through the tube, then through a reservoir, and finally through an unconstricted long pipe whose rheological conductance is small enough to provide the requisite overpressure within the tube to overcome the tendency for the tube to flatten during the shrinkage of its bore.

The concentration of the halide or oxyhalide in relation to the oxygen is such as to compensate at least in part for the volatile oxide loss by volatilization during bore shrinkage.

The preferred method of providing the core material is by codepositing silica with one or more dopants by a vapor phase reaction. An alternative method involves the use of a vapor phase reaction to deposit dopant material upon a silica surface and cause the deposit to diffuse into the silica. In both instances a vapor phase reaction from which hydrogen and hydrogen-containing compounds are excluded is preferred.

The silica tube referred to may be made of a doped silica, and may itself be formed by a deposited layer lining the bore of a tubular substrate.

Referring now to Figure 5.14, in the preferred method of shrinking and collapsing the bore of a silica tube internally lined with a graded index layer of germania doped silica, the tube itself is depicted at 20. A mixture of oxygen and germanium tetrachloride vapor is fed into the tube via delivery pipe 21, and from the tube via a further pipe 22 to a reservoir 23. Connected to the outlet of the reservoir is a long length of tubing 24. Optionally the delivery pipe is provided with a tee 25, one limb of which terminates in a regulating valve 26.

**Figure 5.14: Apparatus for Preform Manufacture
Employing Volatile Dopant**

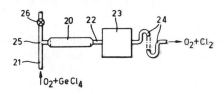

Source: U.S. Patent 4,165,224

The mixture of oxygen and germanium tetrachloride vapor is obtained by entraining the vapor in oxygen bubbled through germanium tetrachloride liquid maintained at a constant temperature. Typically it is maintained at ambient temperature, oxygen is bubbled through at 5 cc/minute and then added to a stream of oxygen providing a total flow rate of 500 cc/minute. The tube is typically 1 m long, has a wall thickness of 1 mm and a bore diameter of 10 mm.

The tube is continually rotated during the process of shrinking and final collapse of its bore. For this purpose the tube is mounted in a kind of lathe (not shown) which is equipped with a pair of chucks (not shown) which are synchronously driven. Rotary gas seals are required to connect the two pipes 21 and 22 with the ends of the tube. The dimensions of the seal between tube and pipe 22 are chosen such that possible blockage by germanium oxide soot swept down from the hot zone is minimized. The reservoir typically has a capacity of the order of 20 ℓ, and can be made of an inert plastic material such as polypropylene. Typically the requisite overpressure in the tube 20 is about 1 mm mercury, and this can be provided by using a length of up to about 50 m of ⅜" bore inert (PVC) plastic tubing for the tubing 24.

In a typical process the bore is completely collapsed after three traverses of the hot zone (not shown). The first traverse takes the bore from 10 mm down to about 4 mm, and the second from about 4 mm down to about 1 mm. For the final traverse it is no longer possible to have a flow of gas down the tube 20, and hence the gas flow through the liquid germanium tetrachloride is shut off. If the apparatus is provided with the tee piece 25 and regulating valve 26 then this valve may be opened and a flow of oxygen maintained during the final traversal in order to provide the requisite overpressure. Under these circumstances the final traversal is from the further end of the tube 20 back up towards the end nearer the tee 25.

B_2O_3- or F-Doped Silica Layer

S. Shiraishi, K. Fujiwara and S. Kurosaki; U.S. Patent 4,161,505; July 17, 1979; assigned to Sumitomo Electric Industries, Ltd., Japan describe a process for producing an optical transmission fiber which comprises feeding highly pure halides, hydrides or organic compounds of Si and B by way of carrier gas on the outer surface of a fused silica rod or a fused silica pipe, or inner surface of a fused silica pipe, oxidizing them and depositing the products to form a pure fused silica layer or a doped fused silica layer containing B_2O_3, melting the pipe and the deposited layer followed by a spinning. The SiO_2 layer can alternatively contain fluorine instead of B_2O_3. A further SiO_2 layer can be deposited thereon to improve the spinning processability and lower the index of refraction of the B_2O_3-containing layer.

Example: The process will be described by way of an experimental example. In Figure 5.15b, Ar gas, selected as a carrier gas, was fed at a flow rate of 2 ℓ/minute, carrying BBr_3 and $SiCl_4$ to the burner while the temperature of the evaporator 5 was kept at 30°C. 60 ℓ/minute of hydrogen gas and 45 ℓ/minute of oxygen gas were fed to the burner shown in Figure 5.15a.

The outer surface of a pure fused silica rod of 10 mm in diameter was contacted with the burner flame and processed for 2 hours to form a rod of about 20 mm diameter. The rod was heated in a vacuum at 1300°C for 2 hours and the rod was then spun by heating the rod in a high frequency furnace to obtain a fiber having a core diameter of 75 μ and a deposition layer diameter of 150 μ. On passing laser light through this fiber, it was found that the light was completely trapped with less scattering losses, and the entire transmission losses were also low.

Characteristics: The optical transmission fiber of this process provides a clad type fiber and an optical O guide of a highly pure fused silica portion in which the optical energy concentrates and a surrounding doped fused silica layer of a lower index of refraction containing B_2O_3 or F and thus it possesses high optical transmission characteristics and extremely low optical transmission losses.

Since completely oxidized SiO_2 or B_2O_3 is deposited on the clean surface of highly pure fused silica body in the doping with B_2O_3 as well as F, the interface is neither contaminated nor are gas bubbles formed (bubbles, if entrapped, can be eliminated by heating in vacuum or by application of supersonic waves under heating) thereby decreasing the scattering losses in the interface between the two fused silica media having a different index of refraction.

In addition, the index of refraction can easily be controlled by varying the amount of B_2O_3 contained in the fused silica. Moreover, the raw materials used in the process such as halides, hydrides or organic compounds of B and Si as well as O_2 gas can be obtained in a highly pure state due to their physical and chemical characteristics thus reducing the impurity content in the fused silica which contains the B_2O_3. This decreases the absorption losses and easily enables the preparation of a fiber with a parabolic index of refraction distribution in which the transmission losses are extremely low.

Figure 5.15: Production of Fused Silica Rod

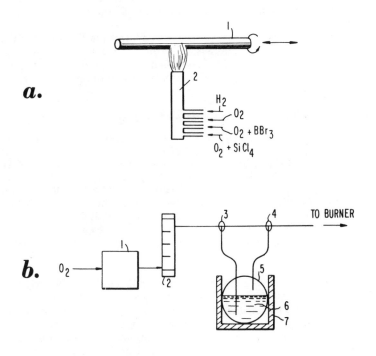

(a) Oxy-hydrogen burner
(b) Apparatus for feeding BBr₃ and SiCl₄ into burner

Source: U.S. Patent 4,161,505

Since the inclusion of F does not substantially affect the light absorption, this process can provide a fiber having a transmission loss as low as that of the fused silica fiber, can provide an easy way to control the index of refraction and can provide a transmission fiber having less overall transmission losses.

In one preferred embodiment, the transmission fiber has a further SiO_2 layer deposited on the outer surface thereof. In the SiO_2 layer containing B_2O_3, its melting temperature is lowered as the content of the B_2O_3 is increased for reducing the index of refraction, which decreases the viscosity of that portion resulting in a deformation in the shape thereof in the melting to spin. In order to avoid this deformation and spin a fiber satisfactorily, an additional SiO_2 layer is preferably deposited on this portion. A further effect is that the index of refraction of the SiO_2 layer is lowered due to the tensile stress exerted thereon, after the spinning, because the coefficient of expansion of the SiO_2–B_2O_3 system is higher than that of fused silica.

In the SiO_2 layer having F incorporated therein, the water-repellent glass layer is stable with respect to atmospheric conditions (primarily for humidity) at room temperature and inhibits water intrusion to the portion 2, which protects the doped fused silica doped with F portion from chemical attack by HF.

Further, the present process comprises a means to control the F content in the SiO_2 and to control uniformly the dispersion of the F therein. It can also prevent the incorporation of hydrogen in the production stage and its effects on the melting to spin and thus the protection of the fiber from destruction can be obtained.

Increase of Refractive Index Without Increase of Doping

In self-focusing fiber optical light conductors a parabolic variation in the refractive index across the radius is obtained by means of a continuous change in the amount of doping. According to a known method internally coated quartz glass tubes are produced by passing a gaseous mixture of silicon tetrachloride, a metal chloride (for example titanium tetrachloride, germanium tetrachloride, boron trichloride or aluminum chloride) and oxygen through the tube. In the heating zone the tube is heated to a temperature between 800° and 1200°C. The deposition is done at a pressure of between 1 and 100 torrs. Up to a thousand layers are deposited on top of each other comprising an increasing amount of doping material.

J. Koenings and D. Küppers; U.S. Patent 4,145,458; March 20, 1979; assigned to U.S. Philips Corporation describe a process to obtain an increase in the refractive index in the plasma-activated chemical vapor deposition method without the necessity of increasing the doping. To enable this, the flow of oxygen which is necessary for a complete reaction is reduced during the deposition reaction $SiCl_4 + O_2 \rightarrow SiO_2 + 2Cl_2$. It then appears that the deposited coatings have an increased refractive index which is the greater the more the oxygen supply is reduced. Presumably chloride is taken up in the coatings in the case of an incomplete reaction.

The same effect can be obtained by continuously reducing the wall temperature. It then appears that the refractive index of the deposited SiO_2 coatings depends on the wall temperature. The lower this temperature the higher the refractive index of the deposited coatings.

In a preferred embodiment of the method the oxygen flow is decreased at a constant silicon halogenide flow in the range from 1 to 50 Ncm^3/min (normal cubic cm per minute) and at a total pressure in the tube of between 1 and 100 torrs from an originally 5- to 10-fold stoichiometrical excess during the production method to values near or below the stoichiometrically required flow.

With a further preferred embodiment of the method the temperature of the tube wall in the range between 800° and 1200°C is reduced during the procedure by 100° to 200°C, the oxygen flow being 5- to 20-fold the silicon halogenide flow.

By accurately adjusting the oxygen reduction in the time and/or the time temperature variation in the time it is possible to produce "graded index" profiles having a parabolic variation of the refractive index, which is important for self-focusing fiber optical light conductors.

Figure 5.16 shows diagrammatically an arrangement for performing the method.

**Figure 5.16: Apparatus for Internal Coating of Tubes by
Plasma-Activated Chemical Vapor Deposition**

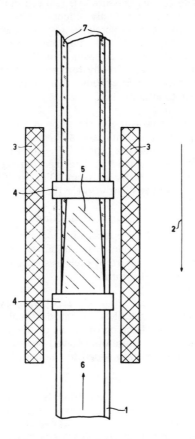

Source: U.S. Patent 4,145,458

A tube **1**, for example of quartz, is moved through a heating device **3**, for example an electric heating coil, in the direction indicated by the arrow **2**. Ring-
shaped electrodes **4** are disposed within the heating device by means of which a
plasma can be produced between the electrodes by applying an ac or dc voltage
in the gas mixture moving in the direction indicated by arrow **6** through the
quartz tube **1**. A microwave resonator may be used instead of the ring-shaped
electrodes.

In the reactive deposition a coating **7** is directly formed on the inner wall of the
tube **1**.

Example 1: A constant $SiCl_4$ current of 10 Ncm^3/min and an oxygen current which decreases linearly versus the time from 200 Ncm^3/min to 30 Ncm^3/min is passed for 1 hour through a quartz tube **1** (length 100 cm, outside diameter 8 mm, inside diameter 6 mm) in the direction of the arrow **6**. The total pressure in tube **1** was 20 torrs; this pressure was reduced during the process to 12 torrs as the oxygen current was gradually reduced. The wall temperature was kept at 1050°C. An oven **3** and a microwave resonator were reciprocated at a speed of 5 m/min along the tube **1**, a plasma **5** being produced and a power of 200 W was picked up by the resonator from a main device. A SiO_2 coating **7**, approximately 300 μm thick, was directly formed on the tube wall. The relative increase obtained in the refractive index relative to the jacket of pure quartz glass was some parts per thousand.

Example 2: Example 1 was repeated in the same arrangement, however, a constant oxygen current of 100 Ncm^3/min and a constant $SiCl_4$ current of 10 Ncm^3 per minute were passed through the tube **1**. The total pressure in the tube was 20 torrs. During the process, which lasted for approximately 1 hour, the oven temperature was continuously decreased from 1050° to 950°C. The change in the relative refractive index was slightly lower than in Example 1.

High Purity Glass Using Chemical Vapor Deposition

For some applications, such as for optical fiber wave-guide manufacture, very high purity glasses are required in which the impurity levels of less than 0.1 ppm are desired. The normal manufacture of such glasses requires the use of high purity source materials in powder form.

P.W. Black; U.S. Patent 4,202,682; May 13, 1980; assigned to International Standard Electric Corporation describes a method of making a glass wherein a melt is made of one or more of the constituents of the glass, or their precursors, which melt may include one or more undissolved phases, and wherein the remaining constituents of the glass are added to the melt by chemical vapor deposition.

In accordance with this method those constituents of a desired glass composition that are readily available in an adequately pure oxide, or oxide precursor, (e.g., nitrate or carbonate) powder form of suitable grain size are melted together, while one or more other constituents that cannot readily be obtained as oxides, or their precursors, of suitable purity and grain size are added to the melt by chemical vapor deposition, using suitably pure volatile compounds.

In the case of a small addition, typically not more than 5 mol %, by chemical vapor deposition, it is generally possible to achieve satisfactory results by arranging for the chemical vapor reactions to take place at, or just above, the surface of the melt. Diffusion of the deposited material into the melt will provide a measure of homogenization, but this may need to be assisted by stirring. For larger additions it may be preferred to perform the chemical vapor reaction in stages interspersed with additions to the melt of batches of the powder constituent or constituents.

It is to be noted that the initial melt does not itself have to be a glass. Thus, for instance, it is possible to make a soda-lime-silicate glass by starting with a melt formed by heating sodium and calcium carbonate, and then adding the required silica by reacting silicon tetrachloride with oxygen. Neither is it necessary in cases where the vapor-deposited constituent or constituents have the property of promoting homogeneity for the initial melt to be homogeneous, or even a single phase.

One example of an application of the process concerns the addition of small amounts, typically up to 2 mol %, of alumina to sodium borosilicate glasses in order to reduce the tendency for phase separation. The alumina may be deposited by entraining an aluminum halide, such as aluminum trichloride, in oxygen, and promoting the oxidation of the halide, either thermally, or with the aid of a plasma flame. The alumina produced by this reaction is in a form that is readily incorporated into the host glass.

Another example concerns the addition of small amounts, typically not more than 1 mol %, of arsenic oxide to modify the valence states of certain residual impurities in sodium borosilicate glasses. The arsenic oxide may conveniently be deposited by a thermally induced halide oxidation reaction using arsenic trichloride entrained in oxygen.

A further example involves the formation of sodium borosilicate glasses themselves. A feature of such glasses prepared from powdered constituents including boric oxide powder is the hydroxyl ion contamination of the product that is attributable mainly to the affinity of boric oxide for atmospheric moisture. By the use of direct oxidation of a boron halide above a sodium silicate melt, the boric oxide may be added in a pure hydroxyl-free form. Elimination of hydroxyl groups is required for many optical applications where the presence of such groups in the final product would cause undesirable absorption.

When two or more constituents are to be added by chemical vapor deposition to a melt, it may not be convenient to deposit them simultaneously having regard in particular to their vapor pressures, reaction rates, and their solubilities in the melt, all of which are significantly temperature dependent. On the other hand it is found in some instances the codeposition of different oxides may increase a rate of reaction. For example the rate of formation and incorporation of silica can be enhanced by codepositing it with a number of different oxides including those of arsenic, phosphorus, and boron.

In certain instances, such as where silica is the sole vapor deposited constituent, the rate of reaction can be enhanced by introducing reagents through a plasma flame to provide additional energy for the reaction. Where the presence of hydroxyl groups is not a disadvantage the rate of reaction can alternatively be enhanced by substituting a flame hydrolysis reaction for the direct oxidation reaction.

Fluorine-Doped, Synthetic Quartz Glass

K. Rau, F. Simmat, A. Mühlich and N. Treber; U.S. Patent 4,162,908; July 31, 1979; assigned to Heraeus Quarzschmelze GmbH, Germany describe a process for the production of synthetic, hydroxyl-ion-free quartz glass by oxidizing a hydrogen-free silicon compound in a hydrogen-free gas stream containing elemental and/or bound oxygen and depositing the oxidation product as a vitreous mass on a refractory support, the gas stream being passed through an induction-coupled plasma burner.

For the achievement of a prescribed reduction of the refractive index of synthetic quartz glass, a hydrogen-free, heat-decomposable fluorine compound in vapor form, especially dichlorodifluoromethane (CCl_2F_2), is introduced into the flame of the plasma burner in the amount of at least 500 g/kg of synthesized SiO_2.

At the same time it has been found advantageous to add the fluorine compound in vapor form to the oxygen being supplied for maintaining the flame in the plasma burner. To obtain a deposition product whose refractive index is to vary in a prescribed manner, it is advantageous to increase or decrease the amount of fluorine compound added during the depositing procedure. In such a manner there is produced a glass whose refractive index is 1.457 to 1.435.

If, in the case of increasing the amount of fluorine compound, a rod of synthetic, hydroxyl-ion-free quartz glass is used as the refractory support and is set in motion relative to the plasma burner, by rotating it for example, during the depositing of the fluorine-doped, synthetic, hydroxyl-ion-free quartz glass, one obtains a foreproduct for the manufacture of light-conducting fibers, which consists of a core of the refractory support material and a covering of fluorine-doped synthetic quartz glass.

A parabolic decrease of the refractive index in the covering is obtained if the amount of fluorine compound added is increased as the thickness of the covering increases. A light-conducting fiber is then produced by drawing a foreproduct of this kind.

Instead of the rod of synthetic, hydroxyl-ion-free quartz glass, a rod of synthetic quartz glass can advantageously be used whose refractive index is increased by the addition of refractive-index-increasing metal ions. It is advantageous to use a doped rod of synthetic quartz glass whose refractive index decreases with distance from the rod axis. Thermally decomposable hydrogen-free fluorine compounds which can be used include $CClF_3$ and CF_4.

As sources of hydrogen-free silicon compounds which can be used to provide hydroxyl-ion-free quartz glass one can employ the following: $SiCl_4$, $SiCl_3F$, $SiCl_2F_2$, $SiClF_3$.

Generally speaking the amount of hydrogen-free thermally decomposable fluorine compound relative to the hydrogen-free silicon compound would depend upon the nature of the glass desired and in particular the relative amount of the fluorine in the compound. Broadly speaking, however, the weight relationship of the fluorine in the thermally decomposable fluorine compound to the silicon in the hydrogen-free silicon compound will be in the range of 50 to 800 g fluorine per kg of silicon, preferably between 150 and 300 g fluorine per kg of silicon.

The process is conducted by heating the hydrogen-free silicon compound in the presence of the hydrogen-free gas stream containing elemental and/or bound oxygen and thermally decomposable fluorine compound at temperatures generally in the range of 1800° to 2600°C, preferably between 1850° and 2000°C. The temperatures are employed in an amount sufficient to deposit the so-heated mass upon a refractory support as a vitreous mass. The fluorine is present together with the silicon compound for a residence time of between 0.02 and 0.3 second, preferably between 0.04 and 0.15 second.

It has been found advantageous to use a burner having three quartz glass tubes disposed concentrically at some distance from one another, the outer tube over-reaching the middle and innermost tube and the middle tube overreaching the innermost tube. The working gas and the silicon compound including the fluorine compound present in vapor form are fed through the innermost tube, and a separating gas, preferably oxygen, is fed through the interstice between the innermost and the middle tube and between the middle tube and outermost tube.

The process differs from the state of the art particularly in that the fluorine doping of the synthetic quartz glass is no longer subject to caprice, but takes place at a specific, predetermined rate. Reductions of the index of refraction to values of 1.4532 can easily be achieved in the synthetic quartz glass produced by the process, thereby providing the assurance that this quartz glass will be suitable also for the manufacture of light-conducting fibers, especially for those light-conducting fibers whose core consists of quartz glass of high purity.

Fluorine-Doped, Synthetic Quartz Glass Preform

In an extension of the previous process, *K. Rau, A. Mühlich, F. Simmat and N. Treber; U.S. Patent 4,165,915; August 28, 1979; assigned to Heraeus Quarzschmelze GmbH, Germany* describe a semiproduct for use in the manufacture of light-conducting fibers.

The semiproduct comprises a core of synthetic quartz glass having a thickness of 6 to 400 mm, especially synthetic quartz glass obtained from gaseous silicon halide, the quartz glass containing less than 10 ppm hydroxyl ions and having, in the near infrared spectral range, an optical loss totaling less than 4 dB/km, measured in the mass, the core fused with a jacket having a wall thickness of 2 to 20 mm and consisting essentially of synthetic quartz glass, especially synthetic quartz glass derived from gaseous silicon halide, containing more than 4,000 ppm of fluorine, the jacket having a length of at least 200 mm.

A semiproduct in accordance with the process is prepared in a preferred manner as follows: a cylinder provided with a bore and made from synthetic quartz glass containing less than 10 ppm of hydroxyl ions and more than 4,000 ppm of fluorine ions, whose outside surface is finely ground and whose finely ground inside surface, after cleansing in a cleaning agent containing at least 30% hydrofluoric acid, is fused in a tubular furnace to a solid cylinder of synthetic quartz glass containing less than 10 ppm of hydroxyl ions and having an optical loss total, in the near infrared spectral range, of less than 4 dB/km, measured in the mass, whose surface has been finely ground and cleansed with a cleaning agent containing at least 30% hydrofluoric acid. The fusion product thus obtained is then immediately drawn to a diameter of more than 8 mm but less than 60 mm.

It has been found desirable to polish the inside surface of the bored cylinder before introducing the solid cylinder. It has proven advantageous, before the insertion of the solid cylinder, to subject the bored cylinder to an ultrasonic cleaning process. It is also advantageous to polish the surface of the solid cylinder prior to insertion into the bored cylinder. Flame polishing has proven an especially good method for this polishing operation and that of the inside surface of the bored cylinder.

The polished surface of the solid cylinder is also advantageously subjected to an ultrasonic cleaning before the solid cylinder is inserted into the bored cylinder.

Ga_2O_3-P_2O_5-GeO_2 Optical Transmission Line Glass

K. Inoue, J. Goto and Y. Kawabata; U.S. Patent 4,197,136; April 8, 1980; assigned to Fujitsu Limited, Japan describe an optical transmission line glass which always includes gallium oxide (Ga_2O_3) and includes phosphoric oxide (P_2O_5) and germanium oxide (GeO_2) as the principal components. The glass is sufficiently vitrified at 1400°C and has a nonalkaline nature. Therefore, it may be easily vitrified in a silica vessel without corroding the vessel.

The glass is easily soluble in water when it includes a very small amount of Ga_2O_3, but has excellent waterproof characteristics when Ga_2O_3 of about 10 wt % is added.

In order to manufacture the glass, a halogenide of phosphorus and a halogenide of germanium are mixed in the gas phase and the mixed gas is heated with oxygen for the purpose of oxidization. The glass-forming oxide is produced by such oxidization. Since the volatile P_2O_5 is combined with germanium (Ge) and formed into a nonvolatile compound in such oxidized reaction, the loss of phosphorus due to vaporization of P_2O_5 may be reduced drastically.

When Ga_2O_3 is added to the mixed oxide, or is mixed in the form of $GaCl_3$ to a gaseous phosphorus compound such as, for example, phosphoric trichloride (PCl_3), and a germanium compound such as, for example, germanium tetrachloride ($GeCl_4$), and formed into glass soot and these are vitrified, transparent glass including Ga_2O_3 is obtained. The addition of Ga_2O_3 insures sufficient waterproof characteristics. The refractive index of the glass material formed is lowered as the content of Ga_2O_3 increases. Thus, by controlling the mixed content of Ga_2O_3, a glass material having an adequate refractive index may be obtained for use as a core and as cladding.

In accordance with the process, an optical transmission line glass essentially includes 10 to 58 wt % of phosphoric pentoxide P_2O_5, 15 to 85 wt % of germanium dioxide GeO_2, and 5 to 40 wt % of gallium trioxide Ga_2O_3.

The gallium trioxide Ga_2O_3 is provided as a network modifier for increasing water resistance.

The phosphoric pentoxide and germanium dioxide are the principal components and are in a weight ratio of 4:6, and the gallium trioxide is the remainder in an amount of 10 to 25 wt %.

Fiber Optics Fused Array with Improved Blemish Quality

In the manufacture of a fused fiber optics array such as a faceplate certain aspects of the process result in blemish defects within the plate. One type of blemish is commonly referred to as "chicken wire." It is characterized by reduced transmission in those fibers at or near the boundaries of the component building blocks known as multifibers or multi-multifibers. Quite often only the outermost row of fibers is affected but in severe cases the effect may extend into the next row, rarely deeper.

Since a major cause of chicken wire blemishing is the increased exposure to heat and contamination of the outer rows of fibers in the multi- or multi-multifiber drawing process, *W.P. Siegmund; U.S. Patent 4,175,940; November 27, 1979; assigned to American Optical Corporation* describes a process to provide for protection against excessive heating or contamination of such outermost fibers of component building blocks for fiber optics arrays during the process of drawing the building blocks to desired cross-sectional sizes.

Another object is to afford protection against excessive surface heating and contamination of multi- or multi-multifiber components in a manner compatible with usual optical fiber processing and with minimal complication and/or additional operational expenditure.

These objects are accomplished by providing a removable layer of protective material, e.g., leachable glass, around a multi- or multi-multifiber preform prior to its drawing to reduced cross-sectional size. In so doing, both excessive surface heating and contamination during drawing may be greatly reduced to preserve the light-conducting capabilities of outermost fibers of the resultant multi- or multi-multifibers. Removal of the protective layer from the fibers after drawing renders them adaptable to use as component building blocks for fiber optics arrays. The improved transmission properties at edges of the multi- or multi-multifibers minimizes chicken wire blemishing in fused assemblies thereof.

The aforesaid protective layers can be applied to the multi- or multi-multifiber preform as a row of square fibers, series of plates or strips of material, a frit or a specially formed sleeve or tube.

Protective Coating

R.D. Maurer; U.S. Patent 4,173,393; November 6, 1979; assigned to Corning Glass Works describes a protective coating for optical waveguides. In accordance with this process, a coating of metallic glass is applied around the periphery of an optical waveguide. The coating is applied as a liquid which solidifies when it cools to room temperature. When the metallic glass coating cools, it contracts thereby placing the surface of the waveguide under compression. When a waveguide is placed under compression, it is strengthened considerably against fracture. In order to fracture the waveguide, the compression force must be overcome. Only after the waveguide is placed under tension can it be fractured.

The metallic glasses are particularly suitable for use as protective coatings because they are compatible with the required high cooling rate, because they form a good moisture barrier, because they have a high yield point and because they possess the necessary toughness.

In general, simple metallic glasses are made from one noble metal element (or one transition metal element) and one metalloid (B, C, Si, N, P, Ge, Sb). There are advantages in including many constituents of each type.

One example of a metallic glass suitable for use is $Ni_{36}Fe_{32}Cr_{14}P_{12}B_6$ (Allied Chemical Company). Other metallic glasses which are suitable for use include $Fe_{803}P_{163}C_{39}B$ and $Fe_{40}Ni_{40}P_{14}B_6$ (Allied Chemical) or $Pd_{77}Au_5Si_{18}$; $Pd_{77}Ag_5Si_{18}$; $Pd_{77}Cu_5Si_{18}$.

Manufacture of Continuous Optical Preform

H.F. Sterling and J. Irven; U.S. Patent 4,195,980; April 1, 1980; assigned to International Standard Electric Corporation describe a method of manufacturing a continuous preform for an optical fiber of the type produced from a silica tube via a rod preform stage. The method includes collapsing inwardly lengths of silica tube each into a rod preform, removing the end portions of each rod preform, arranging the preforms end to end in a rectilinear array, placing silica sleeving tubes end to end over the rods, each sleeving tube being a sliding fit on the rods, and fusion sealing the assembly into a continuous rod, and in which the sleeving tubes are so arranged that the joints therebetween are not coincident with the joints between the rods.

The continuous preforms prepared by the process may be drawn into fiber by conventional drawing techniques. In a particularly advantageous arrangement the preform manufacture and fiber drawing processes may be operated in tandem so that a continuous production of fiber is achieved.

Joining of Optical Fibers with a Link Piece

M. Chown, A.W. Horsley and D.G. Dalgoutte; U.S. Patent 4,183,737; Jan. 15, 1980; assigned to International Standard Electric Corporation describe a method of making an optical fiber coupler for coupling first and second optical fibers of the plastic clad silica or similar type both of which have the same cross section.

The method includes the steps of providing a glass sleeve whose bore is a clearance fit around the fibers; of positioning a link piece of glass optical fiber having the same cross section as the two fibers to be joined and a length that is less than that of the sleeve in the central section of the sleeve bore to leave the bores of both of the end sections of the sleeve unobstructed; of securing the link piece in position within the sleeve by heating the central section of the sleeve so as to cause it to collapse around the link piece over substantially the whole of its length while leaving the portions of the sleeve bore beyond the link piece ends large enough to accommodate the ends of the first and second fibers; and in which the link piece is a single material fiber whose refractive index is higher than that of the collapsed tube, the link piece and the collapsed tube forming a single fiber coupling element.

Coupling of Glass Fibers Using an Etchant

W.J. Stewart; U.S. Patent 4,159,863; July 3, 1979; assigned to Plessey Handel und Investments AG, Switzerland describes a method of treating an end of an optical glass fiber with a view to facilitating the establishment of a good efficiency optical coupling of the fiber end to a small-dimension source or receiver of light, for example to a light-emissive diode.

According to the process, the end of a glass fiber of the graded-, or at least stepped-index type, or in the case of a clad fiber, the end of its core, is etched with an etchant which varies in etching power according to the composition, and thus in a graded- or stepped-index fiber according to the index, of the fiber-core material, thereby producing at the end of the fiber a recess or "well" whose depth increases from the edge towards the axis of the fiber.

In one form of the process the fiber is made of glass having a high melting point, with approximately parabolic grading of the index, and a predetermined small quantity of a readily fusible glass, which will melt at a substantially lower temperature than the glass constituting the fiber, for example a chalcogenide, is placed in the well while the fiber is held in a vertical disposition, and the end of the fiber with the fusible glass in the well is then heated. When the fusible glass becomes liquefied, it will wet the material of the fiber and take up the shape of the fiber end, or form a somewhat convex surface determined by surface tension, and will thereby form a good quality lens, which enables the fiber end to be optically coupled with good efficiency, for example to a small source of light such as a light-emissive diode.

The fiber, or the fiber core in the case of a clad fiber, which is to be subjected to the treatment, may be made of doped silica by chemical vapor deposition, and buffered hydrofluoric acid has been found to be a suitable etchant for most fibers thus produced. The depth of etching may be varied according to requirements; a depth of 50 microns has been found useful in fibers of about 120 micron diameter.

Connection of Optical Fibers Using Vibration

J. Eldin; U.S. Patent 4,159,900; July 3, 1979; assigned to Compagnie Industrielle des Telecommunications Cit-Alcatel SA, France describes a method of connecting optical fibers by using a capillary tube which is flared at both ends. The capillary tube is formed so that its bore is cylindrical along at least half of its length, the fibers are inserted into the bore by vibration until their respective end faces come into contact with each other and the capillary tube is then heated to its softening point (at least at its ends) and allowed to cool to obtain a tightening of the ends of these fibers.

By way of an example, in the case of optical fibers having an outside diameter of 100 μm, the capillary tube is chosen with a length of about 5 mm, the bore is cylindrical along a length of about 4 mm and has a diameter close to 110 μm along this cylindrical part; the two ends of the bore, each having a length of about 500 μm, are conical, the diameter of their opening is about 1 mm. The capillary tube has an outside diameter of about 2 mm.

The optical fibers are inserted in the capillary tube by guiding the fibers into the passage and vibrating them while they are being guided until the end faces of the fibers come into contact with each other at the central part of the passage. Prior to the insertion of the fibers in the tube a protective covering is stripped from the ends of the fibers and their end faces are prepared, e.g., by clean breakage.

Once the fibers are in place in the tube, at least the ends of the capillary tube are heated to their softening temperature. This is done by heating the entire capillary tube (or merely its ends) to a regulated temperature between 400° and 700°C according to the type of glass of which the capillary tube is made.

The ends of the tube having reached the softening point, the capillary tension of the glass produces a gripping effect at the ends of the tube on the fibers. The cooling of this clamping can, of course, be controlled by nipping the ends of the capillary tube.

PHOTOCHROMIC GLASS

Lithium Boroaluminosilicate Sheet Glass Compositions

The basic patent in the field of photochromic glass, U.S. Patent 3,208,860, describes a broad range of silicate glass compositions exhibiting reversible phototropic properties. These glasses are rendered reversibly phototropic through the inclusion of specified quantities of silver halide in the glass composition, and through appropriate heat treatment of the glass after forming to cause the precipitation and growth of silver halide crystallites in the glass. These crystallites are small enough to be invisible, yet are darkenable under the action of light to reduce the light-transmitting capabilities of the glass. Upon shielding from light, the crystallites fade to the colorless state, restoring the original light transmission characteristics of the glass. In photochromic glasses these darkening and fading cycles may be repeated indefinitely without fatigue.

A major use for photochromic glasses exhibiting reversible darkening under the action of visible light has been in the manufacture of photochromic ophthalmic lenses.

D.J. Kerko and T.P. Seward, III; U.S. Patent 4,148,661; April 10, 1979; assigned to Corning Glass Works describe a region of glass compositions which can be used in processes for forming drawn sheet, and which yet provides haze-free optical quality, good chemical durability, chemical strengthenability, and photochromic properties which are sufficiently developed to be useful in thin sheet form.

Broadly, glass compositions in accordance with the process consist essentially, in weight percent on the oxide basis as calculated from the batch, of about 54 to 66% SiO_2, 7 to 15% Al_2O_3, 10 to 25% B_2O_3, 0.5 to 4.0% Li_2O, 3.5 to 15% Na_2O, 0 to 10% K_2O, 6 to 16% total of $Li_2O + Na_2O + K_2O$, 0 to 3.0% PbO, 0.1 to 1.0% Ag, 0.1 to 1.0% Cl, 0 to 3% Br, 0.008 to 0.12% CuO, and 0 to 2.5% F. The glass may optionally additionally contain colorant oxides selected in the indicated proportions from the group consisting of 0 to 1% total of transition metal oxide colorants and 0 to 5% total of rare earth oxide colorants.

Glasses provided from the abovedescribed compositions exhibit viscosities of at least about 10^4 poises at the liquidus temperature thereof, a liquidus-viscosity relationship which permits forming by direct sheet drawing from the melt.

Glasses of these compositions also exhibit long-term stability against devitrification in contact with platinum at temperatures corresponding to glass viscosities in the range of 10^4 to 10^6 poises, and can thus be drawn from the melt at those viscosities using platinum or platinum-clad drawbars, downdraw pipes, or other sheet-forming means to provide glass sheet of optical quality.

The best combination of photochromic and physical properties utilizing conventional heat treating and strengthening processes are found within a preferred group of glasses having compositions consisting essentially, in weight percent on the oxide basis as calculated from the batch, of about 57.1 to 65.3% SiO_2, 9.6 to 13.9% Al_2O_3, 12.0 to 22.0% B_2O_3, 1.0 to 3.5% Li_2O, 3.7 to 12.0% Na_2O, 0 to 5.8% K_2O, 6 to 15% total of $Li_2O + Na_2O + K_2O$, a ratio of Li_2O content to $Na_2O + K_2O$ content not exceeding about 2:3, 0.7 to 3.0% PbO, 0.1 to 1.0% Ag,

0.15 to 1.0% Cl, 0 to 3.0% Br, 0 to 2.5% F, 0.008 to 0.12% CuO, 0 to 1.0% total of transition metal oxides selected in the indicated proportions from the group consisting of 0 to 0.5% CoO, 0 to 1.0% NiO, and 0 to 1.0% Cr_2O_3, and 0 to 5.0% total of rare earth metal oxides selected from the group consisting of Er_2O_3, Pr_2O_3, Ho_2O_3, and Nd_2O_3.

Example: A glass batch is compounded and melted in a batch melter at a temperature of about 1400°C to provide a molten glass having a composition, in parts by weight as calculated from the batch, of about 58.1 parts SiO_2, 16.9 parts B_2O_3, 11.3 parts Al_2O_3, 3.8 parts Na_2O, 5.8 parts K_2O, 1.8 parts Li_2O, 0.24 part Ag, 0.30 part Cl, 0.26 part Br, 0.27 part F, and 0.018 part CuO.

The molten glass is fed into a refractory overflow downdrawn fusion pipe at a viscosity of about 10^4 poises and is delivered from the pipe as drawn glass sheet about 1.5 mm in thickness. This drawn sheet is cooled below the glass softening temperature and separated into sections of glass sheet from which small samples of drawn sheet glass are cut.

The drawn sheet glass articles prepared as described are then subjected to a heat treatment to develop the photochromic properties thereof. The heat treatment comprises heating the articles in a lehr, supporting the article in a manner to prevent surface damage, at a rate of about 600°C per hour to a temperature of 630°C, holding the articles at this temperature for ½ hour, cooling the articles at a rate of about 600°C per hour to a temperature at least below 350°C, and finally removing the articles from the lehr.

The photochromic drawn sheet glass articles prepared as described are then subjected to an ion-exchange strengthening treatment comprising immersion in a molten $NaNO_3$ bath at a temperature of 390°C for 16 hours. After treatment the samples are removed from the salt bath, cooled, washed to remove excess salt, and tested for strength and photochromic properties. The unabraded modulus of rupture strengths of the articles averages about 50,200 psi with depths of surface compression being about 3.5 mils.

The faded luminous transmittance (Y) of a typical 1.5 mm thick photochromic drawn sheet glass article produced as described above is about 90.9%. When darkened by a 20 minute exposure to a pair of 15 watt black-light blue fluorescent UV lamp bulbs spaced a distance of about 3¾ inches from the glass sheet, a darkened luminous transmittance of about 22.8% is obtained. The darkened glass typically fades about 15.6 percentage points to a luminous transmittance of about 38.4% in a 5 minute fading interval.

Ophthalmic Glass Exhibiting Rapid Darkening and Fading Properties

G.B. Hares, D.L. Morse, T.P. Seward, III, and D.W. Smith; U.S. Patent 4,190,451; February 26, 1980; assigned to Corning Glass Works describe the production of transparent photochromic glass which, in the preferred embodiment, will be suitable for ophthalmic applications and which, in 2 mm thickness, will exhibit the following photochromic behavior:

(a) At about 20°C, the glasses will darken to below 40% transmittance in the presence of actinic radiation, e.g., bright outdoor sunlight; the glasses will fade at least 30 percentage units of transmittance after 5 minutes' removal from the actinic radiation;

and the glasses will fade to a transmittance in excess of 80% in no more than 2 hours after being removed from the actinic radiation;

(b) At about 40°C, the glasses will darken to below 55% transmittance in the presence of actinic radiation, e.g., bright outdoor sunlight; the glasses will fade at least 25 percentage units of transmittance after 5 minutes' removal from the actinic radiation; and the glasses will fade to a transmittance in excess of 80% in no more than 2 hours after being removed from the actinic radiation;

(c) At about –18°C, the glasses will not darken below 15% transmittance in the presence of actinic radiation;

(d) The glasses are capable of being strengthened via thermal tempering or chemical strengthening while maintaining the desired photochromic properties; and

(e) The glasses have compositions susceptible to refractive index adjustment without loss of the desired photochromic properties.

These objectives can be achieved in glass compositions which in their broadest terms consist essentially, in weight percent on the oxide basis, of 0 to 2.5% Li_2O, 0 to 9% Na_2O, 0 to 17% K_2O, 0 to 6% Cs_2O, 14 to 23% B_2O_3, 5 to 25% Al_2O_3, 8 to 20% $Li_2O + Na_2O + K_2O + Cs_2O$, 0 to 25% P_2O_5, 20 to 65% SiO_2, 0.004 to 0.02% CuO, 0.15 to 0.3% Ag, 0.1 to 0.25% Cl, and 0.1 to 0.2% Br. When less than about 5% P_2O_5 is present in the composition, the minimum SiO_2 content will range about 45%.

Various compatible metal oxides such as those given below in the indicated amounts may be included to improve the melting and forming capabilities of the glass and/or to modify the physical and optical properties thereof: 0 to 6% ZrO_2, 0 to 3% TiO_2, 0 to 0.5% PbO, 0 to 7% BaO, 0 to 4% CaO, 0 to 3% MgO, 0 to 6% Nb_2O_5, and 0 to 4% La_2O_3. Up to about 2% F may also be included to assist melting of the glass. Finally, colorant oxides may optionally be included in the glass compositions. In general, such additions may consist of up to 1% total of transition metal coloring oxides, e.g., CoO, NiO, and Cr_2O_3 and/or up to 5% total of rare earth metal oxides, e.g., Er_2O_3, Pr_2O_3, Ho_2O_3, and Nd_2O_3. In general, the sum of all extraneous additions to the base glass will not exceed about 10%.

The optimum photochromic properties are normally secured where the "photochromic elements" are maintained within the following ranges; i.e., where the CuO is included in amounts between 0.005 and 0.011%, the Ag is held between 0.175 and 0.225%, the Cl is maintained between 0.12 and 0.225%, and the Br is present between 0.1 and 0.15%.

Furthermore, in the preferred compositions, the molar ratio of alkali metal oxide:B_2O_3 will preferably be maintained between about 0.55 and 0.85, if the glass is essentially free from divalent metal oxides other than CuO, the weight ratio of Ag:(Cl + Br) will preferably be held at values between about 0.65 and 0.95, and the molar ratio of (alkali metal oxide minus Al_2O_3):B_2O_3, i.e., $(R_2O - Al_2O_3)$:B_2O_3, preferably ranges between about 0.25 and 0.4. Where ZrO_2 and/or Nb_2O_5 is present in the glass, the molar ratio of (alkali metal oxide minus Al_2O_3 minus ZrO_2 and/or minus Nb_2O_5):B_2O_3, i.e., $(R_2O - Al_2O_3 - ZrO_2$ and/or

Nb_2O_5):B_2O_3, will preferably range between about 0.25 and 0.4. These latter two ranges may not be applicable when the P_2O_5 content is about 5% or greater.

The actual batch ingredients can comprise any materials, either the oxide or other compound, which, when melted together with the other ingredients, will be converted into the desired oxide in the proper proportions.

Photochromic Microsheet for Use in Glass-Plastic Composite Lenses

There has been considerable interest in glass-plastic composite articles and, particularly in the ophthalmic industry, for composite glass-plastic lenses. There are a number of plastics having densities substantially less than glass. As a result, in both the prescription and nonprescription sunglass markets, plastics have seen increasing service since the lightness thereof causes less discomfort to the wearer and has permitted the merchandising of lenses of larger area since the weight, when compared to glass, is much less. Nevertheless, plastic lenses have one drawback which has limited their universal acceptance; that is, the surfaces of plastic lenses do not possess the surface hardness of glass and, hence, are susceptible to being scratched. Therefore, care must be exercised in handling such lenses. Also, but less importantly, plastics do not exhibit the heat resistance and chemical durability of glass.

Accordingly, efforts have been undertaken to produce transparent glass-plastic composite lenses wherein the body of the lens would consist of plastic but at least one surface thereof, conventionally the surface of the lens away from the wearer's face, would have a thin skin of glass laminated thereto. Such a glass surface can provide the desired resistance to surface abrasion, heat resistance, and chemical durability lacking in the plastic. And the resulting composite article would be lighter than a lens formed from glass alone.

D.J. Kerko, J.-P. Odile, C.J. Quinn and P.A. Tick; U.S. Patent 4,168,339; September 18, 1979; assigned to Corning Glass Works describe the production of photochromic glass microsheet which is particularly useful in the fabrication of transparent glass-plastic composite lenses. In such lenses, the photochromic microsheet will either be buried within the plastic or will act as a surface layer thereon. The microsheet consists essentially, in weight percent on the oxide basis as calculated from the batch, of 54 to 66% SiO_2, 7 to 15% Al_2O_3, 10 to 30% B_2O_3, 3 to 15% Na_2O, 0.4 to 1.5% PbO, 0.2 to 0.5% Br, 0.5 to 1.2% Cl, 0.2 to 0.5% F, 0.008 to 0.03% CuO, and >0.03% to 1% Ag.

Various optional ingredients can be added including 0 to 4% Li_2O and 0 to 10% K_2O, but the sum of $Li_2O + Na_2O + K_2O$ will not exceed about 15%, 0 to 3% P_2O_5, 0 to 1% I, and 0 to 0.5% CdO. Well-known glass colorants may also be included, and are selected from the group consisting of 0 to 1% total of transition metal oxide colorants and 0 to 5% total of rare earth oxide colorants.

The method comprises the following steps:

(a) A glass-forming batch of the proper composition is melted;

(b) The melt is adjusted in temperature to impart a viscosity thereto of about 10^4 to 10^5 poises, and then

(c) The melt within that range of viscosities is drawn past refractory forming means in a downdraw process such as is described

in U.S. Patents 3,338,696 and 3,682,609 to produce potentially photochromic glass microsheet having a thickness between about 0.25 and 0.50 mm.

Potentially photochromic glass is defined as glass including silver halide-containing crystals and sensitizing agents or activators, such as copper oxide, which can
be rendered photochromic by means of a predetermined heat treatment subsequent to the forming step.

Gradient Photochromic Glass Containing Unnucleated Portions

Photochromic or phototropic lenses have the disadvantage that recovery of high
transmissivity takes several minutes. This has been noticed with discomfort and
dislike by wearers under such conditions as driving an automobile where low
levels of illumination exist inside the car and high levels of illumination may exist outside the vehicle. While it is desirable to reduce the light intensity to the
driver's eyes while observing road and traffic conditions, the driver must be permitted to clearly view information presented by instruments on the vehicle instrument panel where a low level of illumination normally exists. Indeed, it
may be dangerous to prevent this.

A similar type of problem may be found in occupations where sudden changes
in the level of illumination from bright to dim occur either (1) by rapid changes
in the intensity of the light source, or (2) by movement of the wearer of the
spectacles from high level of intensity to a darker environment.

*A.F. Menyhart; U.S. Patent 4,154,590; May 15, 1979; and D.A. Krohn; U.S.
Patent 4,155,734; May 22, 1979; both assigned to American Optical Corporation*
describe a process whereby glass lenses or lens blanks, containing all the ingredients necessary to produce phototropic or photochromic behavior, are treated in
a conventional production furnace to produce a locally variable heat treatment,
wherein at least one portion thereof is raised to a temperature exceeding the
glass strain point but not the softening point, and other portions are heated to
variable temperatures decreasing from the strain point.

The treatment causes development of phototropic or photochromic behavior
only in those portions of the lenses or lens blanks exposed to the temperature
above the strain point. The remaining portions of the lenses which do not exhibit photochromic behavior are referred to as "unnucleated" photochromic
glass.

By example, illustrated in Figure 5.17, a stainless steel fusing tray 30, of dimensions approximately 12" long, 6" wide and 2½" high, has its inside bottom preferably covered with a 1" asbestos sheet 32. A series of lenses 34, each mounted
on a chromite fusing block 36, may be positioned on the asbestos. A strip of
asbestos 38 is mounted on each lens or series of lenses in such a manner that
the strip 38 forms a dam at about the middle of the lenses. Thus, this strip 38
includes cutouts 38a, 38b and 38c which conform closely to the curvature of
the lens blanks so as to rest on the surface of the lens blanks.

Between the series of lenses 31 and the series 33 and their positioned or overlying asbestos strip 38 is a bed of conventional glass makers silica 39 of the usual
coarse type and of conventional high purity. The high purity is required to ensure low iron content to avoid contamination of the lens blanks, e.g., 200 ppm

or less. The bed should be of sufficient depth to blanket that portion of each
of the lenses which is to be heated to a temperature below the strain point.

Figure 5.17: Apparatus for Variable Temperature Treatment
of Lens Blanks

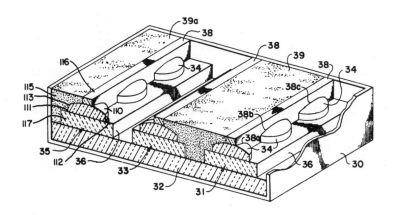

Source: U.S. Patent 4,154,590

A second bed of sand 39a is formed between a wall of the tray and the dam 38
in the series 35. Thus, about half of each lens in each series will be exposed to
full temperature of the atmosphere of the oven, while a lens under the sand is
raised to a lesser temperature. With the lenses, dams, and masking silica mounted
in the tray, the tray is passed through a conventional production furnace, providing heating as described above.

The tray with the silica and the mounted lenses and blocks is placed in a conventional production photochromic bifocal furnace, and the temperature is raised
to about 1220°F over a period of 90 minutes. Once the 1220°F temperature is
reached, the assembly with the lenses is permitted to soak at that temperature
for about 20 minutes. The lenses are thereafter cooled over a period of about
30 minutes to a temperature where they can be handled, that is about 100°F.

The glasses specified by the letters A, B, C, D and E of the following table, may be
used in the process. To use the glasses, the lens or lens blanks are prepared. The
strain point and the softening point of respective glasses are noted and the furnace is operated to permit an appropriate temperature above the strain point
but below the softening point.

The masking provides an appropriate temperature gradient along the lens. The
appropriate temperature gradient permits a potential upper portion of the lens
to have well-developed silver halide crystals, with a controlled progression to substantial freedom from nucleation at the bottom or potential bottom of the lenses
or lens blanks. There is thus established a graded thermal masking of the lenses,
or lens blanks.

Compositions in Weight Percent of Unnucleated Glasses

	A	B	C	D	E
SiO_2	53.0	21.4	58.8	57.3	0.0
Al_2O_3	10.5	37.7	22.9	9.1	8.3
ZrO_2	2.0	0.0	0.0	0.0	1.3
Li_2O	2.1	0.0	4.5	0.0	0.0
BaO	6.0	5.5	0.0	0.0	3.3
SrO	0.2	0.0	0.0	0.0	0.0
Na_2O	0.6	3.8	1.5	6.5	16.2
NaF	1.0	1.0	4.7	3.1	0.0
NaCl	1.0	1.0	1.8	2.6	1.0
Ag_2O	0.4	0.5	0.4	0.5	0.6
PbO	5.1	0.0	0.0	1.0	0.0
CuO	0.1	0.1	0.02	0.02	0.02
P_2O_5	0.0	15.6	0.0	0.0	7.5
B_2O_3	18.0	4.8	2.5	18.6	61.8
K_2O	0.0	8.6	0.0	0.0	0.0
NaBr	0.0	0.0	0.8	1.3	0.0
MgO	0.0	0.0	2.1	0.0	0.0

The maximum masking is over that area which is to serve as the reading, or bottom, portion of the lens, or lens blank, when it is glazed in a frame. There is substantially no masking over that area which is to serve as the distance portion and thus there is thermally introduced maximum nucleation.

Over-Nucleation of Selected Lens Portions

P.I. Kingsbury, Jr. and T.P. Seward, III; U.S. Patent 4,160,655; July 10, 1979; assigned to Corning Glass Works describe a process whereby gradient photochromic glass articles are provided through a process involving the nonuniform nucleation of potentially photochromic glass. This process is based on the fact that extensively nucleated potentially photochromic glass of the silver halide type does not readily develop a strong photochromic response characteristic during a subsequent regular heat treatment. Thus it was found that the development of photochromic properties in selected regions of a potentially photochromic glass article can be suppressed by, in effect, "over-nucleating" those regions prior to the normal heat treatment of the glass to develop photochromic properties therein.

The exact mechanism by which selective over-nucleation operates to provide a photochromic gradient in accordance with the process is not fully understood. However, it is presumed that extensive nucleation provides many sites on which photochromic particles may grow. During the subsequent heat treatment, the available photochromic material in the glass is used up early, resulting in many particles too small in size to be effectively photochromic. Thereafter only relatively long heating, which results in particle modification by an Ostwald-type ripening process, can produce strong photochromic characteristics in the over-nucleated regions of the glass.

Broadly, the process of producing a gradient photochromic glass article comprises the initial step of subjecting a selected limited portion of a potentially photochromic glass article to a nucleation heat treatment. The potentially photochromic glass is one containing silver and halogens, and the nucleation heat treatment is carried out at a temperature above the minimum temperature required for nucleation but below the minimum temperature required for the significant growth of photochromic silver halide particles for a time sufficient to achieve a degree of nucleation which will inhibit the subsequent development of photochromic properties in the selected heated portion of the glass. During this nucleation heat treatment, the portions of the glass which are not to be nucleated are maintained at a temperature below that required for effective nucleation. The selective nucleation step may be carried out using any of the conventional selective heating or insulating devices known in the art.

Following the completion of the nucleation heat treatment, the entire article is heat treated at a temperature sufficient to promote the growth of silver halide-containing particles in the glass, for a time at least sufficient to develop photochromic properties in the regions of the article other than the selected portion exposed to the prior nucleation heat treatment. In general, it is found that substantial photochromic development will not occur in the over-nucleated portions of the glass during a standard heat treatment. Thus excellent control over the photochromic properties of the photochromically developed regions of the article may be exercised, and the simultaneous heat treating and sagging of the article during gradient production is possible.

Example: A potentially photochromic glass article having a composition, in parts by weight as calculated from the batch, of about 56.2 parts SiO_2, 16.0 parts B_2O_3, 8.9 parts Al_2O_3, 1.8 parts Na_2O, 2.65 parts Li_2O, 5.0 parts PbO, 6.6 parts BaO, 2.17 parts ZrO_2, 0.2 part Ag, 0.32 part Cl, 0.63 part Br, 0.2 part F, 0.019 part CuO, 0.032 part NiO and 0.019 part CoO is provided for treatment. The glass article is a glass blank for an ophthalmic lens, having the form of a glass disc about 65 mm in diameter and 3 mm in thickness.

The lens blank described is positioned on the floor of a small refractory brick cavity having a top port through which an operating radiant electric heating element directs heat onto approximately one-half of the top surface of the lens blank. The remainder of the lens blank is shielded from radiant heat. The distance between the radiant heater and the heated portion of the lens blank is about 2 inches.

The lens blank is kept in the cavity for a heating interval of 15 minutes to over-nucleate the heated portion of the glass. During this interval a thermocouple positioned between the heating element and the heated glass indicates a temperature of about 580°C.

At the end of the heating interval the glass with over-nucleated and shielded portions is removed from the cavity and positioned in an air-circulating chest furnace where it is uniformly heated to the furnace operating temperature of 625°C. The glass is kept in the furnace for 15 minutes and then removed from the furnace, cooled, and examined for photochromic properties.

The undarkened lens blank has a uniform visible transmittance across its diameter, a typical value in 3 mm thickness being about 67%. After photochromic darkening of the lens by exposure to a black-light blue fluorescent lamp for an interval of 20 minutes, transmittance measurements are taken across a diameter of the lens running through both the photochromic (originally shielded) and non-photochromic (over-nucleated) regions in a direction at right angles to a line dividing these regions. The following data is recorded:

Measurement Position	Visible Transmittance (%)
Over-nucleated edge	45
¼ Diameter	45
Center	45
¾ Diameter	16
Photochromic edge	11

Manufacture of Gradient Lenses Using Heat Sink Material

E.W. Deeg; U.S. Patent 4,149,868; April 17, 1979; assigned to American Optical Corporation describes a method of treating ophthalmic quality lenses or lens blanks that produces a reversible progressive local variation in phototropic behavior with a continuous variation in transmissivity. The lens or lens blank is composed of a potentially phototropic glass containing all the necessary ingredients including uniformly dispersed silver halide particles therein to develop a phototropic behavior during the heat treatment of the process.

The steps of the method include mounting the lens or lens blank in carrier means, heating the lens or lens blank thus mounted in a heat treatment furnace at a temperature sufficient to develop the phototropic behavior of the potentially phototropic glass. Characteristically the required heat treatment temperature to develop this phototropic behavior is one which exceeds the strain point of the glass but not the softening point thereof. During the heat treatment a further step is maintaining selected portions of the lens or lens blank at a temperature low enough to substantially inhibit the development of the phototropic behavior and thus produce a progressive gradient of transmissivity.

The particular improvement of the process includes heat sinking the selected portion of the lens to be maintained at a temperature which substantially inhibits the development of phototropic behavior by utilizing the latent heat of transformation of a material which has at least one phase transformation temperature below the temperature required for development of phototropic behavior. This heat sinking is accomplished by positioning the heat sinking material proximate to the portion of the lens or lens blank in which the temperature is to be maintained to substantially inhibit the development of phototropic behavior. A path of heat conduction between the heat sinking material and the select portion of the lens or lens blank is established.

Table 1 below is a selection of suggested compositions of potentially photochromic glass useable in the process.

Table 1: Compositions of Potentially Photochromic or Phototropic Glasses

	A	B	C	D	E
SiO_2	53.0	21.4	58.8	57.3	0.0
Al_2O_3	10.5	37.7	22.9	9.1	8.3
ZrO_2	2.0	0.0	0.0	0.0	1.3
Li_2O	2.1	0.0	4.5	0.0	0.0
BaO	6.0	5.5	0.0	0.0	3.3
SrO	0.2	0.0	0.0	0.0	0.0
Na_2O	0.6	3.8	1.5	6.5	16.2
NaF	1.0	1.0	4.7	3.1	0.0
NaCl	1.0	1.0	1.8	2.6	1.0
Ag_2O	0.4	0.5	0.4	0.5	0.6
PbO	5.1	0.0	0.0	1.0	0.0

(continued)

Table 1: (continued)

	A	B	C	D	E
CuO	0.1	0.1	0.02	0.02	0.02
P_2O_5	0.0	15.6	0.0	0.0	7.5
B_2O_3	18.0	4.8	2.5	18.6	61.8
K_2O	0.0	8.6	0.0	0.0	0.0
NaBr	0.0	0.0	0.8	1.3	0.0
MgO	0.0	0.0	2.1	0.0	0.0

Table 2 below is a selection of suggested compounds useable as the heat sink material exhibiting a change of phase transition in the appropriate temperature range. In the table, T is temperature expressed in degrees centigrade, H is latent heat of transformation, expressed in kilojoules per mol, C is specific heat expressed in joules per mol per degree centrigrade, M is molecular weight, P is specific gravity expressed in grams per cubic centimeter, Q is the approximate energy per unit volume absorbed by the heat sink material in the temperature range from 25° to 520°C expressed in kilojoules per cubic centimeter and F is a figure of merit expressed in joules per cubic centimeter per degree centigrade.

Table 2: Example of Active Heat Sink Materials

Substance	M	P	C	T_1	H_1	T_2	H_2	T_3	H_3	T_4	Q	F
H_2O	18.0	1.00	75.2	100	40.7	−	−	−	−	−	4.6	46.0
$Na_2SO_4 \cdot 10H_2O$	322.2	1.46	575.2	32	69.0	..... see H_2O and Na_2SO_4					2.2	9.1
NaOH	40.0	2.13	59.5	239	6.4	319	6.4	1,390	−	−	2.5	7.8
Sn	118.7	7.29	26.4	13	2.1	232	7.1	2,690	−	−	1.5	6.5
Na_2SO_4	142.0	2.70	127.7	177	3.1	241	7.0	884	−	−	1.5	6.2
Zn	65.4	7.13	25.4	420	7.3	907	−	−	−	−	2.3	5.5
Se	78.9	4.79	25.4	217	5.4	685	−	−	−	−	1.2	5.5
Cd	112.4	8.64	26.0	321	6.4	765	−	−	−	−	1.6	5.0
Bi	209.0	9.79	25.6	271	10.9	1,560	−	−	−	−	1.2	4.4
$ZnCl_2$	136.3	2.91	76.5	318	23.0	721	−	−	−	−	1.4	4.4
$K_2Cr_2O_7$	294.2	2.69	220.0	242	−	398	37	−	−	−	1.5	73.8
AgCl	143.3	5.56	55.8	455	12.7	1,547	−	−	−	−	1.7	3.7
Te	127.6	6.25	25.6	450	17.5	990	−	−	−	−	1.6	3.6
Tl	204.4	11.85	26.4	234	0.3	304	4.2	1,457	−	−	1.1	3.6
$CdCl_2$	183.3	4.05	76.6	564	48.6	960	−	−	−	−	2.0	3.5
Pb	207.2	11.34	26.8	327	4.8	1,751	−	−	−	−	1.1	3.4
BrO_3	69.6	1.84	63.0	450	23.0	2,247	−	−	−	−	1.5	3.3
CuBr	143.5	4.72	54.7	380	5.9	465	2.9	488	9.6	1,318	1.6	3.3
AgBr	187.8	6.47	52.4	259	−	430	9.2	1,533	−	−	1.3	3.0
$Ca(NO_3)_2$	164.1	2.47	149.2	561	21.3	−	−	−	−	−	1.6	2.8
$PbBr_2$	367.0	6.67	80.1	488	18.5	892	−	−	−	−	1.2	2.5
$PbCl_2$	278.1	5.85	77.8	498	23.6	951	−	−	−	−	1.4	2.8
Li	133.8	4.06	54.4	449	5.9	1,170	−	−	−	−	1.1	2.4
CdI_2	366.2	5.67	77.5	387	15.3	742	−	−	−	−	0.9	2.3
PbI_2	461.0	6.06	81.2	412	25.1	872	−	−	−	−	0.9	2.2

Copper-Cadmium Halide Glasses

R.J. Araujo and P.A. Tick; U.S. Patent 4,166,745; September 4, 1979; assigned to Corning Glass Works describe compositions for refractive index-corrected copper-cadmium halide photochromic glasses which provide not only the required refractive index, optical clarity and photochromic response necessary for ophthalmic use, but also good chemical durability, a liquidus-viscosity relationship satisfactory for conventional forming processes, and excellent ion-exchange strengthening characteristics. Compositions exhibiting these combined properties include those consisting essentially, in weight percent by analysis, of about 45 to 56% SiO_2, 14.5 to 21% B_2O_3, 9.0 to 15% Al_2O_3, 1.4 to 2.4% Li_2O,

2 to 12% Na_2O, 0 to 6.0% MgO, 0 to 3.5% BaO, 0 to 2.5% TiO_2, 0 to 1.4% ZrO_2, 0 to 1.5% La_2O_3, at least 8% total of MgO + BaO + TiO_2 + ZrO_2 + La_2O_3, 0.1 to 0.5% CuO, 0.3 to 1.5% CdO, 0.3 to 0.9% Cl, 0 to 0.6% SnO and 0 to 2% F.

Compositions within the abovedescribed range exhibit satisfactory forming behavior, having a viscosity at the liquidus of at least about 3,000 poises. They also provide glass products which are low in haze and which exhibit a refractive index (n_D) in the 1.52 to 1.54 range (e.g., 1.523). These products typically possess excellent chemical durability, characterized by freedom from visible surface attack following a 10 minute exposure to 10 wt % HCl at 25°C, and are heat-treatable to provide photochromic properties including, in 2 mm thickness, a transmittance not exceeding about 55% in the fully darkened state and fading of at least about 12 percentage points of transmittance in a 5 minute fading interval from the fully darkened state.

Finally the products are chemically strengthenable by known sodium-for-lithium ion exchange processes to provide unabraded modulus of rupture strengths of at least about 35,000 psi, with an ion-exchange layer depth of at least about 3 mils as determined by conventional stress layer examination techniques utilizing, for example, a polarizing microscope with a Babinet compensator. These strength and compression layer characteristics, which are readily obtainable by sodium-for-lithium salt bath ion-exchange processes at normal ion-exchange temperatures (300° to 450°C), permit ophthalmic lenses of 2 mm thickness to routinely pass standard ball drop impact tests.

Particular examples of glass compositions within the scope of the process are set forth below. Compositions are reported in parts by weight on the oxide basis, except for the halogens which are reported on an elemental basis in accordance with prior reporting practice.

	. . . . Compositions (pbw)							
	1	2	3	4	5	6	7	8
SiO_2	50.5	51.3	49.3	51.0	49.1	54.9	50.0	55.0
Al_2O_3	10.2	10.2	10.1	10.4	13.3	9.3	10.5	8.5
B_2O_3	19.6	20.6	19.6	20.2	20.3	17.9	20.6	16.5
Na_2O	3.1	3.6	4.1	2.9	3.1	3.1	3.7	3.5
Li_2O	1.8	1.4	1.8	1.9	2.5	1.8	2.6	1.5
K_2O	–	–	–	–	0.5	0.5	–	–
MgO	5.1	5.1	6.1	6.4	3.5	4.1	6.1	4.0
BaO	2.9	2.9	2.9	2.9	2.2	2.2	2.2	3.5
TiO_2	1.4	1.65	1.4	1.3	1.8	2.0	2.1	2.0
ZrO_2	0.8	1.1	0.8	0.9	–	0.8	0.6	1.1
La_2O_3	1.0	1.0	1.0	1.2	0.9	1.0	–	1.0
CuO	0.19	0.25	0.29	0.28	0.35	0.30	0.34	0.30
CdO	0.55	1.0	1.0	0.9	1.0	1.1	1.0	1.0
Cl	0.6	0.7	0.7	0.7	0.7	0.7	0.55	0.7
F	1.06	1.2	1.06	1.0	1.28	1.1	2.0	1.1
SnO	0.12	0.10	0.25	0.09	0.07	0.06	0.28	0.10
Cr_2O_3	–	–	0.041	–	–	–	–	0.015

To prepare glass articles of the above compositions, glass batches may be compounded utilizing conventional glass batch constituents and the batches then melted in crucibles or optical glass melting units by heating to temperatures of about 1300°C for a melting interval of about 3 hours. The resulting melts may then be cast into glass patties and placed in an annealer.

Colored Glasses Exhibiting Photoanisotropic Effects

C.-K. Wu; U.S. Patent 4,191,547; March 4, 1980; assigned to Corning Glass Works describes a photosensitive colored glass exhibiting photoanisotropic effects consisting of a body portion and an integral surface layer thereon having a thickness of about 1 to 500 microns, at least a portion of which exhibits photodichroic and birefringent properties. The method consists of the following general procedure:

An anhydrous glass body consisting essentially in mol % on the oxide basis, of about 70 to 82% SiO_2, 10 to 17% Na_2O and/or K_2O, 5 to 15% ZnO, 0.5 to 5% Al_2O_3, and 0.1 to 3% Cl is contacted with an aqueous solution containing Ag^+ ions and acidified with a mineral acid to a pH less than about 2, this contact being made at a temperature in excess of 200°C and at a pressure in excess of 225 psig for a period of time sufficient to hydrate a surface layer thereon and to cause the replacement of Na^+ and/or K^+ ions with Ag^+ ions in the hydrated glass, the proportion of Na^+ and/or K^+ ions in the hydrated glass being less with a corresponding increase in Ag^+ and/or H^+ (or H_3O^+) ions. The Ag^+ ions react with Cl^- ions in the hydrated glass to effect the formation of Ag-AgCl-containing crystals, thereby rendering the glass photosensitive.

Up to about 25% by wt of silver can be incorporated into the hydrated glass structure, the amount so incorporated being a function of the Na_2O and/or K_2O level in the glass composition. Generally, the silver present in the surface layer will range up to about 25% by wt, with contents of about 3 to 20% appearing to produce the maximum photoanisotropic effects.

Thereafter, the hydrated photosensitive surface layer was exposed to ultraviolet radiation at an intensity and for a time sufficient to impart photoanisotropic properties thereto. Depending upon the thickness of the surface layer and the depth desired, this exposure time can vary from as little as 0.25 hour up to, perhaps, 120 hours.

The ultraviolet-darkened, thin surface layer developed by the method can display various shades of blue coloration ranging from greenish-blue, through saturated blue, to purple due to the strong absorption in the red portion of the spectrum. This absorption peaks at wavelengths between 650 and 800 nm, depending upon glass composition and processing parameters. Furthermore, the intensity of the blue coloration, i.e., the optical density in the red portion of the spectrum, is proportional to the energy density of the ultraviolet exposure.

Inasmuch as the glasses are not darkened under conventional indoor lighting and are resistant to thermal fading, the thin ion exchanged layer may be darkened to any desired degree and will remain at that value of optical density for an indefinite period of time. A further exposure to intense ultraviolet radiation, however, can alter the optical density of the glass.

Bleaching with polarized light induces an optically anisotropic state which is manifested via the phenomena of birefringence and dichroism. This optically-induced anisotropy, coupled with a high resolution capability resulting from the fine-grained nature of the photodichroic layer developed by the method, suggest the utility of these materials for optical recording. The bleaching resistivity, or the energy density demanded to create a sufficient dichroic absorption difference for reading, is an important parameter which must be considered for such an application.

Introduction of Silver Ions into Hydrated Glass

In a similar process, *R.F. Bartholomew, J.F. Mach and C.-K. Wu; U.S. Patent 4,160,654; July 10, 1979; assigned to Corning Glass Works* have found that silver ions (Ag^+) can be introduced into an alkali metal silicate glass either (1) concurrently with or (2) subsequently to hydration of the glass.

In the first embodiment of the process, the alkali metal silicate glass, in the form of bodies having thickness dimensions less than about 5 mm, is contacted with aqueous silver salt solutions having a pH no greater than 4 at temperatures of at least about 100°C and at pressures of at least 20 psig. This practice results in the simultaneous occurrence of glass hydration and the exchange of Ag^+ ions for alkali metal ions in the glass. Acidification of the silver salt solution with very minor additions of acid, e.g., less than 1% by wt of a mineral acid such as HNO_3, will frequently yield a product of greater optical clarity and one displaying less of the amber coloration characteristically exhibited by the presence of reduced silver in glass.

In the second embodiment of the process, the alkali metal silicate glass is subjected to the steam hydration-dehydration procedure set forth in U.S. Patent 3,912,481, and Ag^+ ions are thereafter introduced into the hydrated glass by means of contacting the glass with an aqueous solution of a silver salt having a pH not exceeding about 5 at a temperature in excess of about 100°C. The exchange of Ag^+ ions for alkali metal ions in the glass will occur at ambient atmospheric pressure, although the use of added pressure or, conversely, a partial vacuum does not adversely affect the rate of exchange. Again, acidification of the aqueous silver salt solution appears to yield a more satisfactory final product. Normally, the ion exchange reaction will be conducted at temperatures between about 150° and 250°C. However, treatments up to the critical temperature of water, i.e., 374°C, can be contemplated.

Both process embodiments can be applied equally satisfactorily with glass compositions operable in U.S. Patents 3,912,481 and 3,948,629, such glasses consisting essentially, in mol % on the oxide basis, of about 3 to 25% Na_2O and/or K_2O and 50 to 95% SiO_2, the sum of those components constituting at least 55% of the total composition. Other constituents can be included to modify the chemical and physical properties of the hydrated glass and/or the parent anhydrous glass. As illustrative of such, Al_2O_3, B_2O_3, BaO, CaO, CdO, MgO, PbO, and ZnO can be useful in altering the melting and forming characteristics of the base glass and/or enhancing the chemical durability of both the base glass and the hydrated glass. BaO, B_2O_3, CaO, PbO, and ZnO can be included in amounts up to about 25%; MgO can be useful up to about 35%; and Al_2O_3 can be operable up to about 20%.

Where other optional components are utilized, it is preferred that individual additions thereof not exceed about 10%. Li_2O appears to inhibit hydration such that, if present at all, the quantity will be held below about 10%. The inclusion of CaO frequently yields a translucent or opaque hydrated body. Therefore, its use will be avoided where a transparent body is demanded. The common glass colorants such as CdS–Se, CoO, Cr_2O_3, Fe_2O_3, MnO_2, NiO, and V_2O_5 may also be included in customary amounts up to a few percent. Self-evidently, where the function of these ingredients is not restricted to coloring, individual additions up to 10% can be tolerated. Finally, conventional fining agents can be employed in usual amounts.

Although concentrated aqueous solutions of silver salts can be satisfactorily utilized, the properties exhibited by the final product are not significantly different from those obtained utilizing dilute solutions. Hence, a 10% by wt silver ion concentration seems to be a practical maximum with a preferred range being about 0.1 to 2% by wt. Also, increased Ag^+ ion concentration causes the glass to develop a dark amber tint.

As a matter of convenience, hydration of the base glass in either embodiment will preferably be undertaken in an autoclave, because such apparatus permits relatively easy control of the contacting temperature, pressure, and atmosphere. Also, again as a matter of convenience, the maximum hydration temperature employed will be limited to 374°C, the critical temperature of water.

Bifocal Lens System

M. Faulstich and G. Gliemeroth; U.S. Patent 4,149,896; April 17, 1979; assigned to Jenaer Glaswerk Schott & Gen, Germany describe a multifocal, phototropic glass system formed by fusing two preformed phototropic glasses together, one glass in such system having a higher index of refraction than the other, both glasses in such system having similar thermal coefficients of linear expansion.

Preferred phototropic glass compositions which are particularly well suited for use in combination with prior art remote portion glasses, are formed from the following starting components in the respective amounts indicated:

	Percent by Weight
SiO_2	~5-30
B_2O_3	~7-35
PbO	~6-26
ZnO	0-~15
La_2O_3	~12-30
Al_2O_3	12-25
ZrO_2	0-~6
TiO_2	0-~3
K_2O	0-~2
Na_2O	0-~2
Li_2O	0-~4
Ag_2O	~0.1-1.8
CuO	0-~0.05
CoO	0-~0.01

In such preferred compositions, the following anion-portions have replaced the oxygen: Cl, from about 0.2 to 4.5% by wt; Br + I, from 0 to about 4.0% by wt; and F, from 0 to about 4.0% by wt.

In such preferred compositions, the following compositional conditions are maintained:

The total content of alkali oxides ranges from about 0.2 to 8% by wt,

The total content of Al_2O_3 and La_2O_3 ranges from about 24 to 54% by wt,

The total content of alkaline earth oxides ranges from 0 to about 6.51% by wt with the provision that the MgO is kept below about 5.0% by wt,

The total content of ZnO plus ZrO_2 and plus TiO_2 ranges from 0 to about 25% by wt.

The total amount of oxides of Bi, Ta, Nb and W ranges from 0 to about 5% by wt.

The production of phototropy in the glasses of this process takes place by means of an annealing process which is conducted so that separations rich in silver and halogen can form in this glass. This annealing process can be coupled with the melting (initial melt fusing) process, or it can be carried out separately and subsequently after a melting operation. Thus, either a melt or a previously formed solid glass is annealed at temperatures in the range of from about 550° to 675°C for times sufficient to segregate silver halide-rich discontinuities typically ranging from about 50 to 200 Å in size dispersed uniformly in a matrix phase of the glass composition.

This phototropic glass is useful in lens manufacture (as for single strength spectacle glasses or the like). This glass is also useful as near portion material (that means segment glass) in the manufacture of phototropic multifocal glass and such constitutes a preferred use therefor.

It was found in glasses of this process that, with the aid of a combination of Li_2O with other alkali oxides, the inclination to turbidity or opacity of the annealed glass, is influenced in the visible spectral range. While utilization of Li_2O with unadapted selection of the further glass components may easily lead to undesired turbidity or opacity phenomena with too high annealing temperatures, this combination of Li_2O with other alkali oxides brings a clear improvement. The reason may be (and there is no attempt to be bound by theory herein) that the separating maximum of the particles causing turbidity or opacity apparently shifts to higher temperatures above the annealing temperature, so that the range of the annealing process no longer overlaps so strongly upon the temperature range where the opacifying process occurs.

Such a shifting of this region where opacifying particles occur to higher temperatures, and such a narrowing of the temperature range over which opacification occurs, was necessary for the solution of the problem of providing a phototropic glass suitable for melting with a far portion glass material. The glass of the process has such a range shift and such a temperature narrowing.

Example: This example illustrates preparation of a multifocal phototropic lens.

There are weighed in for the production of a near-portion glass suitable for fusion with commercial, phototropic, remote-portion glasses the following components. Each component is substantially pure and has an average grain size below about 0.5 mm in maximum particle dimension (0).

	Grams
High purity powdered quartz	2,350.25
H_3BO_3	16,233.50
Zinc oxide	1,524.00
Red lead	4,597.50
Aluminum trihydrate	8,428.50
Lanthanum oxide	5,281.50
Zirconium oxide	954.00
Rutile powder	138.50
Silver nitrate	44.00
Arsenic oxide	153.00
Lithium carbonate	500.00
Lithium bromide	1,327.50
Cupric oxide	6.00
Nickel oxide	12.44

These components are intimately mixed and 25 ml water is added. The mixture thus moistened and homogeneous is inserted discontinuously in charges of 1,000 grams into a platinum crucible which is heated in advance to 1260°C. The introduction process in this connection continues for 8 to 10 hours. When all of the mixture has been introduced into the crucible, the temperature is increased to 1285°C; after attainment of this temperature, the glass is thoroughly agitated for 10 minutes with a platinum rod 8 mm thick and at a rate of revolution of the agitator of 22 rpm.

The molten glass or glass melt thus homogenized is cooled to 1215°C and subsequently poured into rod-molds of 20 mm width, 30 mm height and 1,000 mm length. The melt rigidifies upon pouring or casting and cools below 500°C in the molds. The molds before attainment of room temperature are placed after pouring in an annealing oven, there again heated to 517°C, left for 27 minutes at this temperature and then cooled with a cooling speed of 7°C per hour to room temperature. Subsequently the rods are sawed out of the phototropic near-portion glass in discs 3.5 mm thick; out of these discs are then finished by means of grinding and polishing, the generally half-moon-shaped near-portions required in their known form for the fusing process.

These near-portions are heated at 40°C per minute to 280°C in an electrically heated chamber-oven or -furnace, inserted with tongs wound about with asbestos in the pressure-head (upper part) of a hydraulic 500 kp-press. In the lower part of the press-head of this press is inserted a remote-portion-glass-blank of commercial size 60 mm x 0.6 mm, which is disposed in a pure aluminum basin whose floor or bottom is curved concavely with a 90° radius and which corresponds to the shape of the bifocal glass later desired with a tolerance of $\pm{}^2/_{10}$ mm, and which is heated with the remote-portion glass to 975°C with a heating speed of 75°C per minute in a second chamber-oven or -furnace.

After pressing-in of the harder near-portion glass into the softer remote-portion glass, the two glasses are combined into a bifocal lens system which is permitted to cool in air at a temperature below the cooling temperature (this temperature is 512°C in the near-portion glass or 547°C in the remote-portion glass); accordingly, the cooling temperature is less than 512°C. Subsequently, the product bifocal lens system is heated at 605°C in a chamber-oven or chamber-furnace, for 65 minutes to produce the phototropy in each of the near- and remote-portions and subsequently the product lens is cooled to room temperature at a speed of

8.5°C per minute. The phototropic bifocal lens blank thus fused is subsequently subjected to the usual grinding- and polishing-processes to produce a product finished lens.

OTHER PROCESSES

Glass Microspheres with High Refractive Index

Ever since the 1940s, when retroreflective sheeting based on a monolayer of embedded microsphere lens elements was introduced and came into widespread use, there has been a desire for glass microspheres having increased indexes of refraction.

Despite the desire for high-index microspheres, the art has never, insofar as known, provided commercially useful glass microspheres having a refractive index higher than about 2.7. The highest-index, prior-art microspheres have been based on large amounts of Bi_2O_3 or PbO, usually in combination with rather large amounts of TiO_2 (see U.S. Patents 2,726,161 and 2,853,393, which describe some PbO-Bi_2O_3-based examples having indices up to 2.59). None of these prior-art microspheres have been commercialized for one reason: because the index of refraction obtained is not high enough to eliminate or sufficiently reduce the thickness of the spacing layer. Also, the PbO-containing products are now considered especially unsuited to commercial use for pollution reasons.

U.S. Patent 3,560,074 states that glass microspheres made from 95 to 100 wt % TiO_2 would have a refractive index of 2.9, but insofar as known, glass microspheres of such composition or index have never been made available to the retroreflective sheeting art. British Patent 1,472,431 shows microspheres that comprise over 50 wt % polycrystalline titanium dioxide in anatase form, but no commercial microspheres of this type have been supplied.

A different prior-art approach to increased refractive index (U.S. Patent 3,149,016) is a process for heat-treating glass microspheres to cause a structural rearrangement within the microspheres. The structural rearrangement, which is suggested in the patent to be development of crystal nuclei, increases the index of refraction of the microspheres. The highest-index microspheres exemplified in the patent, having an index raised from 2.47 to 2.7 by the heat-treatment process, comprise 67.5 parts PbO, and, as noted above, this dependence on PbO limits the utility of the microspheres.

C.F. Tung and J.A. Laird; U.S. Patent 4,192,576; March 11, 1980; assigned to Minnesota Mining and Manufacturing Company describe a process which provides for the first time, insofar as known, solid nonporous transparent glass microspheres free of any substantial proportion of PbO and having a refractive index of 2.7 or higher. These microspheres principally comprise 65 to 85 wt % Bi_2O_3 and 5 to 35 wt % TiO_2, though other ingredients may be included in amounts totaling up to about one-fourth of the composition by weight, preferably less than 10%.

It has been found that, when microspheres of this composition are heat-treated in the manner taught in the abovediscussed U.S. Patent 3,149,016, their index of refraction is increased to at least 2.7, and in preferred compositions to 2.75

or more. Insofar as known, these are the highest-index glass microspheres ever prepared, and they open the way to retroreflective sheeting and other retroreflective products of simplified construction and improved properties.

For example, space coats can be eliminated from retroreflective sheeting of the embedded-microsphere-type by use of microspheres of the process and top coats having a refractive index of 1.45 or less; and the latter can be obtained by mixing small amounts of lower-index polymeric ingredients such as fluorinated polymers with more conventional-index polymeric ingredients such as acrylic-based resins.

Conventional space coats can also be eliminated, even with 1.5 index top coats, by vapor-depositing or otherwise forming very thin transparent layers on the microspheres. For example, low-index materials such as cryolite (Na_3AlF_6) vapor-deposited in thicknesses of about 2 μm or less are useful with microspheres of the process. Such thin coatings can be vapor-deposited in uniform thicknesses, and they avoid many of the limitations of conventional, thicker space coats.

Microspheres of the process are also useful in many other constructions besides retroreflective sheeting. For example, the microspheres can be hemispherically reflectorized in the manner taught in U.S. Patent 2,963,378 (e.g., the microspheres are partially embedded into a carrier such as a polyethylene film; then vapor-coated with a specular reflector such as aluminum; and then removed from the carrier), to form reflective lens elements useful as an additive to transparent films to make the films retroreflective. Typically such films are formed by dispersing the reflective lens elements into a variety of liquid compositions that comprise a film-forming binder material and then coating, casting, or extruding the composition as either a coating on a substrate or an unsupported film. The reflective lens elements are useful when totally embedded in the film, and, as a result, will provide retroreflection whether wet or dry.

If the transparent films in which reflective lens elements are incorporated comprise a low-index binder material, the hemispherical reflector of the reflective lens elements can be in direct contact with the microspheres; if more conventional index materials are used, a thin transparent layer can be vapor-deposited onto the microspheres before coating of the hemispherical reflector. Although the noted U.S. Patent 2,963,378 contemplates a similar use of vapor-deposited space coats, the large thickness needed with the comparatively low-index microspheres available at the time of that teaching prevented effective commercial practice of such a technique. With microspheres of this process, the space coat can be thin enough, generally about 2 μm or less, to provide effective, adequately bonded space coats, thereby forming reflective lens elements useful to reflectorize a wide variety of transparent films.

Microspheres of the process can be prepared by known processes; see U.S. Patent 3,493,403, columns 5 and 6, which describe a method for first pulverizing and then intimately blending and sintering raw materials together; then crushing the sintered blend and screening to appropriate size; and then passing the screened particles through a flame, where they fuse and become spherulized.

The microspheres are then heated, e.g., by placing a tray of the microspheres in a furnace, or by moving them through the furnace on a conveyor belt. As taught in U.S. Patent 3,149,016, the heat-treatment operation is understood as causing a structural rearrangement within the microspheres, such as a crystal nucleation.

Bronze-Tinted Windshield Glass

K. Fischer and R. Keul; U.S. Patent 4,190,452; February 26, 1980; assigned to Saint-Gobain Industries, France describe a neutral bronze glass consisting essentially of the following ingredients in percentages by weight:

	Percent by Weight
SiO_2	60–75
B_2O_3	0–7
Al_2O_3	0–5
Na_2O	10–20
K_2O	0–10
CaO	0–16
MgO	0–10
Fe_2O_3	0.20–0.5
CoO	0.0025–0
Se	0.0025–0.0005

The glass has, at a thickness of 3 mm over the range of wavelengths from 400 to 750 nm, a luminous transmittance above 70%, an excitation purity between 2 and 6%, a monochromatic transmittance to which the best straight line fit has at 400 nm a value between 5 and 15% lower than its value at 750 nm, and a dominant wavelength between 570 to 580 nm.

The process provides tinted glass which has improved properties from the point of view of physiological optics for the persons receiving light and seeing through such windows. Thus, automobile windows and windshields made of the glass of this process assure the transmission therethrough of light in the visible range essential to the needs of automobile drivers, while making it possible to distinguish easily among all colors under various conditions of light and weather. The glass has excellent antiglare and improved visibility properties generally.

ELECTRICAL GLASS

PASSIVATION OF SEMICONDUCTORS

Germanate Glass Coating

H.J.L. Trap; U.S. Patent 4,156,250; May 22, 1979; assigned to U.S. Philips Corporation describes glass which is suitable for passivating semiconductor devices and devices in which such a passivating glass coating is used.

Such a coating is intended for protecting the semiconductor, the p-n junctions in such a device in particular, from atmospheric influences such as water, water-vapor and from those of migrating ions. These migrating ions may emanate as contaminants from the ambient medium or from the casing, which consists, for example, of synthetic resin material.

The glass is characterized in that its composition in mol percent is in the following ranges: SiO_2, 50-70; Al_2O_3, 3.5-20; PbO, 7.5-40; GeO_2, 3-20; and $PbO/Al_2O_3 \geqslant 1$, preferably approximately 2. The glass contains water, expressed in cubic centimeters water vapor, reduced to 0°C and 76 cm Hg per cubic centimeter of glass, between 10×10^{-4} and 100×10^{-4}.

The glass, according to the process, differs from the prior art glass in that it has a much higher breakdown voltage and it gives the semiconductor device passivated herewith a much longer life.

Addition of Cordierite

G. Muller; U.S. Patent 4,133,690; January 9, 1979; assigned to Jenaer Glaswerk Schott & Gen., Germany, describes glass compositions useful for passivating silicon semiconductor elements or bodies and semiconductor elements or bodies coated by such glass compositions.

The glass compositions are applied onto the surface of the semiconductor in the form of a finely-ground powder and thereafter fused on at least a portion of the semiconductor element.

The composition comprises: (a) glass; and (b) powdered cordierite in an amount effective to provide a coefficient of thermal expansion of up to 40 x 10^{-7}/°C for the composition within the temperature range of 20° to 300°C. The compositions are compatible with the thermal expansion of the silicon semiconductor and capable of adhering to the silicon in layers greater than 10 μm without the formation of cracks.

Example: Starting with the raw materials of quartz sand (SiO_2), $Al(OH)_3$, H_3BO_3 and minium (Pb_3O_4), a molten glass was produced with the following synthesis composition: 42% by weight SiO_2, 3% by weight Al_2O_3, 10% by weight B_2O_3 and 45% by weight PbO.

This glass was ground and sifted through a sieve with a 60 μm mesh width. 20% by weight cordierite, in proportion to the total mixture, was then added to the glass. The passivating glass produced in this manner was applied in the form of an aqueous suspension in a layer of about 250 μm thickness on the surface of Si-rectifiers with mesa configuration, dried and subsequently melted on at 700°C for 10 minutes. (The term "passivating glass" was retained in accordance with the customary technical terminology, even though the material is in fact a mixture of a glass and a filler.)

After cooling, the surface (circumference) of the mesa rectifiers was covered with a gap-free, porcelain-type, white glass layer of about 200 μm thickness, which exhibited no cracks after it had been treated repeatedly by temperature shocks resulting from ice water and boiling water. The rectifiers which had been coated in this manner showed stable electric properties.

The passivating glass was sintered into rods in order to determine the thermal expansion. An expansion of 38 x 10^{-7}/°C was measured on these rods in comparison to an expansion of 46 x 10^{-7}/°C for the pure glass without any cordierite modification.

OTHER ELECTRICAL USES

Electrical Coating on Inner Surface of Glass Tubing

C.C. Lagos, R.A. Lanio, J.L. Crowley and N.A. Moreau; U.S. Patent 4,175,941; November 27, 1979; assigned to GTE Sylvania Incorporated describe the deposition of a transparent electrically conductive tin oxide film on the inner surface of glassware, for example, tubing to be used, for example, in the manufacture of fluorescent lamps. The use of such a film is disclosed in U.S. Patent 4,020,385.

In the prior art, such tin oxide films were generally deposited by means of liquid sprays or liquid-gas sprays or by chemical vapor deposition techniques. The temperature of the glass during deposition was less than its softening temperature.

In this process, the tin oxide film is deposited on the glass while the glass is at a temperature above its softening point temperature. The film is deposited from an aqueous solution that is applied to the softened glass in the form of a liquid stream. The solution contains tin chloride and hydrofluoric acid. Figure 6.1 is a diagrammatic representation of apparatus that can be used in the process.

Figure 6.1: Apparatus for Internal Coating of Glass Tubing

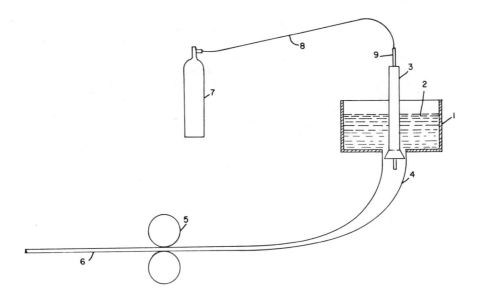

Source: U.S. Patent 4,175,941

As shown in the figure, a tank **1** contains molten glass **2** which is drawn down around mandrel **3** to form a bag **4** of softened glass which is drawn by tractor **5** to form tubing **6**. A container **7** contains a solution as per this process, which is carried by line **8** into tube **9**. Tube **9** is disposed within mandrel **3**, and the bottom of tube **9** projects into bag **4** where the solution can be disposed onto the softened glass.

Because the solution impinges on molten glass bag **4** as a liquid stream, the viscosity and the specific gravity must be controlled, so that soft molten glass bag **4** and the first few feet of soft glass tubing **6** are not distorted or otherwise physically affected by the liquid stream.

The viscosity should be between about 2 to 8 cs and the specific gravity should be no heavier than about 1.65. The viscosity and specific gravity are dependent on the concentration of $SnCl_4$ in the solution which also controls the conductivity of the film formed on the glass tubing. The rate of flow of the solution should be at least enough to maintain a continuous stream of liquid at the exit end of tube **9**.

In one embodiment, the solution comprises tin chloride ($SnCl_4$) and hydrofluoric acid (HF) in water (H_2O). The $SnCl_4$ can be added as $SnCl_4$ or $SnCl_4 \cdot 5H_2O$ or any other form that will give Sn^{+4} and Cl^{-1} in solution. The preferred concentration of $SnCl_4$ is between 560 and 900 g/ℓ. The HF concentration may be varied from about 0.5 to 10% by weight, based on the equivalent SnO_2 concentration. Lower percentages of HF (~1.5%) are preferred because of the corrosiveness of the higher HF concentrations.

In order to increase the electrical conductivity of the tin oxide film, a water-soluble alcohol such as methyl, ethyl or propyl may be included with the water in an amount up to about 75% by volume of the total solvent.

Sodium-Ion-Conducting Sodium Aluminum Borate Glasses

C.L. Booth; U.S. Patent 4,190,500; February 26, 1980; assigned to E.I. DuPont de Nemours and Company has found that sodium aluminum borate glasses with compositions corresponding to points that lie within the quadrilateral ABCD region of the Na_2O, B_2O_3 and Al_2O_3 ternary diagram of Figure 6.2 where the mol percent of the compositions corresponding to the corners of the quadrilateral (shown in the table) and containing additions of about 2 to 11 mol % of at least one of ZrO_2, MgO, P_2O_5 and SiO_2 are good ionic conductors and are resistant to corrosion by molten sodium and sulfur at about 300°C.

	Na_2O	B_2O_3	Al_2O_3
A	52	43	5
B	42	33	25
C	33	42	25
D	43	52	5

Figure 6.2: Ternary Diagram of Compositions of Na_2O-B_2O_3-Al_2O_3 Glasses

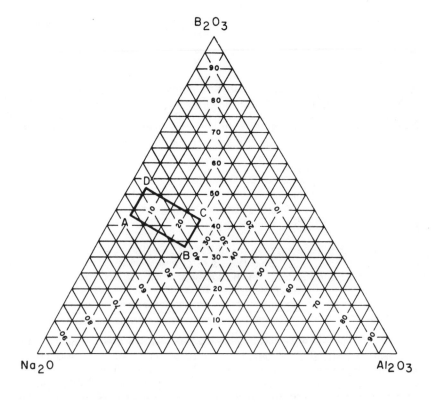

Also described is an electrochemical device having:

(a)   Two electrodes, one of which is a sodium-containing electrode capable of supplying sodium ions;

(b)   A solid electrolyte separating the two electrodes; and

(c)   An inert connecting electrical connector to complete an electrical circuit between the two electrodes wherein the solid electrolyte consists essentially of the aforementioned sodium-ion-conducting glasses.

As used in the battery cell, the glass electrolyte can be in the form of thin membranes fabricated in various orientations. These membranes can be in the form of flat plates, corrugated sheets, spirals or other designs which, during operation, will provide for anode metal ion transfer, but will keep separate the liquid anode and cathode materials.

A preferred form for the electrolyte is fine, hollow glass fibers wherein the individual fibers have an outside diameter wall thickness ratio of at least 3, ordinarily from about 3 to 20, and preferably from about 4 to 10. Usually within these ratios fibers having an outside diameter from about 20 to 1,000 μ and a wall thickness of from about 5 to 100 μ are used. Such hollow fibers provide a high strength, thin-walled membrane and give a high ion conductivity. They also provide a very large surface area to volume ratio.

Although less advantageous in the latter respect, fibers as large as 5,000 μ outside diameter and having walls as thick as 1,000 μ can be used.

For use in a battery cell, the hollow fibers can be fabricated into bundles of circular, rectangular prismatic or other geometric cross sectional shapes which provide for a controlled orientation and substantially uniform spacing between fibers.

The actual fabrication of the electrolyte fibers into a predetermined configuration readily can be carried out by one skilled in the art using known handling, packing and fabricating techniques. To illustrate, bundles of the fibers each having one end closed, can be prepared wherein the open ends of the fibers are passed through and sealed into a common header which, in turn, either serves as or communicates with a reservoir for anode metal.

These hollow fibers can be sealed in place as a bundle in a header, for example, by adhesives such as glazing or potting compounds, solder glass, high temperature thermosetting resins and the like materials. Preferred are Na-S batteries in which the solid electrolyte consists of sealed hollow fibers which contain sodium.

Glass Electrode

Electrodes are known in the art for measuring the oxidative potentials in liquid media, comprising sensing elements made of noble metals, such as platinum and gold. These electrodes have the following disadvantages:

(a)   Sensing elements made of noble metals are susceptible to a comparatively easy contamination by catalyst poisons (for instance, by SO_2 and other sulfurous compounds);

(b)   The presence of gaseous oxygen or hydrogen in a medium being analyzed affects the electrode potential; and

(c)   Noble metals are capable of catalyzing the decomposition of some redox systems (for instance, of hydrogen peroxide).

M.M. Shults, A.A. Beljustin, A.M. Pisarevsky, L.V. Avramenko, S.E. Volkov, V.N. Lakhtikova, V.A. Dolidze and V.M. Tarasova; U.S. Patent 4,140,612; February 20, 1979 describe a glass electrode wherein the sensing element is made of a material which enables the glass electrode to measure the oxidation potential of liquid media having a pH value of below 3 at a temperature of above 60°C in the presence of dissolved oxygen, hydrogen and catalyst poisons.

The glass electrode (Figure 6.3) for measuring the oxidative potentials of liquid media is made in the form of a cylindrical tube 1 of a high-resistance glass, which tube serves as a casing.

Figure 6.3: Glass Electrode for Measuring Oxidation Potential of Liquid Media

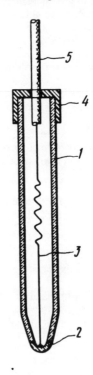

Source: U.S. Patent 4,140,612

A sensing element 2, secured by soldering to one end of the tube 1, is made of a glass featuring electronic conduction, composed of 32 to 45 pbw SiO_2; 7.0 to 26.0 pbw Me_2O, wherein Me is Li, Na, K, 16.0 to 40 pbw TiO_2; 0.8 to 4.2 pbw Ti_2O_3; 2.0 to 32.0 pbw Nb_2O_5 and/or Ta_2O_5.

The sensing element **2** is connected to a metallic current lead **3** disposed within the cylindrical tube **1** and led out of it in the form of a cable **5** through the top of the tube protected by a cap **4**, which ensures hermeticity of the inner cavity of the electrode.

The glass electrode and an auxiliary electrode, for instance, a chloro-silver electrode, are immersed in a solution containing a redox system. Potentials arise in a circuit thus formed, and in the event of the glass electrode, these arise at a sensing element glass solution interphase.

The potential of the glass electrode depends on the redox state of a medium, whereas that of the accessory electrode remains constant. The current leads of these electrodes are connected to a measuring instrument, a high-resistance millivoltmeter which registers the arising potential difference determining the ratio at which the oxidized and reduced forms of elements occur in the solution being analyzed according to the Nernst equation.

The proposed electrode makes it possible to measure the oxidative potentials within the range of from –700 to 1,250 mV at pH values ranging from –0.5 to 14. The electrode can operate within a temperature range of from 0° to +150°C, at an electrical resistance of ≤10 Mohm.

Stem Sealing Method for Assembling Electron Tubes

J.I. Nubani and R.L. Muenkel; U.S. Patent 4,165,227; August 21, 1979; assigned to RCA Corporation describe a stem sealing method for use in assembling electron tubes, including an improved method for removing the excess glass or cullet from the sealing machine.

Stem sealing is a widely used process in which a glass stem is sealed into the glass neck of an electron tube. The stem is a wafer-shaped piece of glass with the exhaust tubulation extending from one side, the stem leads sealed into and extending therethrough, and the electron gun mounted on the stem leads on the other side.

Stem sealing is usually conducted on an automated machine comprising a rotatable turret having a plurality of rotatable supporting means, each including a mount pin for supporting a stem and an envelope holder for supporting an envelope in controlled positions with respect to one another.

The envelope and stem are loaded onto the supporting means at a loading station. The turret then rotates intermittently from one station to the next until the stem is heat-sealed into the neck and the confronting glass, which has been melted to make the seal, is solidified and annealed.

Prior to sealing, the neck (which usually is tubular), is much longer than necessary. During the sealing operation, when the glass is molten, the excess glass is cut off with a sharp flame and the molten edge solidifies and sticks to and around the mount pin.

The operator, upon unloading the sealed envelope, smashes the excess glass or cullet from the mount pin, permitting the fragments to fall in and around a bin adjacent the unloading station.

The practice of smashing the cullet is undesirable because it results in poor housekeeping around the sealing machine. The glass fragments also may cause injuries to nearby personnel. In addition, the practice of smashing the cullet generates very fine glass particles which become airborne and drift through the factory, settling randomly on work surfaces and surfaces of the product. For example, where the factory is making aperture-mask-type color picture tubes, glass particles frequently settle on in-process aperture masks. When these masks are assembled into tubes, the presence of these glass particles may cause the finished tubes to be defective.

This method follows the prior procedure described above, except that, after the cullet is cut off, the cullet is released from the mount pin without substantially fracturing it. The cullet is then removed from the vicinity of the mount pin again without fracturing the cullet.

One method for releasing the cullet from the mount pin is to rotate the mount pin and cullet and simultaneously apply heat to the cullet to expand it away from the mount pin. Preferably, a platform is positioned under and close to the cullet so that, when the cullet is released, it is supported on the platform.

With the platform supporting the cullet, the cullet may be raised or the mount pin lowered so that the mount pin is essentially clear of the cullet. The supporting means is moved to the next station, which is equipped with a chute on one side. Means such as an air jet, on the other side, drives the cullet from the platform into the chute without fracturing the cullet, at least until it is in the chute.

By practicing this method, all of the cullet is transferred substantially unfractured into the chute and may be collected in a closed system connected thereto. Thus, the housekeeping and safety around the sealing machine are markedly improved. Also, little or no fine glass particles are generated, resulting in fewer rejected finished electron tubes that result from airborne fine particles depositing on product or work surfaces.

Glass-Crystalline Material for Microwave Circuits

P.I. Litvinov, V.M. Firsov and G.B. Knyazher; U.S. Patent 4,162,921; July 31, 1979 describe a material which can be used as the bases or supports for integrated circuits in microwave equipment, consisting of a glass crystalline material with dielectric properties close to those of alumina ceramics (dielectric constant of 10 and dielectric loss factor of $3\text{-}5 \times 10^{-4}$ at a frequency of 10^{10} Hz) containing SiO_2, Al_2O_3, MgO and TiO_2 and characterized in that it also contains oxides of rare earth elements selected from the group consisting of CeO_2, La_2O_3, Pr_2O_3 and Nd_2O_3, the oxides of the rare earth elements being present in combined amounts from 14.0 to 17.0 wt %.

The glass crystalline material is characterized in that it has the following composition in weight percent: 31 to 45 SiO_2, 20 to 30 Al_2O_3, 5 to 15 MgO, 15 to 25 TiO_2, 7.0 to 8.5 CeO_2, 3.0 to 4.6 La_2O_3 and other oxides of rare earth elements besides those mentioned above, 3.0 to 4.6.

A method is described of producing the glass crystalline material, depending on the initial glass composition being heated to the maximum crystallization temperature and held at this heat treatment temperature. The method is characterized in that heating is performed at a rate of 60° to 300°C per hour, up to 1200°C and exposure at this temperature lasts for 3 to 6 hours.

TECHNICAL
AND SPECIALTY GLASS

SEALING OR BONDING GLASS

Sealing Glass Preform

Integrated circuits, formed on a small chip of silicon, are often enclosed in a dual-in-line package which is a rectangular enclosure having electrically conducting pins extending from two parallel sides of the package for making electrical connections to the enclosed integrated circuit. When such packages are made of ceramic they comprise a ceramic substrate (for example, about 7 mm wide x 20 mm long x 1.9 mm thick) having a depression in the center of one of the large flat surfaces. The bottom of the depression is usually coated with gold to serve as an electrical contact. The silicon chip containing the integrated circuit is bonded to the bottom of the depression. The lead pins are then located along the long sides of the substrate in a lead frame and are connected to the integrated circuit.

A ceramic cover, having about the same length and width as the substrate but a smaller thickness, e.g., about 1.3 mm, and optionally also having a central depression, is then placed on top of the substrate. Either the substrate and/or the cover is provided with a layer of low-melting sealing glass disposed generally around the edges of a mating face. After assembly the package is heated to a temperature at which the sealing glass softens and forms a bond and seal between the cover and the substrate.

Heretofore, the layer of sealing glass on the substrate or cover has been formed by a method involving screen printing. According to the screen printing process, a paste comprising a mixture of powdered low-melting sealing glass and a binder, such as a synthetic resin dissolved in a suitable solvent, is applied to the substrate and/or cover by screen printing and dried. The steps of application and drying are repeated several times in order to build up the layer to the proper thickness (0.35 to 0.5 mm). The layer is then fixed by baking, i.e., by raising the temperature of the coated element to 410°C over a period of 1.5 hours.

This method of preparing the element having a layer of sealing glass is very inefficient. It requires three to six screen printing and drying operations to build up the required film thickness of 0.35 to 0.5 mm. Furthermore, since the printed film shrinks about 30% when it is baked, the printed layer must be made oversize to account for the shrinkage.

Since the printed paste thus extends over the sides of the substrate, the process has the further drawback that the sealing glass may adhere to the sides of the substrate or damage the gold plating of the central depression where the silicon chip is mounted. In addition, proper registration of the successive printing steps is difficult to attain and hence it is difficult to obtain glass layers of exact dimensions, even with careful work.

Another problem with the printing technique is the frequent presence of air bubbles which are formed in the glass layer and cause pinholes which prevent the seal from being perfectly gastight.

K. Kita; U.S. Patent 4,165,226; August 21, 1979; assigned to Narumi China Corporation, Japan describes a simpler and more economical method of preparing an element of a dual-in-line package having a layer of sealing glass.

In one embodiment, fine powder, 100 mesh or finer, of glass having a softening point of about 350°C is mixed with a binder and formed into a blank of the proper shape by compression molding in a die, e.g., a metal die.

The binder may be any material which will retain the glass particles in the proper shape when molded at relatively low temperatures such as room temperature, but which will decompose, or burn away at the temperatures used (300° to 390°C) for sintering the molded preform subsequently.

Suitable binders include synthetic polymeric resins dissolved in a solvent. A preferred binder comprises a 10% solution of an acrylic resin dissolved in a solvent such as toluene or xylene. Since the binder coats each particle of glass when it is mixed with the powder, less binder can be used than is required in the process using screen printing. The amount of resin can also be reduced to about 75% of that used in the case of screen printed glass sealing layers.

After the glass powder is mixed with the resinous binder the powder is molded into the desired shape for the sealing glass preform. This molding is carried out by conventional compression molding using dies of the proper shape and size. Since some shrinkage occurs in the subsequent sintering operation, allowance must be made for this in molding the powder preform. The shrinkage is generally about 10%, which compares favorably with the 30% shrinkage commonly observed with the conventional screen printing process.

It has been found that the use of a water-insoluble resin and solvent as the binder for the glass powder prevents adhesion of the powder to the die during the molding steps. Hence such binders and solvents are preferred. Most preferred binders and solvents are acrylic resins and aromatic solvents such as toluene or xylene.

The molding pressure used in molding the powder into properly shaped preforms may be any pressure which suffices to compress the powder into a self-coherent mass which will retain its shape through the handling needed to prepare the final sintered glass preform. Typical pressures range from 450 to 1,400 kg/cm². The pressure used will have some effect on the properties of the preforms. Higher pressures generally produce a molded powder which shrinks somewhat less during sintering. The relation between molding pressure and shrinkage during sintering is shown below.

Pressing Pressure (kg/cm²)	Shrinkage (%)	
	Sintering	Baking
450	10	2
650	9	1
850	9	1
1,000	9	1
1,200	9	1
1,400	8	0.8

It may be seen, for example, that the shrinkage will be about 9% when a molding pressure of 650 kg/cm² is used and the die can be designed to allow for this.

Pinholes in the glass preform may be formed by residual solvent or resin which escapes later during the sealing of the ceramic dual-in-line package. However, the most common cause of pinholes is air bubbles which remain between the particles of glass during sintering. Higher molding pressures decrease the formation of pinholes. Pressures over 600 kg/cm² are generally effective to prevent pinholes. Pressures over 1,000 kg/cm² do not generally further decrease the likelihood of pinhole formation. Hence the preferred molding pressures are 600 kg/cm² or greater and most preferably between 600 and 1,000 kg/cm².

The molded powder is then sintered. During the sintering the binder is burned away or volatilized and the glass powder becomes a cohesive glass preform which can be handled easily. The sintered preform can be stored indefinitely if necessary.

The sintering is carried out by gradually heating the molded powder, keeping the blank in the temperature range of 300° to 340°C for a long enough time to burn away any residual carbon from the binder which might cause pinholes later when the package is sealed. The sintering is completed by raising the temperature to 390°C after about 1.5 hours. The highest temperature of the sintering step must be carefully controlled because the glass may begin to melt at higher temperatures and adhere to the support.

In another embodiment, no binder is used but the glass powder is directly placed into molding dies heated to a temperature higher than the glass transition temperature but lower than the softening temperature and the powder is formed by heat and pressure in the die into a solid glass preform of the desired size and shape. Since this method uses no binder, it eliminates the step of mixing the glass powder with the binder as well as the sintering procedure to remove the binder. This procedure also reduces the number of pinholes caused by the presence of the binder.

In this embodiment, the dies which have the shape and thickness of the preform desired for a particular application are heated to a temperature higher than the glass transition temperature of the sealing glass but lower than its softening temperature. This temperature will depend upon the particular sealing glass being used but will generally be in the range of 350° to 380°C for the glasses in common use. The glass powder is then loaded into the heated die and compressed at a pressure of 500 kg/cm² or more. The powder is thereby formed into a thin solid glass preform of the required shape and thickness (0.4 to 0.5 mm).

Finally the sintered glass preform is mounted on a surface of the substrate and/or cover and is baked at a temperature of 405° to 410°C to fuse the glass to the material to be sealed.

This process prepares sealing glass preforms having dimensional accuracy, a minimum of deviation from the proper thickness, and smooth surfaces, all of which make it easier to control the quality in the subsequent steps in the sealing process than when the screen printing process is used. Moreover, since this process does not require as much skilled labor as the screen printing process and has fewer steps, it is more suitable for large-scale factory production and increases the efficiency of the manufacturing process with accompanying reduction of production costs.

Sealing Glass Not Requiring Devitrification

M.E. Dumesnil and U. Schreier; U.S. Patent 4,186,023; January 29, 1980; assigned to Technology Glass Corporation describe a sealing glass which is non-devitrifiable in the sense that essentially no crystallization occurs when the glass is exposed to a temperature of 430°C for a period of 15 minutes. Under the sealing conditions required to produce a seal with this glass which are commonly about 400°C and 5 to 8 minutes, no crystallization is observed and the glass in the completed seal is vitreous.

The sealing mixture contains glasses capable of being intimately mixed with large amounts of a wide variety of fillers with minimal effect on glass flow at the sealing temperature while effecting a substantial decrease in the coefficient of thermal expansion of the mixture. Thus vitreous glass seals characterized by high mechanical strength, good chemical stability, low coefficients of expansion and capable of withstanding repeated thermal shocks (MIL-STD-883 specifications) are achieved with these low temperature sealing glass compositions (360° to 430°C).

These very fluid, very low melting glasses are produced in the lead-borate and lead-zinc-borate glass-forming systems by the joint addition of cuprous oxide and fluorine (Cu_2O + F) in concentrations of 0.1 to 10% by weight, the molar ratio of cuprous oxide to fluorine being in the range of 1:0.25 to 1:10, preferably in the range of 1:1 to 1:5. The resulting glass can be made even more fluid by the addition of up to 5% by weight bismuth oxide (Bi_2O_3).

The preferred glass compositions are characterized by a DTA (differential thermal analysis) softening point in the range of 260° to 280°C, linear thermal expansion coefficients of the order of 105 to 110 x 10^{-7}/°C, remarkable water-insolubility, resistance to steam, good glass flow even in a nitrogen atmosphere and the ability of being admixed with large quantities of filler(s) in particulate form while retaining high fluidity at low temperatures.

In accordance with this process, a particulate filler is admixed in amounts reaching 56% by volume, preferably 5 to 50% by volume, to tailor the thermal expansion of the resulting sealing glass to a value as low as 50 x 10^{-7}/°C. The particulate fillers are refractory or semirefractory powders which are any such well-known materials, synthetic or natural, conventional in the art and also include materials prepared from glass, recrystallized glass, glass-ceramics, coprecipitated or sintered materials.

The glasses of this process contain lead oxide, optionally zinc oxide, boron oxide, optionally bismuth oxide, silicon dioxide, cuprous oxide and a minor proportion of at least one solid nonvolatile fluoride, the proportions being lead oxide 75 to 85% by weight, zinc oxide when present up to 10% by weight and preferably below 8% by weight, boron oxide 8 to 15% by weight, silicon dioxide 0.75 to 2.5% by weight, bismuth oxide when present 1 to 5% by weight, cuprous oxide 0.5 to 5.5% by weight, and nonvolatile metal fluoride in amounts such that the mol ratio of cuprous oxide to the fluoride content of the metal fluoride is in the range of 1:0.25 to 1:10.

The completed seal obtained with these mixtures consists of finely divided refractory particles dispersed in a vitreous glass matrix.

Example: A base glass was prepared by mixing 4,150 g of red lead oxide (Pb_3O_4), 150 g lead fluoride, 350 g zinc oxide, 800 g boric acid, 50 g silica and 120 g cuprous oxide. After heating the mixture in a platinum crucible at 1000°C for 20 minutes, the melt was poured through cold steel rollers to facilitate subsequent crushing.

The resulting glass flakes had a composition in weight percent of PbO, 80.8; ZnO, 6.75; B_2O_3, 8.68; SiO_2, 0.96; Cu_2O, 2.31 and F, 0.45; and a corresponding molar ratio Cu_2O:F of 1:1.5, a linear thermal expansion (25° to 200°C) = 106 x 10^{-7}/°C, and a DTA softening point = 272°C.

Glass-to-Metal Seal Involving Iron Base Alloys

J.M. Popplewell; U.S. Patent 4,149,910; April 17, 1979; assigned to Olin Corporation has found that iron base alloys can be used in glass or ceramic-to-metal composites or seals, provided that the iron base alloy has certain inherent oxidation characteristics. The characteristics required in the iron base alloy are that it has formed on its surface an oxide consisting essentially of chromium oxide and α-Fe_2O_3 in the form of a compact continuous film. This oxide film must form immediately adjacent to the metal, be strongly adherent to it and comprise from at least 10% up to 100% of the total oxide film thickness.

Suitable iron base alloys for use in the glass or ceramic-to-metal composites or seals contain from 1 to 5% silicon and from 1 to 10% chromium. In a preferred embodiment the alloys consist essentially of from 1 to 3% silicon, from 1 to 3% chromium, and the balance essentially iron.

In addition to the above elements, the alloys may also contain aluminum in an amount from 0.001 to 1%, nickel in an amount from 0.001 to 5%, cobalt in an amount from 0.001 to 5% for additional strength and grain refinement, and carbon in an amount from 0.001 to 1% for some applications requiring increased strength and wear resistance. Impurities may be present in amounts not adversely affecting the properties of the glass or ceramic-to-metal composite or seals. In particular, the impurities may include less than 1% nitrogen, less than 1% sulfur, less than 1% manganese, less than 1% molybdenum and less than 1% phosphorus.

Example: An iron base alloy was provided which possessed the following composition: 3.0% chromium, 2.0% silicon, and the balance iron. The alloy possessed coefficients of thermal expansion of 12.2 x 10^{-6}/°C through the temperature range of 20° to 300°C, and 13.4 x 10^{-6}/°C through the temperature range of 20° to 800°C, respectively.

The alloy was cast into ingot form by the Durville casting method, and was hot and cold rolled to 0.030" gauge. The metal was sheared into specimens measuring 2" x ½", which were then degreased in benzene and rinsed in methanol and acetone. The specimens were then heated at 800°C in air for 3 minutes and cooled rapidly.

Samples were prepared for sealing by the placement of a small quantity of glass powder thereon. Six glass powders were tested which possessed differing coefficients of thermal expansion, as set forth below.

Owens Illinois Glass No.	Thermal Expansion Coefficient in in/°C (0°-300°C)	Sealing Temperature (°C)
00130	40 x 10^{-7}	615
00756	75 x 10^{-7}	450
00564	83 x 10^{-7}	460
00766	87 x 10^{-7}	435
00578	108 x 10^{-7}	380
00583	117 x 10^{-7}	365

The six samples were then heated to the normal temperature of sealing of the particular glass powders disposed thereon. After the sealing operation and sufficient cooling had been accomplished, the samples were visually examined and it was observed that a low contact angle existed between the glass and the metal and excellent flow characteristics were observed. These experiments confirmed that the varous glasses effectively wet the thermally formed oxide film.

Glass-to-Aluminum Seals

J.A. Wilder, Jr.; U.S. Patent 4,202,700; May 13, 1980; assigned to U.S. Department of Energy describes a glass or glass-ceramic composition that forms high strength hermetic seals with aluminum having low gas permeability and high chemical durability.

The glassy composition contains from 10 to 60 mol percent Li_2O, Na_2O, or K_2O with from 35 to 40 mol percent of Na_2O and up to 5 mol percent K_2O being preferable. The composition also contains from 5 to 40 mol percent BaO or CaO with 10 to 15 mol percent BaO being preferable. The composition further contains from 0.1 to 10 mol percent Al_2O_3 with from 0.5 to 3 mol percent being preferable. In addition, the composition still further contains from 40 to 70 mol percent P_2O_5 with 45 to 55 mol percent being preferable.

The glassy composition may either be employed as a glass, or where greater strength or thermal shock resistance is desired, as a glass-ceramic. In order to cause the devitrification needed to produce a glass-ceramic, there may be employed up to 0.1 mol percent of a nucleating agent such as platinum or up to

10 mol percent of nucleating agents such as TiO_2, ZrO_2, Y_2O_3, La_2O_3, or Ta_2O_5, with up to 5 mol percent being preferable.

A satisfactory heat treatment schedule which will cause the glass to form a seal and subsequently devitrify into a glass-ceramic is to heat the glass in contact with the metal to 350° to 400°C for about 2 hours to form the desired seal with the the metal, thereafter increasing the temperature to 400° to 450°C for about 2 hours for nucleation of the crystal phase, subsequently increasing the temperature to 450° to 525°C for about 2 hours for growth of the crystal phase, and then cooling to room temperature. A satisfactory heating rate is 5°C per minute and a satisfactory cooling rate is 1°C per minute.

It has been found that the use of up to 10 mol percent but preferably up to 5 mol percent of a wetting agent such as CoO, CuO, SnO, Ag_2O, Cr_2O_3 or NiO will help the glass wet and therefore seal to the aluminum.

In addition, up to 30 mol percent but preferably up to 10 mol percent of other materials such as PbO, WO_3, B_2O_3, SiO_2, GeO_2 or ZnO_2 may be added to modify the mechanical properties of the composition.

A specific example of the glassy composition contains 40 mol percent Na_2O, 10 mol percent BaO, 1 mol percent Al_2O_3 and 49 mol percent P_2O_5. Up to 0.1 mol percent Pt may be added as a nucleating agent and up to 5 mol percent CoO may be added as a wetting agent.

This particular formulation has a thermal expansion coefficient of 203 x 10^{-7}/°C as a glass and 225 x 10^{-7}/°C as a glass-ceramic. The formulation has a glass transition temperature of 303°C with viscosity of 10^{14} poise at 300°C falling to 10^8 poise at 400°C. It exhibited a fracture toughness of 0.423 KSI-in^2, a Young's modulus of 5.8 x 10^6 psi, and a shear modulus of 2.25 x 10^6 psi. Electrical resistivity was measured as 10^{11} ohm-cm at room temperature and 10^6 ohm-cm at 200°C. Chemical durability as measured by the leach rate in deionized water was found to be 10^{-6} g/cm^2/min. The noted presence of Al_2O_3 was found to profoundly affect the chemical durability. The formulation also possesses low helium permeability, being about 10^9 atoms/sec/cm/atm.

It has been found that through appropriately modifying the proportions of the constituents, glassy compositions are obtained having thermal expansion coefficients from 160 to 260 x 10^{-7}/°C. For sealing to aluminum, a composition with a thermal coefficient of expansion of 235 x 10^{-7}/°C would be used.

The glasses may be prepared by mixing powders in the appropriate proportions of the metal oxides or carbonates plus ammonium dihydrogen phosphate. The mixed powders are first calcined at 300°C to drive off water and ammonia and then fused at from 850° to 1000°C. If platinum is to be used as a nucleating agent, it may be added to the melt as a solution of chloroplatinic acid. The glass may then be used or cast into bulk slabs or preforms for later use in manufacturing the desired seals.

Corrosion-Resistant Hermetic Plug Seal

J.A. Topping and P. Mayer; U.S. Patent 4,199,340; April 22, 1980; assigned to the Minister of National Defence, Canada describe glass compositions which can be sequentially fluidized and ceramed in a single rapid heating step to yield hermetic plug seals which exhibit good corrosion resistance in a KOH environment, such as in Ni-Cd batteries.

Figure 7.1a illustrates a hermetic plug seal **1** in which a metal pin electrode **2** is mounted in a metal collar **3** by glass-ceramic seal **4**.

Suitable glass compositions for use in making the corrosion-resistant glass-ceramic seal **4** are from the $ZnO-Al_2O_3-SiO_2$ system, and generally comprise about 25.0 to 32.0% by weight ZnO, about 2.5 to 10.0% by weight Al_2O_3 and about 30.0 to 60.0% by weight SiO_2.

The glass compositions normally include one or more fluxing and/or nucleating agents selected from the group consisting of CaO in an amount of up to about 10.0% by weight, PbO in an amount of up to about 14.5% by weight, P_2O_5 in an amount of up to about 2.5% by weight, LiO_2 in an amount of up to about 10% by weight, Na_2O in an amount of up to about 3.0% by weight, K_2O in an amount of up to about 3.0% by weight and ZrO_2 in an amount of up to about 12.5% by weight.

A particularly preferred glass composition comprises about 30.0% by weight ZnO, about 5.0% by weight Al_2O_3, about 40.0% by weight SiO_2, about 2.5% by weight CaO, about 5.5% by weight PbO, about 2.5% by weight P_2O_5, about 5% by weight LiO_2, about 3.0% by weight Na_2O, about 2.0% by weight K_2O and about 4.5% by weight ZrO_2.

Another particularly preferred glass composition comprises about 30.0% by weight ZnO, about 5.0% by weight Al_2O_3, about 35.0% by weight SiO_2, about 2.5% by weight CaO, about 5.5% by weight PbO, about 2.5% by weight P_2O_5, about 5.0% by weight LiO_2, about 3.0% by weight Na_2O, about 2.0% by weight K_2O and about 9.5% by weight ZrO_2.

Figure 7.1b illustrates a simple apparatus for use in making the plug seals. The apparatus includes an oven body **11** which may, for example, be a length of fused quartz or Vycor tubing, about which is mounted a RF induction coil **12** which is connected to a suitable generator/power source (not shown). Oven body **11** and induction coil **12** are supported by suitable means (not shown) such as laboratory clamps, etc.

In the embodiment illustrated, the upper end of oven body **11** is sealed by a stopper member **16** protected on its underside by a heat shielding layer **17** which, for example, may be a layer of Fibrefax. A glass inlet **18** extends through stopper member **16** and heat shielding layer **17** and is connected to a source of inert gas (not shown).

Within oven body **11** is a jig **13** which is adapted to hold the plug seal components in desired relationship during the sealing and ceraming process. Jig **13** is composed of a material suitable for use at the temperatures reached during the sealing and ceraming process. Satisfactory results have been obtained when jig **13** is made from Varemco 502-1300 machinable ceramic rod. Jig **13** is mounted upon a support rod **14** which in turn is mounted upon a laboratory jack or other suitable means (not shown) in order that jig **13** is movable between two positions.

Figure 7.1:   Glass-Ceramic-to-Metal Seals

a.

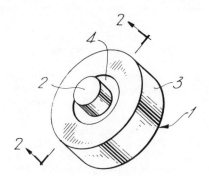

b.

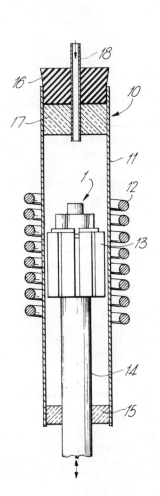

(continued)

Figure 7.1: (continued)

c.

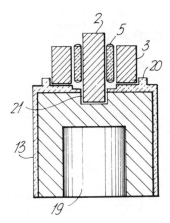

(a) Hermetic plug seal
(b) Apparatus for making plug seal
(c) Jig and plug seal

Source: U.S. Patent 4,199,340

In one position, jig **13** is situated within oven body **11** essentially as illustrated in Figure 7.1b, at which position sealing and ceraming of the glass composition is carried out. In the other position, jig **13** is remote from oven body **11** for jig loading and unloading purposes.

A gas-permeable seal **15** is mounted about support rod **14** at a location such that it seals oven body **11** below jig **13** when jig **13** is operatively positioned to effect sealing and ceraming of a glass composition. Gas-permeable seal **15**, like the heat shielding layer **17**, may be of Fibrefax.

Figure 7.1c illustrates a jig **13** similar to that depicted in Figure 7.1b upon which are mounted the component parts necessary to make a plug seal according to this process. Jig **13** has a lower bore **19** for purposes of mounting upon the support rod **14** of Figure 7.1b. The upper surface of jig **13** is provided with a lip **20** for retaining metal collar **3**, and an axial bore **21** for retaining the lower end of metal pin **2**. Between metal pin **2** and metal collar **3** is a glass composition preform **5** which, upon sealing and ceraming, becomes glass-ceramic seal **4** which is illustrated in Figure 7.1a.

A typical sealing and ceraming operation to produce the plug seal is as follows. Jig **13**, in its position remote from oven body **11**, is loaded with the pin **2**, the collar **3** and the glass composition preform **5**, essentially as illustrated in Figure 7.1c. In order to obtain a good glass-ceramic-to-metal seal each of the components, as will be understood by those skilled in the art, is matched for size, cleaned and dried prior to loading of the jig. The jig is then moved into its position within oven body **11** essentially as illustrated in Figure 7.1b, the lower end of the oven body being sealed by the gas-permeable seal **15**.

The interior of the oven body is next flushed with inert gas, for example, purified argon. The purification can be carried out by passing the gas through columns filled with Drierite before and after passage through copper turnings heated to a temperature of about 600°C.

Next, a RF magnetic field is generated by RF induction coil **12** which is connected to a generator/source of power (not shown). The application of RF power causes the metal parts to heat to the required sealing temperature. It will, of course, be obvious that suitable temperature sensing means can be employed in association with the oven body so that the application of RF power can be controlled to in turn provide temperature control.

The glass composition preform **5** which is located between and in proximity to pin **2** and collar **3** is indirectly heated by conduction and radiation until the glass is sufficiently fluid to first flow and fill in the space between pin **2** and collar **3** and adhere to the metal parts, and then ceram to form the glass-ceramic seal **4** illustrated in Figure 7.1a. The plug seal is then allowed to cool in the sealing atmosphere in the oven body to room temperature, a step typically requiring 15 to 20 minutes.

Glass-to-Metal Seal for Electrochemical Cells

J.J. Decker and D.J. Kantner; U.S. Patent 4,178,164; December 11, 1979; assigned to GTE Sylvania Incorporated describe a glass-to-metal seal for an electrochemical cell terminal which comprises an apertured metal plate having a terminal pin substantially centrally located within the aperture and held therein by a glass seal. The aperture in the plate is surrounded by a concavo-convex relief ring the walls of which define the boundaries of the glass seal on one side of the plate.

On the opposite side of the plate the terminal pin is provided with a broadened area which is spaced from the plate and the area between it and the plate is also filled with the material of the seal. The diameter of the broadened area of the pin is greater than the diameter of the aperture. The seal is formed by subjecting the parts to a melting temperature for the seal material while applying a force thereto while the parts are in a mold.

The seal can be more clearly understood by reference to Figure 7.2. It will be seen that seal **14** is formed in an apertured plate **16** which can be the cover portion of the hollow metal body of the cell. The aperture **18** therein has a given diameter and is surrounded by an annular concavo-convex relief ring **20** with the convex portion interior of the cell. At least the area **22** of plate **16** surrounding aperture **18** is planar.

Terminal pin **12** is positioned substantially centrally of aperture **18** and has a first diametered longitudinally extending portion **24** and a second diametered portion **26** formed at one end thereof external of the cell. The first diametered portion **24** is smaller than aperture **18** and the second diametered portion **26** is larger than aperture **18** and is spaced therefrom.

The sealing material **28** fills the space defined by the relief ring **20**, surrounds terminal pin portion **24**, projects through aperture **18** and encompasses surface **30** of terminal pin portion **26**. The seal is preferably formed in a reducing atmosphere.

Figure 7.2: Cross Section of Electrochemical Seal

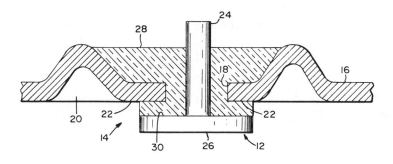

Source: U.S. Patent 4,178,164

This seal provides many advantages over the prior art. The combination of the
relief ring with the final shape of the glass seal, which extends on both sides of
the plate **16**, gives greatly increased strength by allowing a certain amount of
flexure to plate **16**. Further, the additional area of terminal pin portion **26** aids
in this strengthening feature as well as providing increased contact area.

Method of Forming a Lead-Through in a Ceramic Component

*M.A. Monneraye and M.J.C. Monnier; U.S. Patent 4,163,656; August 7, 1979;
assigned to U.S. Phillips Corporation* describe a method of manufacturing a lead-
through of a metal element through a ceramic component by means of sealing.
The metal element may be an alloy of iron, nickel, cobalt—known as fernico
(commercially Dilver P, Vacon 12, Nilo K, etc.).

Such seals are necessary particularly for manufacturing conducting lead-throughs,
which are gastight and vacuum tight, through ceramic materials such as aluminum
oxide or beryllium oxide and which may be used, for example, for the manufacture of connections through a substrate having an output circuit of a housing
accommodating semiconductor elements, output electrodes, high voltage arrangements, connections to the electrodes of photomultiplier tubes having a ceramic
structure, etc.

According to this process, a suspension of glass powder based on zinc-borosilicate
which also comprises lithium oxide in a molar quantity of from 3 to 5% is deposited on the substrate at the area of the lead-through whereafter the assembly
is heated for several minutes in a nitrogen atmosphere at a temperature near
1000°C and is cooled to room temperature.

In this method, the product is to be maintained preferably during cooling at
the value of the annealing point (AP) of the glass for at least **15** minutes so that
the sealing operation can be controlled.

Figure 7.3a shows a metal element **1** referred to as the lead-through of a fernico alloy through an aperture **2** of a ceramic component **3** (for example, in aluminum oxide or beryllium oxide).

Figure 7.3: Two Stages of Manufacturing Fernico Lead-Through

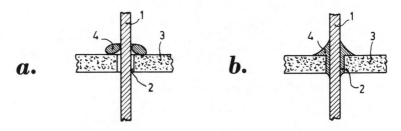

Source: U.S. Patent 4,163,656

The difference between the respective diameters of the apertures and of the metal elements does not exceed 0.15 mm (0.10 to 0.05 mm). A quantity of glass powder having a suitable composition chosen from the range given below and including in suspension a small quantity of lithium oxide in an organic liquid is provided at the area of the aperture **2** and around the lead-through **1**, for example, with the aid of a brush. The bead of glass powder thus obtained is denoted by **4**.

The assembly is heated for 10 minutes at approximately 1000°C in a furnace through which a hydrogen-nitrogen stream saturated with water vapor is passed at room temperature. The molten glass wets the ceramic material and the metal lead-through and fills up the free space in the aperture **2** by capillary action. After cooling, a lead-through **1** as shown in Figure 7.3b sealed in the ceramic material is obtained.

In order to control the sealing operation the product is maintained at the value of the annealing point of the glass (AP) preferably during cooling for 15 to 30 minutes.

The table shows different compositions based on zinc-borosilicate comprising lithium oxide in quantities which are less than or equal to 5%. This table likewise shows for each composition the coefficient of expansion, α, measured between 20° and 300°C, the annealing point and the density, ρ.

	. . . . .Mol Compositions							
	1	2	3	4	5	6	7	8
SiO_2	5	10	15	20	5	10	15	15
B_2O_3	35	30	26.5	20	40	35	30	30
ZnO	56	55	55	55	50	51.5	50	45
Li_2O	4	5	3.5	5	5	3.5	5	5
Al_2O_3	0	0	0	0	0	0	0	5
α $(10^{-7}/°C)$	49.5	49.5	49.5	49	51	50	49	48
AP (°C)	505	520	515	515	520	510	520	515
ρ (kg/m³)	3,630	3,600	3,650	3,695	3,420	3,510	3,570	3,400

Sealing of CRT Faceplate to CRT Envelope

The manufacture of specialized cathode-ray tubes (CRT), for example, high-contrast CRT's using transparent phosphor-black absorption layer technology, has not proved to be easy. In particular, difficulty has been experienced in frit-sealing the high-temperature faceplate required under these circumstances to conventional CRT bodies. While not fully understood, the problem seems to lie in the fact that the thermal coefficient of expansion of the high-temperature faceplate is considerably less than the coefficient of expansion of the envelope. Thus, as the seal cools, the different expansion rates cause stress to be generated within the envelope-faceplate area, resulting in a failure of the seal.

A.G. Hager and P.F. Krzyzkowski; U.S. Patent 4,194,643; March 25, 1980; assigned to the U.S. Secretary of the Army describe a method of joining a first member having a first thermal coefficient of expansion to a second member having a second thermal coefficient of expansion.

This method comprises the step of joining one end of a third member having a third thermal coefficient of expansion to the first member and also joining the other end of the third member to the second member, the third member acting as an intermediate stress-absorber and having a thermal coefficient of expansion which is less than both the first and the second thermal coefficients of expansion.

As shown in Figure 7.4, a high-temperature, alumina-silicate CRT faceplate **10**, illustratively manufactured from Corning type 1720 or 1723 glass, is sealed to a conventional CRT envelope **11**, illustratively manufactured from Corning type 7052 glass, by means of a cylinder **12** of uranium glass, illustratively Corning type 3320. Cylinder **12** is typically ½ inch or greater in depth and has the same outer diameter as the funnel end of the envelope **11**.

As will be explained, cylinder **12** is sealed to faceplate **10** by some suitable vitreous frit **13**, for example, Owens-Illinois SG-7. At the other end, cylinder **12** is flame-sealed to the CRT envelope.

Cylinder **12** acts as an intermediate stress-absorber between faceplate **10** and envelope **11** and is the primary glass of this graded seal approach. One could expect that the intermediate stress-absorber would have a coefficient of expansion falling between those of the faceplate and the envelope. The 3320 uranium glass has a coefficient of 40×10^{-7} which is less than either of the other two glasses, 42×10^{-7} for the 1720 faceplate and 46×10^{-7} for the 7052 envelope, and one would expect that this would aggravate, rather than alleviate, the situation; however, this is not the case and the uranium glass is highly effective as a stress-absorber.

In operation, the cylinder **12** is flame-sealed to, and coaxial with, the envelope **11**. Next, liquid honing is used to roughen the outer ⅛ inch periphery of faceplate **10**. This provides a roughened seal area. The butt edge of the cylinder **12** is likewise ground flat, preparatory for sealing.

Next, the solder glass frit **13** is mixed with the appropriate vehicle and uniformly coated onto the roughened butt edge of the cylinder. The frit is maintained in a liquid state by continuously circling the edge of the tube. The envelope is then centrally positioned on the faceplate and the frit allowed to set-up and dry. Advantageously, only a minimum of frit is used—enough only to build up a small bead on the butt edge of the cylinder. The glass faceplate-envelope combination is then fired in a furnace for 60 minutes at 600°C.

Figure 7.4: Cathode-Ray Tube

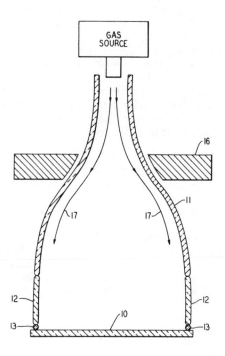

Source: U.S. Patent 4,194,643

Argon gas **17** is piped into the envelope to protect the transparent phosphor-black layer against temperatures over 300° to 400°C. The flow rate of the argon is typically 5 ft³/min and is adjustable. A weight **16** of approximately 200 g is used during the fritting cycle to ensure that the glass frit will compact into a homogenous, voidless layer when it melts.

After the 60 minute fritting cycle, the furnace is allowed to cool by natural means and the CRT removed. The electron-gun, deflection plates, etc., are then added and the tube evacuated and sealed in the conventional manner.

It should be noted that any suitable glass frit can be employed providing, of course, that it has an appropriate thermal coefficient of expansion. Additionally, more than one intermediate glass may be used in combination with other frits. This latter approach is appropriate to larger diameter CRT's and also to other types of faceplate material, e.g., sapphire, etc. Also, inert gases other than argon may be used to bathe the phosphor coating at a small positive pressure, the gas overflowing out into the oven chamber.

Capacitive Pressure Sensor

Capacitive pressure sensors work on the principle that the plates of a parallel plate capacitor form two walls of an enclosure which bounds a volume whose pressure is some reference pressure. Changes in the ambient pressure in relation to the reference pressure cause a flexing of at least one of the capacitor plates. Such a flexing changes the spacing between the capacitor plates and thus the capacitance measured between them. Thus, measuring the capacitance provides a measure of the ambient pressure with respect to the reference pressure.

E.K. Davis, K.W. Hansen, S.W. Taylor and R.M. Wentworth; U.S. Patent 4,184,189; January 15, 1980; assigned to Motorola, Inc. describe a pressure sensor which is comprised of two flat glass plates having conductive electrodes disposed on their surfaces. The electrodes are positioned in a spaced apart, substantially parallel relationship to provide the two plates of a parallel plate capacitor.

A sealing glass mixture is applied to at least one of the glass plates to form a perimeter seal. The space between the plates is then evacuated to a reference pressure and the plates are heated to a temperature sufficient to cause the sealing glass to flow and seal together the two plates. The two plates and the sealing glass then bound a volume within which the pressure is the desired reference pressure.

The sealing glass, which must be compatible with the glass plates, is a mixture which comprises, in weight percent, 8 to 10% SiO_2, 1 to 2% Al_2O_3, 55 to 60% PbO, 7 to 9% PbF_2, 7 to 10% ZnO, 4 to 6% CdO, and 10 to 15% B_2O_3. To provide a hermetic seal under these vacuum sealing conditions, the sealing glass mixture must be vacuum fined before fritting.

Gap Formation in Magnetic Heads

In the data recording and reproduction art, the density of the data recorded on a recording media is in part dictated by the gap length of the magnetic recording head. Also, the same is true for data reproduction from recording media; hence the ability to accurately reproduce densely recorded data depends in part on the gap length of the magnetic playback head. As the data recording and recovery art advances toward more densely recorded data, a corresponding need has developed for magnetic heads with smaller gap lengths.

The gap length of a magnetic head is dependent upon the size of a nonmagnetic space between opposite circuit parts of a magnetic circuit (usually formed by opposite sections of magnetic ferrite or magnetic nonferrite thin film). One early technique for the manufacture of nonmagnetic gaps in magnetic heads involved positioning the opposite portions of the ferrite cores a fixed distance apart and drawing nonmagnetic glass into the space by capillary action to form the gap.

As the state of miniaturization of magnetic heads advanced, problems developed concerning molecular migration between the glass and ferrite. It was discovered, for example, that a mutual diffusion of glass and ferrite occurred at the interface of the glass-ferrite bond. This diffusion caused deteriorations in the surface layer of the ferrite, causing distortion in the shape of the gap between the ferrite cores and a lowering of permeability and maximum magnetic flux of the ferrite. Also, different thermal expansion characteristics between the ferrite and certain glass formulations created residual strain and distortion in the ferrite structure upon cooling.

A. Calderon, Jr., D.J. Hennenfent and A.L. Holmstrand; U.S. Patent 4,182,643; January 8, 1980; assigned to Control Data Corporation describe a method of manufacturing magnetic heads whereby a predetermined layer of spacer material is deposited onto the surface of each confronting magnetic core and a bonding glass is deposited onto the spacer materials.

The cores are then pressed together at an elevated temperature to fuse the bonding glass and to extrude essentially all of the glass, thereby leaving the spacer layers contiguous to each other. The spacer material is selected to act as an isolation barrier between any residual glass and the magnetic cores.

One feature of this process resides in the fact that by isolating the glass from the ferrite cores the glass need be heated only to the temperature at which the opposing glass surfaces fuse and the excess glass is extruded. Thus, the glass need not be heated to such an extreme as to lower its viscosity, as in capillary processes. Hence, this process permits the use of a wider selection of bonding glasses.

Bonding of Bioglass to Metal

Various biologically active glasses have been introduced for the preparation of artificial prostheses. It is known that bone and other biological tissue will bond to or grow on these biologically active glasses. However, the strength characteristics of the glasses are such that it is impossible to construct sufficiently strong orthopedic or dental devices therefrom.

It has been suggested to overcoat metal substrates with biologically active glasses to provide sufficiently strong orthopedic or dental devices capable of bonding to bone tissue. However, there are numerous difficulties associated with bonding such glasses to metal surfaces. For example, the thermal coefficients of expansion of the metal and glasses are so dissimilar at both the melting and softening points of the glasses that cooling the coated metal substrate results in extreme thermomechanical stresses in the glass and metal layers which, when relieved, cause cracks, etc., in the glass coating.

L.L. Hench and P.J. Buscemi; U.S. Patent 4,159,358; June 26, 1979; assigned to Board of Regents, State of Florida describe a method of bonding a bioglass layer to a metal substrate comprising:

(1) Heating a metal substrate having a roughened oxidized surface to about a maximum temperature (T_1) where T_1 is selected such that the total volume expansion of the metal is substantially equal to that of the bioglass at the temperature (T_s) at which the temperature dependence of the volume of the bioglass becomes nonlinear;

(2) Providing a body of molten bioglass at temperature T_w where T_w is sufficiently high that the bioglass is sufficiently

fluid to allow immersion of the metal heated to temperature T_1;

(3) Immersing the metal surface in the molten bioglass for the minimum time required to permit a layer of desired thickness of the bioglass to adhere to the surface upon termination of the immersion, the time of the immersion being of such duration that the temperature of the metal surface does not rise substantially above T_1;

(4) Terminating the immersion;

(5) Allowing the coating to cool rapidly from T_W to about T_3, whereby the thermomechanical stresses in the bioglass layer are rapidly relieved; and

(6) Allowing the coated substrate to further cool to a temperature below about T_s whereby the thermomechanical stresses in the bioglass coating and the metal surface are relieved at a substantially equal rate due to the substantially linear thermal expansions thereof, the bioglass coating being bonded to the metal surface by ion diffusion through the oxidized surface.

Any suitable bioglass capable of bonding to bone or other living tissue may be employed. Suitable bioglasses include those having the following composition, by weight: SiO_2, 40 to 60%; Na_2O, 10 to 32%; CaO, 10 to 32%; P_2O_5, 0 to 12%; CaF_2, 0 to 18%; and B_2O_3, 0 to 20%.

Specific bioglasses include those having the following compositions:

Bioglass A

	Percent
SiO_2	45.0
Na_2O	24.5
CaO	24.5
P_2O_5	6.0

Bioglass B

	Percent
SiO_2	42.94
Na_2O	23.37
CaO	11.69
P_2O_5	5.72
CaF_2	16.26

Bioglass C

	Percent
SiO_2	40.0
Na_2O	24.5
CaO	24.5
P_2O_5	6.0
B_2O_3	5.0

Suitable metals include any metal from which an artificial prosthesis or orthopedic or dental device may be fabricated. Suitable metals include steels such as surgical stainless steel and carbon steel; cobalt-chrome alloys; titanium and titanium alloys; noble metals such as platinum; and noble metal alloys such as platinum-rhodium (90% and 10% by weight, respectively) and molybdenum-nickel-cobalt-chrome alloys.

Example: A structure designed as a replacement for a total hip joint in a monkey composed of stainless steel having the composition:

	Percent by Weight
C	0.03
Mn	1.5
Si	0.5
Cr	18
Ni	13
Mo	2.25
Impurities (P, S)	<0.3
Fe	Balance

was thoroughly cleaned by sandblasting with 180 grit alumina at 80 psi to remove foreign scale and roughen the surface to about a 150 μm finish. The roughening increases the surface area of the metal substrate thereby providing more area for a diffusional bond between the glass and metal.

The device was then thoroughly cleaned ultrasonically in acetone three times (at least 10 minute cycle). The device was then suspended in the center of a tubular oxidizing furnace open to the atmosphere and maintained at 800°C (T_1). The device was allowed to remain in the furnace for 20 minutes to allow for complete linear expansion and to provide an oxide finish in the roughened metal surface of about 1 to 2 μm in thickness.

A biologically active glass having the composition of Bioglass A above was melted in a platinum crucible for a period of 1 hour at 1325°C. The volume of molten glass was sufficient to allow complete immersion of the steel device. The glass was very fluid at this temperature and had a viscosity of about 2 poises.

The metal device and crucible containing the bioglass were simultaneously withdrawn from their respective furnaces. The metal device was immediately immersed in the molten bioglass with a quick, smooth motion and withdrawn at a rate of about 2 cm/sec. This produced a fluid coating of glass of about 1 mm in thickness in the surface of the device. The entire procedure required about 3 to 5 seconds. Obviously, variations in the thickness of the bioglass layer may be achieved by controlling the viscosity of the glass, the length of time of residence of the device in the molten glass, and the rate of withdrawal of the device from the glass.

The glass coated device was held in the air for 20 to 30 seconds to allow the surface temperature of the glass to reach about 800°C. During this period the glass flowed thereby relieving any induced stresses. Also during this period, diffusion of metal from the thin oxide layer into the first few (5) μm of glass occurred.

After the temperature of the surface of the glass had cooled to about 700°C, the coated device was placed in a cooling furnace and allowed to cool to room temperature thereby permitting uniform contraction of the glass and metal.

Optionally, the coated device may be reheated to 500° to 700°C or allowed to remain at 500° to 700°C after coating for a predetermined period of time to allow for partial or full crystallization of the glass.

Halogen Cycle Incandescent Lamps

R. Kiesel, G. Maier and H. Grahmann; U.S. Patent 4,178,050; December 11, 1979; assigned to Patent-Treuhand-Gesellschaft fur Elektrische Gluhlampen mbH, Germany describe a method of economic manufacture of halogen cycle incandescent lamps provided with bulbs of hard glass.

The method consists of the following steps. A bulb with a rounded-off bowl and integrally formed exhaust tube is shaped from a length of hard glass tubing, the ends of which are placed in clamps and are drawn apart after a respective section of the length of tubing has been heated.

The transition area between bulb bowl and exhaust tube is additionally constricted in a second processing step. A subassembly of a mount of one or more individually connectible filaments, two or more lead-in wires, preferably having preoxidized portions and, if required, metal shielding caps, is then pinch-sealed into the bulb, sealing being effected so that the preferably preoxidized portions of the lead wires are positioned within the pinch-seal area.

The pinch-seal is then subjected to additional heating, and additionally subjected to another pinching-and-forming or molding operation. The lamp is finished by filling with inert gas, such as nitrogen, argon, krypton, or xenon, with halogen additive, and then tipping off the exhaust tube.

Alternatively, the lead wires with their preoxidized portions are preassembled by sealing them into a bead member of hard glass which, together with the wires, is then pinch-sealed into the bulb of another hard glass having a higher transformation temperature. The pinch-seal is heated again to obtain a hermetic seal between the bead member of hard glass and the bulb of hard glass.

The transformation point of the bead member of hard glass is lower than the transformation point of the bulb material so that sealing of the lead wires into the bead member of hard glass is possible at a lower temperature than when sealing them directly into the bulb. During the succeeding pinching operation, heating of the hard glass of the bulb is not required to the same extent as when sealing the lead wires directly into the bulb.

During heating of the bulb to form the pinch-seal, around the bead member, the abovementioned hermetic seal is established between the hard glass bead member and hard glass bulb without any noticeable deformation of the pinch-seal. For stabilization of the mount into a discrete subassembly, bonding of the lead wires to a wire bridge, e.g., by welding, is possible. This bridge is provided externally of the sealing area and is removed when the pinch-seal has been established.

This process is particularly suitable to make halogen lamps for use in automotive headlights, in which two separately switchable incandescent filaments are used, respectively, for high beam and low beam. This process is particularly appropriate for automatic manufacture of such lamps, in which rotary or turret-type lamp manufacturing machines are used. Such machines have a plurality of work stations, arranged around the circumference of a circle, at which the respective elements or parts are worked on.

STAINED GLASS PHOTOMASKS

Electron Bombardment Method

Photomasks are used in the art of photolithography for printing microelectronic circuits and other precision photofabricated parts, such as television shadow masks.

In a photolithographic process a substrate is covered with a layer of photoresist in which a pattern is photographically developed by superimposing over the photoresist a photomask having patterned transparent and opaque areas, and then passing actinic radiation, especially ultraviolet light, through the transparent areas of the photomask. A pattern is then developed in the photoresist as a relief image by means of the differential solubilities of the exposed and unexposed portions.

The resulting image may be either a negative or a positive image of the photomask, depending upon whether the photoresist is negative-working or positive-working. Etching or other treatments may then be carried out on the underlying exposed portions of the substrate.

F.M. Ernsberger; U.S. Patent 4,144,066; March 13, 1979; assigned to PPG Industries, Inc. has found that stain-producing ions such as copper and/or silver can be injected into a surface portion of a glass substrate by means of electron bombardment through a thin film of a source of the stain-producing ion.

The electron bombardment causes the injected ions to immediately become reduced to their elemental state, whereupon the application of heat causes the atoms to agglomerate into colloidal microcrystalline color centers which result in the formation of a stain within the body of the glass substrate. The agglomerating heat may be applied by the electron beam itself or by auxiliary means. The stain may be generated in the pattern desired for a photomask either by employing a photoresist through which the electron bombardment may be carried out or by scanning a focused electron beam along a controlled locus.

In the first step of the process as shown in Figure 7.5a, a glass substrate **10**, such as a sheet of conventional soda-lime-silica glass, has a photoresist layer **11** applied to one surface thereof. The type of glass used is not particularly critical, so long as it includes at least a small amount of mobile cations such as alkali metal ions.

The photoresist shown in Figure 7.5 is a positive-working photoresist, but it should be understood that negative-working photoresists are equally suitable. The photoresist **11** is exposed to ultraviolet radiation through a master mask **12** having patterned apertures **13**. Upon development of the photoresist, patterned openings **14** are produced therein, as shown in Figure 7.5b.

Figure 7.5: Process Steps of Fabricating a Photoresist by Electron Bombardment

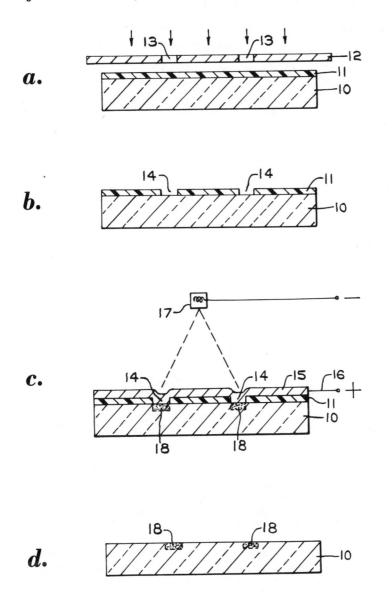

Source: U.S. Patent 4,144,066

Next, a thin conductive film **15** containing a source of stain-producing cations is deposited over the developed photoresist. Film **15** may be a metal or a metal oxide or salt, and the stain-producing ions are preferably silver and/or copper although gold and thallium may also be employed. Preferably, film **15** is a metallic film of silver deposited by sputtering or evaporation. The thickness of film **15** should be sufficient to provide at least the required amount of stain-producing cations, although its thickness is preferably minimized so as to minimize the voltage requirements for the subsequent electron bombardment step.

Referring to Figure 7.5c, the electron bombardment step shown takes place in an evacuated chamber. Film **15** is connected as the anode by means of lead **16** to establish an electrical field between the film **15** and a source of electrons **17**. Film **15** is uniformly flooded with electrons from source **17** with sufficient energy for at least some of the electrons to penetrate through the thickness of film **15** and into the underlying glass.

The energy level of electrons will depend upon a number of factors including the thickness of film **15**, the depth of electron penetration desired, and the degree of temperature rise desired to be generated in the surface portions of the glass. Voltages of at least 5 to 10 kV have been found sufficient for most purposes. Electron penetration of about 1 μ has been found to result from the use of about 10 kV.

In zones **18** of the glass underlying the patterned openings **14** through the photoresist, electron penetration establishes a localized negative charge within those zones which induces migration of the stain-producing cations from film **15** into the glass. The remaining portions of photoresist **11** mask ion migration in the other areas of the glass surface. Because of the availability of excess electrons, reduction of the injected cations takes place immediately within zones **18**, and if the heating effect of the electron beam is sufficiently intense, agglomeration of the atoms will also take place simultaneously.

However, if a lower energy electron beam is employed, agglomeration may be carried out by heating the glass substrate **10** from an independent source of heat. This independent source of heat may be located within the vacuum chamber so that heating may take place simultaneously with the electron bombardment, or the agglomeration heating may take place in a separate, subsequent operation outside the vacuum chamber.

Upon removal of the photoresist **11** and film **15**, the final product as shown in Figure 7.5d is an all-glass photomask having a pattern of stained areas **18** within the glass. The opacity of the stained areas is determined by the amount of stain-producing ion injected into the glass, which, in turn, is controlled by the length and conditions of the electron bombardment step.

Typically, glass stains of the type disclosed here can readily achieve a transmittance of less than 1% in the ultraviolet range of wavelengths (3,000 to 4,000 angstroms) most commonly used in photolithographic processes. At the same time, these same areas remain sufficiently transparent in the visible wavelength spectrum to advantageously provide enough transparency to enable an operator to align the photomask with a substrate to be printed.

The unstained areas of the photomask retain the radiation transmittance properties of the base glass from which the substrate **10** is made, which may typically be at least an order of magnitude more transparent than the stained areas in the ultraviolet range.

Electromigration Method

F.M. Ernsberger; U.S. Patent 4,155,735; May 22, 1979; assigned to PPG Industries, Inc. has found that a high resolution stained glass photomask may be made without the need for etching a metal film, thereby simplifying the process and reducing the chances for loss of resolution.

In accordance with the process, electromigration under carefully selected conditions permits the use of a continuous layer of a source of stain-producing ions in combination with an organic photoresist to migrate stain-producing ions into predetermined, patterned zones of the glass surface.

During the ion migrating step the use of an electric field of sufficient potential enables the use of temperatures below the point at which the organic photoresist begins to melt or decompose. Electromigration of the stain-producing ions will not take place in the areas of the glass underlying the photoresist due to the low electrical conductivity of the photoresist. Thus, electromigration of the stain-producing ions is limited to the portions of the glass underlying apertures through the photoresist which are photographically developed in the photoresist.

In the exposed zones of the glass, cations from the continuous film (preferably silver and/or copper) are induced to migrate into the glass substantially perpendicularly to the surface of the glass due to the unidirectionally applied electric field. Within the glass, the stain-producing cations replace mobile cations of the base glass composition, especially alkali metal ions, which migrate deeper into the glass toward the cathode side of the electric field. Reducing the stain-producing ions to their elemental state and agglomerating them into submicroscopic crystals within the glass results in a visibly stained pattern within the body of the glass.

Referring to Figure 7.6a, there is shown a glass substrate **10** onto which a film **11**, which serves as a source of stain-producing ions, has been applied. Stain-producing ions which may be used include silver, copper, gold and thallium, with silver being preferred.

Film **11** may be applied as a compound of one or more of the stain-producing cations having relatively low electrical conductivity (e.g., AgCl), employing conventional coating techniques such as evaporation, sputtering, wet chemical deposition, or any other known technique, although it has been found that coatings of desirably higher uniformity can be achieved by sputtering in particular.

In Figure 7.6b, a layer of photoresist **12** has been applied over film **11** and is exposed to radiation through patterned apertures **14** in a master mask **13**. When the photoresist **12** is then developed, a pattern of apertures **15** is produced in the photoresist, as shown in Figure 7.6c.

The photoresist shown in all of the figures is a positive-working photoresist, so that the portions **15** which have been dissolved correspond to the patterned apertures **14** in the master mask while the remainder of the photoresist remains intact. It should be apparent that a negative-working photoresist could be used instead, in which case photoresist would remain only in the light-exposed areas after development.

Figure 7.6: Process Steps of Fabricating a Photoresist by Electromigration

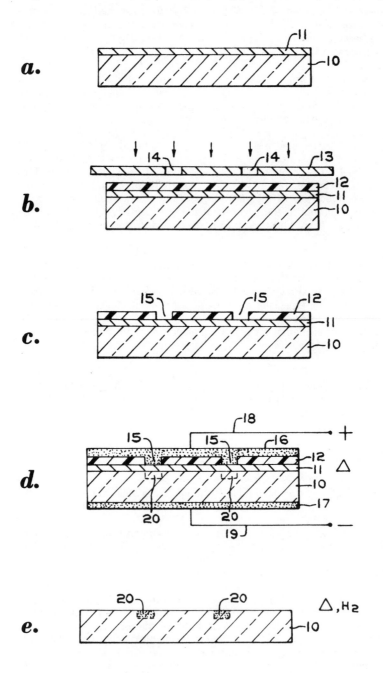

Source: U.S. Patent 4,155,735

Specific examples of commercially available photoresists which may be used include LSI-195 photoresist (Philip A. Hunt Co.), KPR photoresist (Eastman Kodak Co.) and AZ-111 (Shipley Co.).

After the photoresist has been developed, the piece is ready for electromigration of the stain-producing ions into the glass as depicted in Figure 7.6d. Electrically conductive layers 16 and 17 are applied to opposite sides of the glass and serve as anode and cathode, respectively, when they are connected to a source of electrical potential by means of leads 18 and 19.

Because it is easy to apply and remove, colloidal graphite is the preferred material for electrode layers 16 and 17. The colloidal graphite layers may be applied to the substrate in slurry form in which the colloidal graphite is suspended in an aqueous or alcoholic vehicle or the layers may be applied by means of commercially available aerosol sprays of colloidal graphite. Whatever material is used for the electrode layers 16 and 17, it must be applied sufficiently thick to render its resistance insignificant relative to the resistance of the glass (e.g., less than about 10%).

Imposing an electric field between the electrode layers 16 and 17, with layer 16 serving as the anode, causes a migration of mobile cations, especially alkali metal ions, out of zones 20 of the glass which underlie apertures 15 in the photoresist. These mobile cations are repelled by the anode and are thus driven deeper into the glass substrate. At the same time, the electric field causes the stain-producing ions from layer 11 to be injected into the glass in the patterned zones 20 where they take the places vacated by the migrated glass constituent ions. Because of the low electroconductivity of the photoresist, ion migration does not take place in the portions of the glass underlying the photoresist 12.

The rate of ion migration is influenced by temperature and the applied voltage. At room temperature and a potential of only a few volts, the rate of ion migration would be virtually imperceptible. Therefore, elevated temperatures, preferably above about 100°C and a potential of at least a few hundred volts are preferred in order to obtain reasonable treatment times.

Although photoresists are subject to melting and thermal decomposition at elevated temperatures, it has been found that the use of moderately elevated temperatures in combination with an electric field permits a satisfactory electromigration to be carried out without harming the photoresist. Thus, the temperature is maintained above about 100°C but below the temperature at which the photoresist begins to melt or decompose, which is typically around 200°C, depending upon the particular photoresist being used. As an example, temperatures in the range of about 160° to 200°C, together with a potential of about 300 to 400 volts, have been found to yield satisfactory results.

Temperature and voltage are also interdependent since the conductivity of glass increases with increasing temperature, thereby lessening the voltage requirement. Thus, by selecting an appropriate combination of temperature and voltage, a practical rate of ion migration can be attained without requiring the use of deleteriously high temperatures or excessive voltages. It is necessary to avoid voltages so high as to cause arcing through the glass.

After the stain-producing ions have been electromigrated into the glass to the desired depth, all coatings may be removed from the glass substrate, thereby leaving nothing but a latent image within the glass substrate.

As depicted in Figure 7.6e, development of coloration in the ion migrated zones **20** requires the application of heat in a reducing atmosphere to first reduce the stain-producing ions to their elemental state and then to agglomerate the metallic atoms into a submicroscopic crystalline form. These steps can be conveniently carried out, for example, by heating to at least about 200°C, preferably about 400° to 500°C, in an atmosphere containing hydrogen.

At some point in the process following the electromigration step, it may be desirable in some cases to subject the glass to a light etching so as to remove a thin layer of glass into which staining ions may have migrated beneath the photoresist as a result of current leakage through the photoresist. Any known glass etchant, such as dilute, aqueous hydrofluoric acid, may be used.

HEAT-REFLECTING GLASS

Aluminum-Containing Coating

S. Okino, T. Mori, T. Yanai and T. Sawa; U.S. Patent 4,160,061; July 3, 1979; assigned to Central Glass Company, Limited, Japan describe a heat-reflecting glass plate which has a metal oxide film coated at least on one side of the glass plate, and the metal oxide film comprises aluminum oxide and at least one metal oxide selected from the respective oxides of heavy metals such as Cr, Co, Ni, Fe, Zn, Sn, Cu, Mn and Ti.

The amount of aluminum oxide in the metal oxide film is preferably such that the percentage of Al to the total metals in the film is from 1 to 50% by weight, and more preferably from 5 to 35%.

According to a method for forming the metal oxide film on the glass surface, a thermally decomposable compound exemplified by acetyl acetonate of Al is dissolved in an organic solvent together with at least one similarly decomposable metal compound selected from the respective compounds of Cr, Co, Ni, Fe, Zn, Sn, Cu, Mn and Ti, and at least one side of the glass plate is coated with the solution. The coated solution is heated to cause thermal decomposition of the dissolved metal compounds to give the respective oxides.

Example: Acetyl acetonates of Cr, Co, Fe and Al were dissolved in a mixture of dichloromethane (75 vol %) and methanol (25 vol %) with various concentrations of aluminum acetyl acetonate as shown in the following Table 1.

Each solution was sprayed onto the surface of a transparent, 30 cm square and 0.6 cm thick glass plate which had been heated to 630°C in an electric furnace. A resulting metal oxide coating formed on the glass surface was 30 to 80 mμ thick. The tint of the oxide-coated glass plates was neutral and warm grayish. Light transmitted through each specimen was most thickly tinged when a reference solution containing no aluminum salt was used and became gradually pale with increase in the concentration of the aluminum salt in the solution.

After measurement of the transmissivity for visible light and reflectivity for solar energy, the oxide-coated specimens were tempered by heating again for 15 minutes in the electric furnace at 650°C and rapidly chilling by a blast of compressed air. This treatment was carried out as a model of commonly practiced heat treatments of heat-reflecting glasses either for bending works or for strengthening. Then the reflectivity was measured again. The results are presented in Table 2.

Table 1

Sample No.	Composition of Solution: Cr Salt (g)	Co Salt (g)	Fe Salt (g)	Al Salt (g)	Solvent (ml)	Al:Total Metals (wt %)
A*	3.0	3.0	3.0	0	100	0
1	3.0	3.0	3.0	0.25	100	1.5
2	3.0	3.0	3.0	0.5	100	2.8
3	3.0	3.0	3.0	0.7	100	3.9
4	3.0	3.0	3.0	0.9	100	5.0
5	3.0	3.0	3.0	1.0	100	5.5
6	3.0	3.0	3.0	3.0	100	15.0
7	3.0	3.0	3.0	4.5	100	20.9
8	3.0	3.0	3.0	6.0	100	26.0
9	3.0	3.0	3.0	9.0	100	34.5
10	3.0	3.0	3.0	11.0	100	39.2
11	3.0	3.0	3.0	17.0	100	49.9

*Reference specimen.

Table 2

Sample No.	Al:Total Metals (wt %)	Transmissivity (%)	Reflectivity Before Heating (%)	Reflectivity After Heating (%)	Lowering of Reflectivity by Heating (%)
A*	0	43.2	30.3	24.8	5.5
1	1.5	42.5	30.5	26.2	4.3
2	2.8	44.0	30.2	27.1	3.1
3	3.9	43.5	30.4	27.2	3.2
4	5.0	47.5	29.5	28.2	1.3
5	5.5	47.3	29.9	29.0	0.9
6	15.0	48.5	28.5	28.0	0.5
7	20.9	49.0	26.3	25.5	0.8
8	26.0	53.4	24.0	23.3	0.7
9	34.5	60.3	22.7	22.3	0.4
10	39.2	65.0	18.9	18.4	0.5
11	49.9	69.2	15.2	14.9	0.3

*Reference specimen.

As seen in Table 2, the reference specimen A (which was coated with a metal oxide film not containing Al and hence was a known heat-reflecting glass) exhibited such a lowering of the reflectivity by the heat treatment that the heat-treated specimen could easily be distinguished from the untreated specimen with naked eyes.

When the percentage of Al to the total metals was 5% by weight or more, the lowering of the reflectivity was to a negligible extent. Accordingly, no poor appearance will be presented even if either a tempered glass member or a bent-formed glass member is arranged adjacent a nontreated glass member so long as both the heat-treated and nontreated members are of the same heat-reflecting glass the metal oxide coating of which contains at least 5% by weight of Al (according to the above definition).

TiO$_2$ Layer in Rutile Form

Heat-reflecting window glass panes of the kind in which a heat-reflecting TiO$_2$ layer is applied by vapor-coating of a titanium layer in vacuo and subsequent oxidation of such layer at very high temperatures in air are known and have been described in a publication by G. Hass, "Preparation, Properties and Optical Applications of Thin Films of Titanium Dioxide," *Vacuum*, Vol 2, No. 4, 331-345 (1952).

Depending upon the conditions in which the Ti layer is vapor-coated in vacuo, there are two TiO$_2$ forms which may be produced in the subsequent oxidation of the Ti layer. When the titanium is vapor-coated rapidly in a good vacuum, i.e., of $\cong 10^{-5}$ mm Hg or even higher, the rutile form is produced, whereas if vapor-coating proceeds relatively slowly in a poorer vacuum, e.g., of approximately 10^{-4} mm Hg, the anatase form is produced.

TiO$_2$ layers produced in this way have a variety of uses in optics to coat glass panes, e.g., as color filters and as sunshine-reflecting coatings, the thickness of the layer being made such as to act as a highly effective quarter-wave interference layer, over the spectral range for which a modification of the reflection properties of the substrate is required.

One particular purpose for which heat-reflecting glass panes of the kind described are used is in facade elements or infilling panels for buildings. The requirements in the case of infilling panels of this kind are for TiO$_2$-coated glass panels which have a high neutral reflection, possibly with a slight blue or yellow tint, in the visible spectral range. The usual practice with panels of this kind is for the TiO$_2$ interference layer to be disposed on the outside of the building, the back of the glass substrate being treated with an opaque enamel or varnish, so that parts of the building behind the panel cannot be seen through it from the outside.

Films of TiO$_2$ in the rutile form are very advantageous, more particularly for the last mentioned purpose, since such layers have a higher index of refraction than anatase layers and can, therefore, provide higher reflection values which are very desirable in the case of facade elements and infilling panels.

It has also been found that rutile films are much harder and much more resistant to abrasion than anatase films. Consequently, panes of glass where the TiO$_2$ interference film on the outside of the building has a rutile form can be exposed directly to the atmosphere without damage for a prolonged period of time. Also, the usual cleaning agents for outside surfaces of glass can be used to clean such panes or panels.

Theoretically, of course, oxidation of the Ti film to TiO$_2$ should be carried out very rapidly to provide an acceptable layer of TiO$_2$ in the rutile form. It is known from G. Hass, *Vacuum*, Vol 2, No. 4, 335, Figure 3, that oxidation proceeds faster in proportion as the oxidation temperature is higher. Unfortunately, when in the known process the oxidizing temperature is increased to above 550°C in order to promote rapid oxidation, alterations occur in the rutile layers, which turn matt and dull and cause so much light scatter both in transmission and reflection that panes having such layers cannot be used for the purposes mentioned.

R. Groth; U.S. Patent 4,188,452; February 12, 1980; assigned to BFG Glassgroup, France describes a heat-reflecting glass pane of the kind described, of use more particularly as an infilling panel or facade element, and a method for the production of such pane, wherein the TiO_2 layer is at least predominantly in the rutile form and the unwanted changes in the coating which occur when the glass is heated to temperatures above 550°C such as are necessary for rapid oxidation of the Ti layer and such as are essential more particularly for toughening, are effectively inhibited.

The method comprises the steps of providing a substantially transparent glass substrate, applying a silicon oxide layer to the glass substrate, applying a Ti layer to the side of the silicon oxide layer remote from the glass substrate by vapor deposition in vacuo, and subsequently heating the pane to at least 550°C to oxidize the Ti layer to form a layer of TiO_2 at least predominantly in the rutile form, and subsequently cooling the pane.

Example 1: A float glass pane 8 mm thick and having external dimensions of 300 cm x 245 cm was first cleaned by glow discharge in a vacuum-coating plant at a pressure of 3×10^{-2} mm Hg conventionally, whereafter a Ti film was applied to the pane by vapor deposition at a pressure of 3×10^{-5} mm Hg.

The coated pane was then heated in a conventional toughening furnace to 620°C, then chilled. During the heating process the titanium film was oxidized to a TiO_2 film 520 Å thick and having the rutile form. However, the film was turbid and matt and was thus unsuitable for architectural reasons, for use as an infilling panel or facade element, for a building, or the like.

Example 2: The same procedure as in Example 1 was followed but with the difference that, after the glow-discharge cleaning and before the application of the titanium film by vapor deposition, a 130 Å thick Si_2O_3 intermediate film was formed on the glass pane by reactive evaporation of silicon monoxide.

The coated pane was toughened in the same way as in Example 1 after application of the Ti film. The TiO_2 film also had a rutile structure but, unlike the TiO_2 film in Example 1, the film was completely clear, so that the finished pane could be used very satisfactorily as an infilling panel or facade element for a building or the like.

MISCELLANEOUS PROCESSES

Impact-Resistant Safety Glass

W. Triebel, S. Hari, H. Koert and F. Evers; U.S. Patent 4,201,828; May 6, 1980; assigned to Deutsche Gold- und Silber-Scheideanstalt vormals Roessler, Germany describe an impact-resistant safety glass laminate formed of at least one silicate glass panel or layer and at least one polycarbonate panel or layer, and which is further characterized in that one silicate glass panel is of at least 6 mm in thickness and is disposed on the side of the laminate exposed to the potential impact.

It is a further feature of the process that the silicate lamina and the polycarbonate lamina are bonded to each other by means of a cured methacrylic resin, which is derived to at least 50% by weight, based on the total weight of the resin, from 2-ethylhexyl methacrylate. The polycarbonate lamina or panel combined with the silicate glass panel is to be 1.5 to 5 mm, preferably 2 to 4 mm thick.

The silicate glass panel and the polycarbonate panel are bonded to each other by an adhesive layer of effectively at least 0.1 mm, preferably at least 0.5 mm thickness. The upper limit for the thickness of the adhesive layer is governed by economic factors, and is about 5 mm, preferably about 3 mm.

It is particularly advantageous to produce the adhesive layer by curing a methacrylate syrup, which consists of:

(A) 50 to 80% by weight of 2-ethylhexyl methacrylate, 10 to 40% by weight of 2-ethylhexyl acrylate, 0 to 5% by weight of other alkyl esters of the acrylic and/or methacrylic acid, 0.1 to 3% by weight of acrylic and/or methacrylic acid, and possibly 5 to 30% by weight of 2-hydroxypropyl acrylate, whereby at least a part of the monomers exist to such an extent in polymer form that the viscosity of the finished syrup at +20°C lies between 10 and 1,000 cp;

(B) 0.1 to 0.6% by weight of glycol dimercaptoacetate;

(C) 5 to 20% by weight of ethyl tributyl citrate; and

(D) 0.1 to 3% by weight of at least one multivalent ester of the methacrylic acid,

whereby all quantitative data are based on the weight of the finished syrup.

The syrup of methacrylate adhesive is inserted between the two laminae or panels that are to be bonded together in accordance with known techniques and they are cured by a cold hardening catalyst system in a known manner.

A combination of small quantities, for example, 1 ppm of Cu^{++} ions (added as copper naphthenate) and of a hardener liquid, consisting of a solution of at least one alkylacyl peroxide and at least one compound which contains a labile halogen atom, preferably a chlorine atom, in a viscous vehicle such as a plasticizing agent or viscous solvent is particularly preferred as a catalyst system.

Representative hardeners have been described extensively in German Patent 1,146,254. They consist, for example, of a solution of β-phenyl ethyl dibutyl amino hydrochloride and tert-butyl perbenzoate in dimethyl phthalate and are used in quantities of 0.5 to 2% by weight, preferably of about 1% by weight, related to the weight of the methacrylate syrup.

The curing should take place effectively at a temperature between +15° and +35°C, preferably between +20° and +30°C.

In order to protect the polycarbonate panel also on its reverse side against scratching it can be covered with or bonded to a second, generally thinner, silicate glass pane similar to that disposed on the side of potential impact. However, it is likewise possible to provide the reverse side of the polycarbonate panel with a clear-transparent, scratch-resistant coating of only a few micron thickness. This additional protective coating can be accomplished, for example, in a known manner by vaporization of mineral substances, such as silicon monoxide or dioxide, or metallic oxides or fluorides onto the surface of the polycarbonate panel.

As compared to known impact-resistant glass systems, a considerable savings in overall thickness and weight at the same impact-resistance will be achieved by the safety glass according to this process.

Fireproof Glass

W. Kiefer; U.S. Patent 4,172,921; October 30, 1979; assigned to Jenaer Glaswerk Schott & Gen., Germany describes fireproof glass sheets or panes which resist high temperature and which have such a high thermal strength and dimensional stability that when they are used as a room closure, they withstand a fire test according to DIN 4102 for at least 90 minutes without cracking or causing an opening in the room closure.

For the fire test according to DIN 4102 (1970 Edition), the panes with frames have to be installed as a room or space closure into a fired furnace. With the rapid heating in accordance with the standard temperature curve 1, (STC1), sheet 2, section 5.2.4., a temperature gradient is set up between the center of the pane and the rim or margin thereof, because the rim is initially heated up more slowly than the center of the pane, because of the heat-insulating action of the frame.

Due to this temperature gradient or drop, a tensile stress is set up in the marginal zone, and this can lead to the destruction of the pane. The height of the temperature gradient is, inter alia, dependent on the heating-up speed, the heat insulation of the rim or margin of the pane by the frame and the width of the frame.

With a frame which is 2 to 3 cm wide, the temperature difference is generally between 200° and 300°C. With a wider frame, the temperature difference can be even higher. However, since the stresses are reduced again above the transformation temperature (T_g), the temperature difference which leads to the tensile stresses cannot be greater than T_g, i.e., about 550°C for borosilicate glasses.

Since a frame at least about 2 cm wide is generally required for holding the panes, the stressed glass panes must have such a high thermal strength that they withstand a temperature drop between the hotter central portion of the pane and the colder rim of the pane of about $\geqslant 200°$ to $\leqslant 550°C$, without cracking.

The fireproof and high temperature-resistant glass panes are those which have a compressive stress in the marginal region, or in the marginal region and in the surface layer of the center of the pane, and which consist of glasses of which the product of heat expansion (α) and elasticity modulus (E) is 1 to 5 [kp x cm^{-2} x °C^{-1}] and which have such a strong tendency to surface crystallization that they form an at least 5 μm preferably at least 10 μm thick, continuous crystalline surface layer on being heated in accordance with the STC1.

According to one embodiment, the glass panes can be heated to 50° to 150°C above the transformation temperature in order to produce the compressive stress in the marginal zone and thereafter left for only such a length of time in this temperature range above the transformation temperature, while the marginal region of the pane is cooled to a temperature below the transformation temperature, that there has still not been formed any surface crystallization which reduces the mechanical strength of the panel and/or disturbs the transparency of clear glass panes, before the center of the pane has likewise cooled to below the transformation temperature.

According to another embodiment, in order to produce a compressive stress in the marginal region and a compressive stress in the surface of the central region of the pane, the panes can be heated up to a temperature which is within a range of 150°C below the softening temperature (E_W) of the basic glass at a rate fast enough to prevent crystallization and/or deformation. Thereafter the marginal region of the panes, during a time period in which still no surface crystallization has been formed, which would lower the mechanical strength of the glass and/or interfere with the transparency of clear glass panes, can be cooled to a temperature below the transformation temperature, while the central portion of the pane is kept at a temperature just below the softening temperature, until the marginal region has reached the prescribed temperature.

With the following rapid quenching of the central portion of the pane, a compressive stress is established in the marginal region of the panes and in the surface layer of the central portion of the panes. This stress is only so high that the pane is sufficiently resistive on being heated up in accordance with the STC1.

With thick panes, e.g., 5 to 8 mm, it is desirable for these initially to be left for some time, e.g., 5 to 8 minutes, at the transformation temperature, before rapidly heating them to a temperature just below the softening temperature.

Example: A glass pane (500 x 500 x 5 mm), which consists of a glass of the composition in percent by weight: SiO_2, 65.80; Al_2O_3, 18.00; Li_2O, 4.00; Na_2O, 2.00; MgO, 1.00; ZnO, 6.00; CaO, 0.50; BaO, 1.50; TiO_2, 0.60; ZrO_2, 0.60; and has the physical properties: heat expansion (α) (20° to 300°C) = 50.8 x 10^{-7} [$°C^{-1}$], elasticity modulus (E) = 8.9 x 10^5 kp/cm^2, T_g ($\eta \sim 10^{13.5}$ poises) = 600°C; position of the DTA peak: 757°C; upper devitrification limit UDL: 1253°C; activation energy of the crystal growth speed: 65 kcal/mol, is subjected to compressive stress in the marginal region. For this purpose, the glass pane is placed between two ceramic plates of sillimanite with the size 500 x 500 x 50 mm and the composite unit is heated in a chamber furnace to 650°C.

After reaching this temperature, the composite unit is transferred to a second chamber furnace with a temperature of 400°C. The edge of the pane which lies free cools substantially more quickly than the central portion of the pane. The speed at which the temperature falls in the center of the pane depends to a very great extent on the thickness and the material of the ceramic plates.

The composite unit will be left in the second chamber furnace until the entire plate has cooled to a temperature of 400°C and is subsequently cooled to ambient temperature. In this way a compressive stress zone with a width of about 3 cm has been developed in the marginal region of the glass pane. The compressive stress increases from the interior towards the rim of the pane. The compressive stress immediately in the rim is approximately 450 kp/cm^2.

No crystals are to be detected, even with a microscope, on the surface of the pane, which is moreover completely transparent. The reduction in strength of the pane, caused by the compressive stress, amounts to approximately 75 kp/cm^2.

The glass pane with the compressive stress in the marginal region is fitted into a steel window frame using a conventional sealing composition. The marginal portion of the pane is covered to about 20 mm by the steel frame. The steel window frame is installed in a brick wall, which serves as the space closure of a firing furnace. The firing chamber is heated up in accordance with the standard temperature curve (STC1) DIN 4102, sheet 2, 1970 Edition.

The glass pane goes beyond the first 60 minutes of the firing test without cracking or being substantially deformed. After about 70 minutes, the surface of the glass slowly starts to become cloudy. This clouding increases with the duration of the firing test. The firing test according to DIN 4102 is stopped after 130 minutes. No penetration of fire and smoke can be detected in this instance.

Glass for Radio-Photoluminescence Dosimetry

Glass dosimeters for measuring and recording ionizing radiation have been known for many years. Those currently used are silver-bearing phosphate glass dosimeters which are similar to film dosimeters and are generally used for personnel dosimetry. These glass dosimeters, generally comprising two polished surfaces vertical to each other, are exposed to an ultraviolet light source after having been exposed to prior radiation, especially gamma radiation. The fluorescent radiation thus emitted is proportional to the gamma radiation which has been received, and is measured vertically to the direction of the UV-radiation incidence.

Dosimetry by means of silver phosphate glasses is based on the alternating effect of ionizing radiation at energy levels greater than about 10 keV on the silver ions in the glass. The nature of this alternating effect is still not completely clear, but the active centers formed by such radiation are believed to represent interference centers which, on energization with UV light at a frequency of about 365 nm, emit fluorescent light in the visible spectrum (the so-called radio-photoluminescence) which is then measured by means of a photoamplifier. The intensity of the emitted light is proportional to the gamma radiation dose received up to approximately 4 x 10^3 rads. Such silver-activated metaphosphate glasses should desirably show a low natural fluorescence upon UV light stimulation as well as a high sensitivity to radiation, especially gamma radiation.

The relative dose sensitivity of such dosimeter glasses depends on the energy level of the gamma radiation to which they are exposed, and remains substantially constant from about 0.2 MeV up to about 10 MeV. Below about 0.2 MeV, the dose indication rises to a maximum at approximately 50 KeV. A reduction in the effective order of magnitude of the chemical elements in the glass can be used to reduce their energy dependence, which may require compensation by encapsulation with perforated metal filters.

W. Jahn and W. Schumann; U.S. Patent 4,204,976; May 27, 1980; assigned to Jenaer Glaswerk Schott & Gen., Germany describe a lithium-free silver-activated phosphate glass suitable for radio-photoluminesence dosimetry, consisting essen-

tially of, in percent by weight on the total composition: not more than 26% by weight of alkali metal phosphates; at least 16% by weight alkaline earth metaphosphate; a radio-phospholuminescent activating amount of silver metaphosphate; and the balance, aluminum metaphosphate.

The dosimeter glass compositions are characterized in having an alkali metal metaphosphate content of preferably 15 to 26% by weight and especially about 19 to 24% by weight, and an alkaline earth metal metaphosphate content of preferably 16 to 27% by weight and especially about 20 to 24% by weight.

The preferred alkali metal metaphosphate is $NaPO_3$, while the preferred alkaline earth metaphosphate is $Mg(PO_3)_2$. The remainder of the glass composition preferably consists essentially of $Al(PO_3)_3$ and an activating amount of a suitable radio-photoluminescent activating compound, e.g., about 2 to 8% by weight, preferably about 3 to 4% by weight, $AgPO_3$.

Example: 1,000 grams of a composition consisting of about 23.9% $NaPO_3$, about 24.4% $Mg(PO_3)_2$, about 48.2% $Al(PO_3)_3$, and about 3.5% $AgPO_3$, are melted in a ceramic crucible over a period of approximately 2 hours at 1250°C. The temperature is then raised to 1450°C and refining continued for 65 minutes to remove the bubbles. After standing at a reducing temperature, the glass is cast into iron molds at 1000°C and cooled slowly (from T_g at approximately 6°C per hour). Glass samples treated for 50 days in a conditioning cabinet at 50°C and 100% relative humidity showed only limited flecking.

Stationary Phase Surface for Chromatography

V. Pretorius and J.D. Schieke; U.S. Patent 4,169,790; October 2, 1979; assigned to South African Inventions Development Corporation, South Africa describe a stationary phase for chromatography having a surface configuration comprising a plurality of microvilli projecting from the surface of a substrate.

In a preferred embodiment, the microvilli have a dendritically branched microstructure or texture. The siliceous surface may be silica, silicate, or glass. Preferably, the surface is glass.

The process comprises subjecting a siliceous surface to the action of the vapor of a silica-dissolving medium at a concentration higher than required for an etching effect and lower than that at which coarse crystal growth predominates.

The silica-dissolving substance is preferably a fluorine-containing substance, for example, hydrofluoric acid or methyltrifluorochloroethyl ether or any chemically-related substance known to liberate hydrofluoric acid on heating. Examples include, but are not limited to, 2-chloro-1,1,2-trifluoroethylmethyl ether, 2-chloro-1,1,2-trifluoroethylethyl ether, 2-chloro-1,1,2-trifluoroethylallyl ether, fluorotrichloromethane, 1-fluoro-1,1,2,2-tetrachloroethane and p-fluorophenetole.

Methyltrifluorochloroethyl ether (MTFEE) is a preferred reagent. If the partial pressure of this reagent is too low, an etching effect accompanied by slight surface roughening, but not by the growth of the aforementioned microvilli, results. If the partial pressure is too high, coarse, felted, irregular and uncontrolled textures are formed instead of controlled, substantially regular microvilli textures.

The concentration of MTFEE should be at least 20 ml of reagent in liquid form per liter of vapor space in which the siliceous surface being treated is contained but not more than about 300 ml per liter of such space. Preferred limits are 20 and 150 ml per liter of space, most preferably between 25 and 120 ml, especially 25 and 100 ml per liter of space.

Treatment is generally conducted at a minimum temperature of 240°C, preferably 300°C, whereas the maximum temperature is generally about 500°C, preferably 400° to 450°C, e.g., 420°C. Thus, suitable temperature ranges for the treatment include but are not limited to 240° to 500°C, 250° to 450°C, and 320° to 400°C. A temperature of about 400°C is especially suitable. Optimum growth of micro-villi is attained between 350° and 420°C, more particularly 400°C, using between 25 and 75 ml of liquid reagent per liter of vapor space. The temperature should be kept constant over the length of a tubular column or substrate being treated as should vapor concentration.

In a preferred procedure, the required volume of fluorine-containing reagent is introduced into a capillary to be treated, e.g., a capillary in the form of a coil, and is distributed along the length of the coil so that each turn of the coil contains an approximately equal volume of reagent. The ends of the capillary are sealed and the tube is heated to the required temperature.

The period of heating applied and the temperature used determine the length to which the microvilli grow. A minimum of about 1 hour is required, but the preferred range is 1 to 6 hours. Heating has been continued for up to 24 hours, but after about 6 hours little additional growth is observed. The preferred range is between 2 and 5 hours, e.g., 3.4 hours. The treatment is done in a nonoxidizing atmosphere, e.g., N_2.

From the foregoing it follows that there are mainly three parameters, i.e., reagent concentration, temperature and time by means of which, for a given type of surface, the process can be regulated to control not only the size, but also the density (number of microvilli per unit of surface area) of the microvilli.

Within the general limits stated, a comparatively low temperature and high concentration of reagent will yield a high density of comparatively thick, but (at least initially) short microvilli. A high temperature, but low reagent concentration will yield a high density of thin and (at least initially) short microvilli. Near the lower limits of concentration and temperature, there is obtained a low density of small, at least initially short whiskers.

In all of the foregoing cases the length of the microvilli may be increased by lengthening the time of exposure but, as stated above, after about 6 hours the further growth usually becomes slow and no noticeable further growth was observed after 24 hours.

After heat treatment, the ends of the tube are opened and the interior of the tube is purged with dry nitrogen gas, e.g., at room temperature, or preferably heated, e.g., to 100° to 300°C, most preferably 200°C. If this treatment is omitted, an imperfect microvilli texture, interspersed with crystalline growths, may result. The treatment cycle may be repeated, if desired. Microvilli having a dendritic microtexture have been obtained this way. These have also sometimes formed spontaneously.

The surface of the microvilli thus produced is highly active and the treated body can be used for gas chromatography or, if desired, liquid chromatography.

Example: A coiled glass capillary (Pyrex) having an inner radius of 0.0170 cm and 50 meters long was filled to 7.5% of its volume with methyltrifluorochloro-ethyl ether, which was evenly distributed throughout the length of the column. The ends of the column were sealed and the entire column heated evenly for 3.5 hours at 350°C. The column was opened at both ends and thoroughly purged with nitrogen.

Inspection by scanning electron microscopy indicated that the entire column surface was evenly coated with microvilli approximately 10^{-3} cm long, most of which projected at right angles to the surface towards the column interior and which had an average thickness of approximately $\frac{1}{15}$ the length of the micro-villi. In addition there was evidence of a layer of graphite adhering to the surfaces of the microvilli and their substrate. The graphite apparently was a decomposition product of the reagent. This graphite can be removed by heating the tube for 6 to 12 hours at 400° to 450°C under a gentle flow of oxygen.

Microwave-Safe Vacuum-Insulated Bottle

H.M. Stewart and W.J. Tanner; U.S. Patent 4,184,601; January 22, 1980; assigned to Aladdin Industries, Incorporated describe a class of vacuum-insulated containers which are safe for use in microwave ovens.

Referring to Figure 7.7, a thermos bottle or vacuum-insulated container suitable for carrying liquids in the insulated interior portion thereof is illustrated. The bottle consists of a vacuum filler **10** comprising outer wall **12** and inner wall **14**. An annular space **16** is defined between the inner and outer walls and air is evacuated therefrom through an opening in the bottom of the outer wall which is thereafter sealed with a plug and protective cap **20**.

In conventional constructions the filler **10** is formed of glass or metal. According to this process glass is preferable inasmuch as metal is not suitable for use in microwave ovens. Other electrically nonconducting materials, such as plastic, may be used, if desired, although glass is preferred.

In conventional glass filler constructions the inner surfaces of walls **12** and **14** are coated with a thin layer of silver. The purpose of the silver coating is to reduce the transmission of an infrared wavelength radiation through the vacuum filler. As known in the art, the insulating properties of a vacuum filler can be greatly improved by preventing the transmission of infrared wavelength radiation. Thus, a coating which has a high reflectivity or, inversely, a low emissivity characteristic is desirable for this purpose.

The presence of a silver coating is highly detrimental where the bottle is intended for use in a microwave oven. The silver coating results in joule heating and possible implosion of the filler when subjected to microwaves.

Accordingly, it is desired to insulate the bottle against infrared radiation loss without the use of silver or other metallic coatings to permit its use in a microwave oven.

Figure 7.7: Microwave-Safe Vacuum-Insulated Bottle

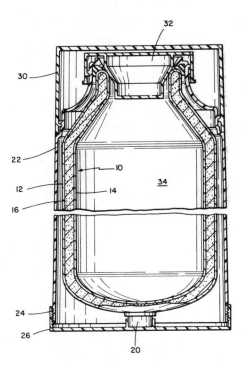

Source: U.S. Patent 4,184,601

According to this process the silver coating is omitted and in its place the annular space between the walls **12** and **14** is substantially filled with finely divided materials which are neither electrically conductive nor adsorbent at microwave frequencies. Examples of such materials include finely divided silica and calcium carbonate.

It is necessary that these materials be electrically nonconductive since microwave heating is accomplished by induced currents in conductive materials when subjected to microwaves. The infrared insulating property of these compounds is, however, the matter of greater moment. These compounds, when provided in the evacuated space **16** between walls **12** and **14**, provide insulation against infrared energy transmission of approximately the same magnitude as the metallic coatings known in the prior art.

The manufacture of the filler **10** is similar to the manufacture of glass fillers for conventional use. The fillers are formed on automatic machinery well-known in the art. Instead of applying a silver coating, however, the granular material is placed into the annular space **16** prior to evacuation. During the evacuation process, which is usually accomplished by means of a vacuum pump, fiber glass batting or other air permeable material may be placed across the filler opening to prevent the vacuum pump from sucking the granular material out of the space **16**. After the space **16** has been evacuated, plug **20** is applied to seal the filler.

The filler is then ready for insertion into a standard vacuum bottle jacket, such as jacket **22** in Figure 7.7. The bottom of the jacket is provided with threads **24** to receive a base **26**. The upper portion of the jacket is adapted to threadingly receive a cup **30**. A stopper **32** engages the opening in the upper portion of the filler **10** to seal the insulating interior **34** of the filler. If desired, of course, the filler may be provided with an inner liner for added protection against possible breakage of the liner and contamination of the foodstuffs contained therein.

Thermos bottles formed according to this process may be filled with liquids or semiliquids and then placed in a microwave oven for rapid heating of the contents of the bottle without damage to the bottle or heating of the bottle by the microwaves or the heated liquids.

The heated contents are then sealed by placing the stopper **32** in position and the contents will remain at approximately the initial temperature for substantial periods of time as in the case with conventionally available thermos constructions.

Glass for Faraday Rotation Element

Previously, glasses containing large amounts of paramagnetic rare earth element ions have been used as Faraday rotation elements because of the large Verdet constant of these glasses. Generally, rare earth element oxides other than cerium oxide are very expensive, and there is a great demand for relatively inexpensive glasses containing Ce^{3+} for Faraday rotation glasses. However, conventional Ce_2O_3-P_2O_5 type glasses are very unstable, and tend to devitrify at the time of melt-molding. Furthermore, since the volatilization of P_2O_5 from this glass composition is marked, the glass composition tends to vary. Hence, the yield of such glasses is extremely low.

S. Hirota; U.S. Patent 4,168,176; September 18, 1979; assigned to Hoya Corporation, Japan has found that a glass composition of Ce_2O_3, P_2O_5, B_2O_3, SiO_2 and/or Al_2O_3 exhibits good stability permitting easy melt-molding and reduced volatilization preventing variation in composition and occurrence of striae, and when produced on a large scale results in good yields and uniform quality.

This glass for a Faraday rotation element comprises, in mol percent, 15 to 18% Ce_2O_3, 64 to 76% P_2O_5, 4 to 16% B_2O_3, 0 to 8% SiO_2, and 0 to 10% Al_2O_3 with 2 to 12% SiO_2 + Al_2O_3.

Transparent Insulating Bodies

G. Schöll; U.S. Patent 4,173,969; November 13, 1979 describes a hollow glass body which can be used as part of a light-transmitting thermally insulating wall, adapted to serve as the cover for a solar-energy collector.

The hollow-silicate-glass-body has a wall thickness which can be of a millimeter or less and which is blown on a hollow glass blowing machine, i.e., a machine of the type which has heretofore been used to blow glass bottles. The hollow body is blown from the silicate glass and is shaped in a plurality of steps. Thus, the body is preblown and/or subjected to prepressing and is then given its final shape by a subsequent or finish-blowing operation.

It is preferred to make the configuration of this body like that of a rectangularly prismatic box having a pair of broad sides and four relatively narrow sides extending around the perimeter of the broad sides. All the sides of the box are integral and unitary with one another. According to an important feature of this process, the sides are of curved profile or cross-section, i.e., are convex and/or concave.

When the hollow bodies are produced in the configuration and in the manner described, they advantageously are flat boxes which can be assembled in lateral and end contiguous relationship.

In order that the box-shaped hollow bodies can be disposed contiguously with a minimum of mutual slidability, it is advantageous to provide the opposite ends of the hollow body with mutually complementary male and female formations. More particularly, the male formation can be a neck or mouth through which the hollow body is blown while the female formation can be an inwardly bent portion of an opposite end wall adapted to receive the neck or mouth of the adjacent boxlike body.

The opposite longitudinal walls of the body also can be mutually complementary, i.e., with one being outwardly convex while the other is outwardly concave, so that the convex wall of one body can engage in the concavity formed by the concave wall of an adjacent body and contiguous therewith.

This permits the hollow bodies to be brought together as close as possible and eliminates to a large extent, the clearances existing between the body. Any remaining interstices may be filled in with plugging material, if desired, to minimize convection within the layer of hollow bodies.

The hollow bodies can be disposed in a single layer to form a light-permeable wall, although it is preferable to arrange them in at least two layers or tiers and to cover the layer or layers with a thin sheet or foil of transparent synthetic-resin material for protection from weather influences such as rain, snow and frost. Similar light-permeable sheets or foils can, of course, be disposed between the layers of the hollow bodies.

While the hollow bodies according to the process have been found to be particularly satisfactory for use as the insulating transparent layers of a solar collector, they are also practical for use as transparent insulating windows, doors or the like or between panes of such doors or windows. The hollow bodies form layers with an extremely high heat-lagging effect and a particularly high light transmissivity, while being of low weight and having considerable strength.

Cathode Ray Tube Panel

D.C. Boyd, H.E. Hagy and D.A. Thompson; U.S. Patent 4,179,638; Dec. 18, 1979; assigned to Corning Glass Works describe a glass panel for a cathode ray tube to be employed in a color television receiver, the panel having a low degree of compaction after normal annealing, a strain point over 470°C, a liquidus temperature below 900°C, and the glass composition, as calculated from the batch, being chemically composed essentially of, in addition to silica, 1 to 3% Al_2O_3, 1.5 to 3% MgO, 2.5 to 4.5% CaO, 5 to 10% SrO, 3 to 10% BaO, 1 to 2.5% PbO, 6 to 10% Na_2O, 4 to 8% K_2O, the total content of SrO + BaO + CaO + MgO being 15 to 24%.

Preferably, the ratio of Na_2O to K_2O is at least 1:1, and the ratio of BaO to SrO is also at least 1:1. Up to 0.4% fluorine may be present. Additive oxides normally present in a panel glass for a color television tube include up to 1% each of CeO_2, TiO_2, As_2O_3, Sb_2O_3, and known glass colorant oxides.

The process is illustrated in the table below wherein are shown compositions of several glasses within the scope of the process. The electrical resistivity is shown as the logarithm, base 10, of the value measured at 350°C.

Also shown are several relevant properties measured on these glasses including softening point (S.P.), strain point (St. P.), coefficient of thermal expansion x 10^{-7} (Exp.), and liquidus temperature (Liq.). All temperatures are in degrees Centigrade (°C).

	. . . . Compositions					
	1	2	3	4	5	6
Components, % by wt						
SiO_2	59.5	60.2	59.2	61.3	60.3	59.3
Al_2O_3	2.0	1.3	2.0	2.0	2.0	2.0
MgO	2.6	2.6	2.6	1.9	1.9	2.3
CaO	3.9	3.9	3.9	2.8	2.8	3.8
SrO	7.7	7.7	7.2	7;7	8.7	7.7
BaO	7.7	7.7	8.4	7.7	7.7	7.7
PbO	2.4	2.4	2.4	2.4	2.4	2.4
Na_2O	8.6	8.6	8.6	8.6	8.6	7.9
K_2O	5.6	5.6	5.6	5.6	5.6	6.6
TiO_2	0.5	0.5	0.5	0.5	0.5	0.5
CeO_2	0.16	0.16	0.16	0.16	0.16	0.16
As_2O_3	0.2	0.2	0.2	0.2	0.2	0.2
Sb_2O_3	0.4	0.4	0.4	0.4	0.4	0.4
F	−	−	−	−	−	0.3
Properties						
S.P.	696	693	695	691	695	691
St. P.	479	478	480	472	474	472
Exp.	98.4	98.7	98.1	96.1	98.5	97.4
Liq., °C	884	799	867	805	843	871
Log R (350°C)	7.425	7.430	7.435	7.245	7.315	7.650

The compositions may be melted and worked in accordance with standard practice for panel glass. For example, glasses having the compositions above were melted experimentally by mixing batches from commercial ingredients including sand, feldspar, lime, fluorspar (if fluorine is required), strontium, barium, and sodium carbonates, litharge, potassium carbonate and sodium antimonate.

The properly proportioned batch was mixed intimately and placed in a crucible which was electrically heated at 1550°C and held for 4 hours to thoroughly melt. Each melt was then poured into slab molds, drawn as cane, or otherwise worked in suitable manner.

COMPANY INDEX

The company names listed below are given exactly as they appear in the patents, despite name changes, mergers and acquisitions which have, at times, resulted in the revision of a company name.

INVENTOR INDEX

U.S. PATENT NUMBER INDEX

Copies of U.S. patents are easily obtained
from the U.S. Patent Office at 50¢ a copy.

4,168,959 - 118	4,184,863 - 11	4,194,898 - 75
4,169,182 - 154	4,185,419 - 96	4,195,980 - 244
4,169,790 - 308	4,185,980 - 122	4,195,981 - 17
4,170,459 - 137	4,185,981 - 107	4,195,982 - 52
4,170,460 - 82	4,185,983 - 58	4,196,004 - 35
4,170,461 - 84	4,185,984 - 21	4,197,103 - 135
4,171,212 - 138	4,186,023 - 277	4,197,105 - 175
4,172,712 - 3	4,187,094 - 70	4,197,136 - 242
4,172,921 - 305	4,187,095 - 58	4,197,349 - 158
4,173,393 - 243	4,187,115 - 168	4,198,223 - 221
4,173,459 - 197	4,188,089 - 205	4,198,224 - 39
4,173,460 - 86	4,188,198 - 228	4,198,226 - 75
4,173,461 - 51	4,188,199 - 79	4,198,463 - 75
4,173,969 - 312	4,188,228 - 144	4,198,466 - 85
4,175,939 - 115	4,188,444 - 80	4,198,467 - 85
4,175,940 - 243	4,188,452 - 303	4,199,335 - 202
4,175,941 - 266	4,189,325 - 171	4,199,336 - 149
4,175,942 - 49	4,190,451 - 247	4,199,337 - 215
4,177,077 - 150	4,190,452 - 264	4,199,339 - 178
4,177,319 - 185	4,190,500 - 268	4,199,340 - 280
4,178,050 - 293	4,191,039 - 48	4,199,364 - 147
4,178,082 - 188	4,191,545 - 209	4,200,445 - 27
4,178,162 - 120	4,191,546 - 30	4,200,447 - 56
4,178,163 - 38	4,191,547 - 257	4,200,448 - 14
4,178,164 - 284	4,191,548 - 53	4,200,467 - 182
4,178,414 - 74	4,191,585 - 156	4,200,485 - 145
4,179,189 - 201	4,192,252 - 160	4,200,681 - 33
4,179,300 - 185	4,192,576 - 262	4,201,559 - 176
4,179,638 - 313	4,192,663 - 157	4,201,561 - 97
4,180,409 - 151	4,192,664 - 36	4,201,828 - 303
4,181,403 - 230	4,192,665 - 168	4,202,682 - 238
4,182,437 - 164	4,192,666 - 173	4,202,700 - 279
4,182,619 - 74	4,192,667 - 16	4,203,746 - 101
4,182,643 - 290	4,192,689 - 66	4,203,747 - 101
4,183,620 - 228	4,193,784 - 22	4,203,750 - 22
4,183,737 - 244	4,193,807 - 183	4,203,774 - 101
4,184,189 - 289	4,194,643 - 287	4,204,976 - 307
4,184,601 - 310	4,194,807 - 207	4,205,974 - 89
4,184,859 - 201	4,194,895 - 126	4,205,976 - 71
4,184,860 - 216	4,194,896 - 110	4,206,253 - 67
4,184,861 - 9	4,194,897 - 105	

NOTICE